APPLIED OPTICS

and

OPTICAL ENGINEERING

VOLUME VIII

CONSULTING EDITOR

RUDOLPH KINGSLAKE

APPLIED OPTICS

and

OPTICAL ENGINEERING

EDITED BY

ROBERT R. SHANNON

and

JAMES C. WYANT

Optical Sciences Center
University of Arizona
Tucson, Arizona

VOLUME VIII

1980

ACADEMIC PRESS

A Subsidiary of Harcourt Brace Jovanovich, Publishers

New York London Toronto Sydney San Francisco

ACADEMIC PRESS, INC.
111 Fifth Avenue, New York, New York 10003

United Kingdom Edition published by
ACADEMIC PRESS, INC. (LONDON) LTD.
24/28 Oval Road, London NW1 7DX

Library of Congress Cataloging in Publication Data
Kingslake, Rudolf, ed.
    Applied optics and optical engineering.

    Vol. 7–   edited by R. R. Shannon and J. C. Wyant.
    "Cumulative index" : v. 5 , p.
    Includes bibliographical references.
    PARTIAL CONTENTS.––v. 1. Light, its generation and
modification.––v. 2. The detection of light and infra–
red radiation.––v. 3. Optical components. [etc.]
    1. Optical instruments. 2. Optics. I. Shannon,
Robert Rennie, Date   II. Wyant, James C.
III. Title.
QC371.K5   621.36   65–17761
ISBN 0–12–408608–X (v. 8)

PRINTED IN THE UNITED STATES OF AMERICA

80 81 82 83   9 8 7 6 5 4 3 2 1

Contents

CHAPTER 1

Photographic Lenses

Ellis I. Betensky

CHAPTER 2

Lens Mounting and Centering

Robert E. Hopkins

CHAPTER 3

Aspheric Surfaces

Robert R. Shannon

CHAPTER 4

Automated Lens Design

William G. Peck

CHAPTER 5

Radiometry

William L. Wolfe

CHAPTER 6

The Calculation of Image Quality

William B. Wetherell

CHAPTER 7

Circuits for Detectors of Visible Radiation

William Swindell

CHAPTER 8

Arrays and Charge-Coupled Devices

James A. Hall

List of Contributors

Numbers in parentheses indicate the pages on which the authors' contributions begin.

ELLIS I. BETENSKY, *Opcon Associates, Cincinnati, Ohio* (1)

JAMES A. HALL, *Advanced Technology Laboratories, Westinghouse Defense and Electronic Systems Center, Baltimore, Maryland 21203* (349)

ROBERT E. HOPKINS, *Laboratory for Laser Energetics, College of Engineering and Applied Science, University of Rochester, Rochester, New York 14627* (31)

WILLIAM G. PECK, *Applications Division, Genesee Computer Center, Incorporated, Rochester, New York 14605* (87)

ROBERT R. SHANNON, *Optical Sciences Center, University of Arizona, Tucson, Arizona 85721* (55)

WILLIAM SWINDELL, *Optical Sciences Center, University of Arizona, and Department of Radiology, Arizona Health Sciences Center, Tucson, Arizona 85721* (317)

WILLIAM B. WETHERELL, *Optical Systems Division, Itek Corporation, Lexington, Massachusetts 02173* (171)

WILLIAM L. WOLFE, *Optical Sciences Center, University of Arizona, Tucson, Arizona 85721* (117)

Foreword

It has been about ten years since Volume V completed the initial set of books on this topic which were edited by Rudolph Kingslake. During this period, these books have served as the major reference on the many topics associated with the design and use of optical systems. The treatise was intended for use as a reference by the practicing engineer, and the use of mathematics and derivations was kept to a minimum for this purpose, with lucid explanation serving as the mechanism for communication rather than the exposition of theory.

Many changes have occurred in the field of optical engineering during the past decade. Lasers in various wavelength regions, widespread use of computers and digital processors, new optical materials, and widespread innovations in electro-optical sensors are but a few of the new topics that need to be reviewed. Dr. Kingslake has acknowledged the need for treatment of new topics and updating of some previous topics, and he has collaborated with Brian Thompson on Volume VI, for which the Table of Contents can be found elsewhere in this volume.

This volume is the second of a planned set of successor volumes. We are initiating these with the intention of providing as useful a reference as that provided by the previous volumes. Two efforts are required here. First, much of the reference material on topics such as sources and detectors and materials must be updated because these items have been supplanted by newer devices or materials that were not available ten years ago. Second, many new concepts and topics are of interest, some of which were not even in the conceptual stage at that time. In a few cases, the understanding of some topics, such as the transmission of light through the atmosphere, has been improved to the extent that a new treatment of the subject is in order. We are planning to respond to all of these needs in the subsequent volumes.

Our philosophy is a bit different than that of the previous volumes. The field of applied optics moves so rapidly that it is not possible to establish a grand plan to cover several volumes, with each volume covering a specific topic. We have instead chosen to include a broad sample of topics from each of the above categories in each volume. This generates a continuing series of reviews on topics in applied optics that will be both as up to date and as comprehensive as possible. Some of the newer topics in optics are sufficiently complex that it is not possible to retain the

xi

mathematical level of complexity at as low a level as that which could be maintained in the previous volumes. In such cases, we have endeavored to have the authors provide key formulas and explanations insofar as possible.

We believe that these volumes should serve as valuable general references for the practicing optical engineer. Further, we intend that the choice of topics will be such that an engineer can learn of some of the newer techniques and processes that will be of vital interest if he is to advance in his field. It is our hope to be able to produce about one volume a year which meets this criterion. We are attempting to obtain as authors leaders in the field who will write articles of lasting significance and interest.

ROBERT R. SHANNON
JAMES C. WYANT
1980

Preface

This is the second volume of "Applied Optics and Optical Engineering" done under our editorship. This volume continues the variety of basic and applied topics that were introduced in Volume VII.

The first chapter, by Ellis Betensky, on photographic lenses, consists of a presentation of examples of the state of the art in photographic lens design, including wide angle and zoom lenses. In this chapter, for the first time a representative set of transfer function characteristics is presented from which the reader can obtain some idea of the level of image quality to be expected in modern camera lenses.

The second chapter, by Robert Hopkins, is a discussion of lens mounting and centering techniques. In this chapter, some of the classical as well as modern techniques for determining whether or not the lens is indeed built and aligned as the designer intended are presented.

Aspheric surfaces form the topic of the third chapter in this volume, by Robert R. Shannon. The purpose of the chapter is to discuss some of the characteristics of aspheric surfaces and provide the reader with a general view of the concepts involved in the design and application of nonspherical surfaces in optical systems.

The fourth chapter, by William Peck, on automated lens design, gives a brief look at some of the principles underlying modern lens design programs. From this chapter, it is evident that interplay between the designer and the program is of utmost importance in obtaining a solution. It is also obvious that the lens designer today would be lost without access to an adequate computer program for design.

In the fifth chapter, William Wolfe has examined the concepts of radiometry and presents a unified approach to the problem of units, ranging from photons to footcandles, that might be encountered in radiometric problems. Included in this chapter are detailed formulas as well as rules of thumb for the calculation of radiometric quantities. Special care has been taken to ensure that the definitions of all quantities are consistent and appropriate to the problem.

A very lengthy sixth chapter, on the calculation of image quality, has been prepared by William Wetherell. The author of this chapter examines in detail some of the fundamental relationships among aperture shape, aberrations, and the expected transfer function that may be obtained. The results of this chapter may be used as a guide for both optical designers and system engineers in writing specifications on high quality optical systems.

The final two chapters enter into the concepts of detectors to be used with optical systems. In the seventh chapter, William Swindell discusses circuits applicable for detectors of visible radiation. The purpose of this chapter is to allow the nonspecialist to select the most appropriate circuitry to match a given detector. As such, the chapter may be considered as a "how to do it" section for the optical specialist rather than the electronics specialist and, it is hoped, may reduce the optical engineer's dependence on those whose skills lie primarily in the electronics area.

The eighth chapter, by James Hall, is a review of available and soon to be available detector arrays, including the most recent charge-coupled detectors. The purpose of this chapter is not to explain in detail the methods of operation of the device but to provide the working optical engineer with an overview of the nature of such detectors and the imagery that may be expected from them.

We believe that these articles continue the development of quality discussions of various topics associated with optical engineering in an up-to-date manner. The material ranges from reviews of classical concepts in optics, such as are found in lens design and radiometry, to the discussion of detector concepts that could not even be considered when the first volumes of this treatise were being planned and written.

Contents of Other Volumes

Volume VII

CHAPTER 1

Photographic Lenses

ELLIS I. BETENSKY

Opcon Associates, Cincinnati, Ohio

I. INTRODUCTION

Photographic lens designs have changed considerably since the mid-1960s. New photographic lenses are now designed to function over a large focusing range, typically at a magnification of 0 to −0.1, and are generally smaller than the older types. Particularly, zoom lenses are smaller, have greater focal-length ranges, and are now accepted as versatile, economical, photographic instruments. In many ways, the designs are so unconventional that the practicing engineer, who often uses a photographic lens as part of an instrument or system, must reconsider the simple rules of focusing and thin-lens formulas to take full advantage of the new lenses.

Since these new designs are not based upon new manufacturing technology or glass types, the question may be asked, Why has this change occurred? Perhaps the greatest factors are the maturing of design optimization programs and extensive use of the modulation transfer function (MTF) as a design tool. Fixed-focal-length lenses are designed for optimum performance over a focusing range by specifying the design as a zoom lens having variable spaces. Zoom lenses are optimized for different

1

focusing positions at the same focal length simultaneously. The general use of through-focus MTF evaluation is achieved by describing lens aberrations in terms of through-focus modulation for a selected spatial frequency. Not only is the through-focus MTF a tool for maximizing depth of focus and determining lens manufacturing tolerances, but it is also a very useful analytic tool for the optical system designer who is concerned with film flatness, temperature variations, mechanical alignment, and stability.

In the following discussion, evaluation techniques are described and significant characteristics identified to indicate the direction of recent developments.

II. THROUGH-FOCUS MTF ABERRATIONS

Classical aberrations in terms of series expansions are convenient for lens design optimization programs but not for quantitative descriptions of system performance. Also, the effects of system components other than the lens cannot be readily assessed in these terms. For these reasons the lens designer's method of through-focus MTF analysis has become quite popular among system designers. While the classical aberration names are used for familiarity, the meanings are not intended to be classical in a strict sense.

(a) *Spherical aberration.* Reduction in peak modulus and shift of best focus from paraxial. May have two extremes as a result of zonal variation (Fig. 1a).

(b) *Coma.* Reduction in peak modulus and broadening of region of focus (Fig. 1b).

(c) *Astigmatism.* Lack of coincidence of radial and tangential peak modulations (Fig. 1c).

(d) *Chromatic aberration.* The difference in peak modulation among colors. The polychromatic transfer function is generally used to determine the effects of a lateral shift (a chromatic variation in magnification), while through-focus curves determine the modulation in each color (Fig. 1d).

To predict resolution with a particular film, the threshold modulation for the particular spatial frequency is used to determine the depth of focus (Fig. 2). This depth is then used as a basis for system performance and tolerances determining how factors (such as film curvature) reduce depth directly or how peak modulation can be reduced by a factor such as scattered light. It is a straightforward procedure for devising a rational toler-

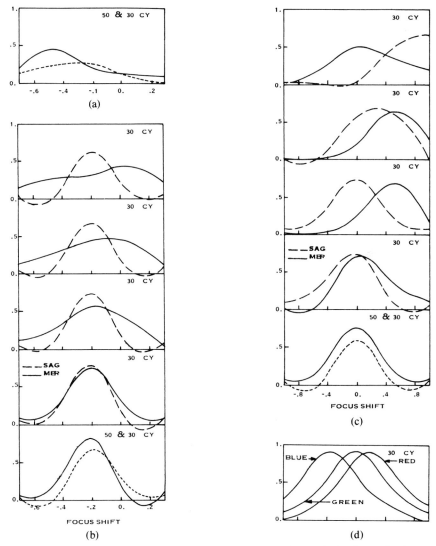

FIG. 1. (a) Spherical aberration, (b) coma, (c) astigmatism, and (d) chromatic aberration.

ance scheme based upon through-focus aberrations, particularly on a system level.

Typically, the nominal depth of focus is allocated for a tilted field, axial astigmatism, and astigmatism due to manufacturing deviation of radii, thicknesses, and indexes of refraction. In addition, the reduction in

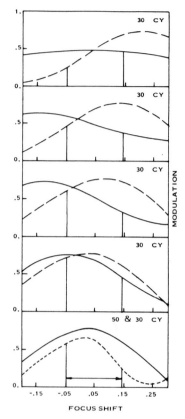

FIG. 2. Depth of focus for threshold modulation at one spatial frequency.

peak modulation is allocated for axial coma due to decentration and for spherical aberration caused by manufacturing deviations of the centered lens. A trial budget for these factors is used for the starting point in the tolerances procedure. A new budget is then determined after the effects upon depth and modulation due to manufacturing deviations are assessed.

III. DISTORTION

Distortion is frequently described in optical design literature as a primary aberration, and thus measurements and standards are often based upon this assumption. However, most inverted telephoto-type lenses and many zoom lenses have a distortion correction based upon secondary and higher orders. The standards for percent radial distortion are thus not an accurate indication of line straightness as demonstrated by the two plots (Fig. 3) which show 2% and less than 5% radial distortion, respectively.

Assessment of distortion for photographic lenses not used for measurement purposes is a subjective interpretation of line straightness. The maximum deviation of the slope of the line image may be a useful design criterion.

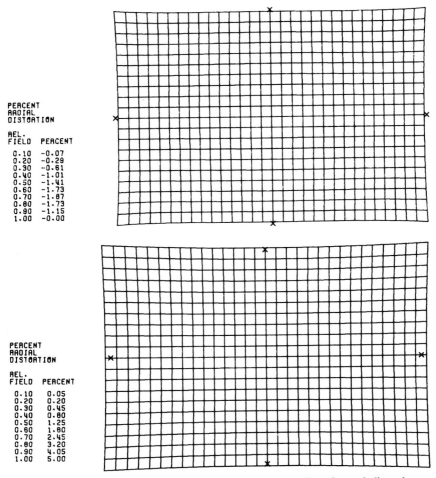

PERCENT
RADIAL
DISTORTION

REL. FIELD	PERCENT
0.10	-0.07
0.20	-0.29
0.30	-0.61
0.40	-1.01
0.50	-1.41
0.60	-1.73
0.70	-1.87
0.80	-1.73
0.90	-1.15
1.00	-0.00

PERCENT
RADIAL
DISTORTION

REL. FIELD	PERCENT
0.10	0.05
0.20	0.20
0.30	0.45
0.40	0.80
0.50	1.25
0.60	1.80
0.70	2.45
0.80	3.20
0.90	4.05
1.00	5.00

FIG. 3. Image of square grid produced by lenses having distortion as indicated.

IV. SPECTRAL WEIGHTING

The polychromatic optical transfer function (OTF) is the weighted vectorial average of the monochromatic OTF over an appropriate range of wavelengths. There does not appear to be a generally accepted standard for choosing weights and wavelengths used in estimating the OTF of photographic lenses. If daylight is combined with average lens transmittance, the spectral distribution shown in Fig. 4 is derived. An optimum set of weights and wavelengths for this distribution is suggested by Crawford (1978). An estimate of the polychromatic OTF using these weights and wavelengths is equivalent to an estimate using nine equally spaced wave lengths and any standard numerical quadrature rule. If the OTF is assumed to be a smooth function of wavelength, this estimate is entirely adequate.

Wavelength (nm)	Weight
433.9	0.1324
482.8	0.2739
557.6	0.2848
633.7	0.2110
686.1	0.0979

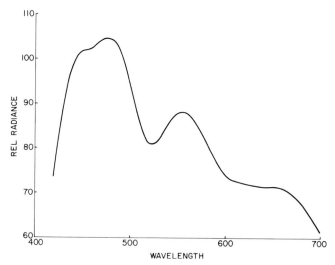

FIG. 4. Spectral distribution of daylight combined with average lens transmittance and film.

V. REVIEW OF CURRENT LENS DESIGNS

The increasing number of applications in photography has resulted in probably as many distinct photographic optics. Some classical standards (triplet, Tessar, Gauss, Sonnar, Plasmat, etc.) are still being designed and produced for new applications. Most, however, are not really new and therefore have been omitted from this review. Other truly new designs for many diverse photographic applications have been omitted for lack of space, including lenses for cinematography and large-aperture, wide-angle objectives representing significant lens design developments. The lens designs chosen for discussion here are primarily intended for use with both miniature and large-format still picture cameras.

In each of the following design analyses, only the polychromatic OTF is shown. The summary indicates angular coverage for approximately 92% of the full field coverage corresponding to a 20-mm image height for a 35-mm film format. The Gaussian optics properties summarized are:

EFL	effective focal length
BRL	vertex-to-vertex barrel length
ENP	front-vertex-to-entrance-pupil distance
BFL	back focal length
FVD	front-vertex-to-focal-plane distance
EXP	rear-vertex-to-exit-pupil distance

A. NORMAL LENSES

The normal lens of large aperture, once represented by many design forms, is typically now a double Gauss form with a broken contact doublet such as the example shown in Fig. 5 (Behrens and Glatzel, 1973). This type of lens is well corrected for astigmatism throughout at least a 20° semifield of view but has increasingly reduced modulation for the radial orientation as the field angle is increased.

The new Gauss-type lens developed by Mandler (1975) is interesting because the glass types chosen are far from typical. More important, the design represents a successful attempt to reduce the cost of a precision lens by using plastic lens mount parts and fewer lens surfaces. Because the design permits a mounting technique which ensures centration, the actual manufactured performance can be very close to the calculated values. It is conceivable that these design procedures will be applied to other lens types in the future (Fig. 6).

An exception to the trend toward novel lenses with special focusing features and unusually high speed or compactness, the normal lens for a

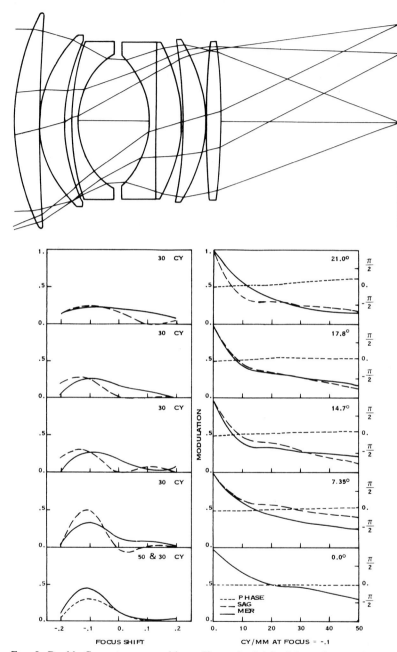

FIG. 5. Double Gauss-type normal lens: 50 mm 1 : 1.4 for 35-mm format; $f/1.42$; $21.0°$ semifield; EFL = 52.0; BFL = 36.9; BRL = 42.6; FVD = 79.6; ENP = 28.7; EXP = −25.3.

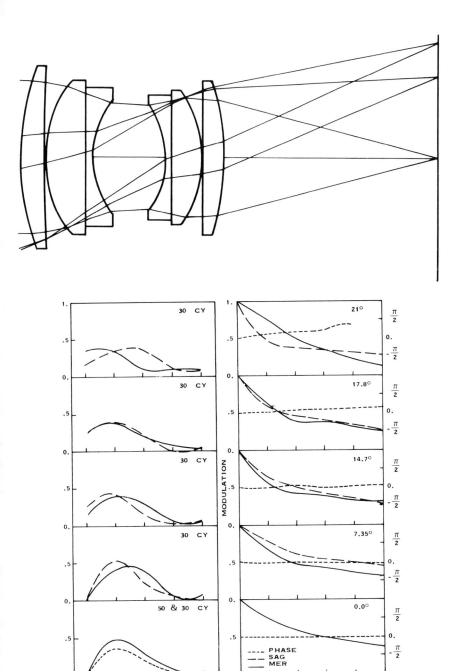

FIG. 6. New double Gauss-type normal lens: 50 mm 1:2 for 35-mm format having three plano surfaces and four radii; f/2.00; 21.0° semifield; EFL = 52.4; BFL = 37.6; BRL = 35.7; FVD = 73.2; ENP = 22.3; EXP = −20.9.

large-format camera is the very old Plasmat type shown here as a Schneider Symmar (Fig. 7). The design in its modern refined form is much superior to the older one, even though none of the newer glass types are needed. It is capable of a very high resolving power and has a stable aberration correction with conjugate change, making it popular in many laboratory applications.

B. WIDE-ANGLE, LARGE-APERTURE, INVERSE TELEPHOTO LENSES

This lens type, in its present compact form, is a very recent development and has advanced rapidly. The size characteristics are summarized in terms of effective focal length f:

Vertex-to-vertex length: $2f$–$2.5f$ for $f/2$; $1.5f$ for $f/2.5$
Front-window diameter: $1.3f$–$1.6f$ for $f/2$; $1.0f$ for $f/2.5$
Relative aperture: to $f/1.4$ with only spherical surfaces; from $f/1.4$ to $f/1.2$ with one aspherical surface

Since the lenses are compact, there is a large concentration of negative power in the front group. To correct for the resulting conjugate shift aberration, most commonly astigmatism, the design compensates through a variable air space formed by moving away from a fixed front group as in the Vivitar Series I 28-mm $f/1.9$. The distortion for this lens type is usually well corrected.

The overall limiting aberration of this lens type is usually astigmatism, the sagittal field tending to have an undercorrected zone and to be overcorrected at the corner. A compact example (Matsubura, 1973) is shown in Fig. 8. In Fig. 9, a form utilizing an aspherical surface is shown (Glatzel, 1974a).

C. ULTRAWIDE-ANGLE, INVERSE TELEPHOTO LENSES

Lenses of this category have a semiangle coverage in excess of 45° and, as a result of efforts to minimize the front lens diameter, the designer finds this problem one of the most difficult. The aberration correction is a direct result of this size restriction, since earlier designs resolved the problem quite adequately where size was not a factor.

Typically the entrance pupil is located approximately one focal length from the front element and the back focal length is often more than $2f$, a combination achieved by a large concentration of negative power in the front group. Thus, simultaneously correcting distortion and astigmatism is a major problem, particularly when considering chromatic variations in

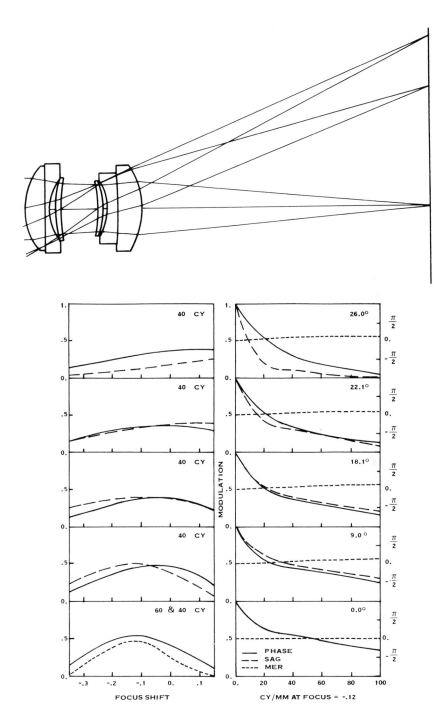

Fig. 7. Normal lens for large format: 150 mm 1 : 5.6 for 4 × 5 in. format; *f*/5.70; 26.0° semifield; EFL = 150.1; BFL = 127; BRL = 52.0; FVD = 179; ENP = 27.0; EXP = −29.4.

11

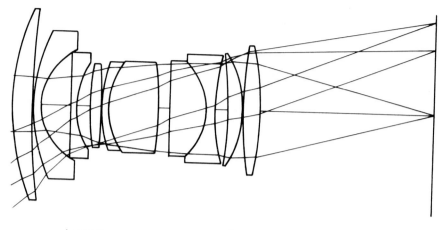

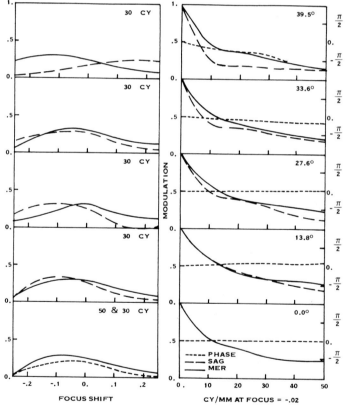

FIG. 8. Wide-angle, large-aperture, inverse telephoto lens: 24 mm 1 : 2 for 35-mm format; $f/2.05$; 39.5° semifield; EFL = 24.6; BFL = 38.3; BRL = 54.8; FVD = 93.1; ENP = 19.6; EXP = −28.5.

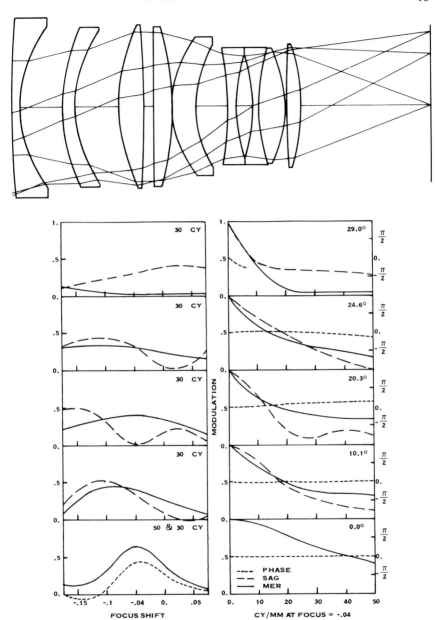

FIG. 9. Wide-angle, large-aperture, inverse telephoto lens with aspherical surface: 35 mm 1:1.4 for 35-mm format; $f/1.44$; 29.0° semifield; EFL = 36.6; BFL = 35.8; BRL = 80.2; FVD = 116.0; ENP = 32.1; EXP = −28.3.

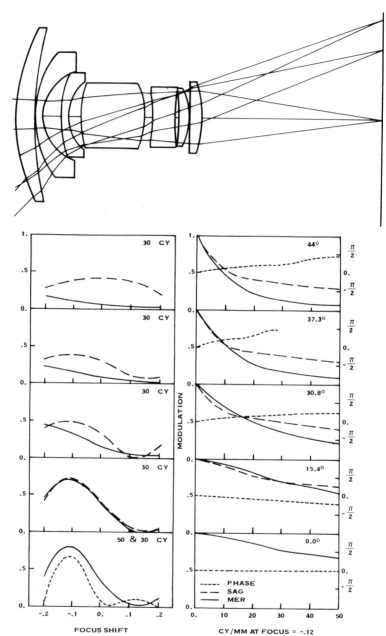

FIG. 10. Ultrawide-angle, inverse telephoto lens: 21 mm 1 : 3.5 for 35-mm format; $f/3.60$; 44.0° semifield; EFL = 21.5; BFL = 37.2; BRL = 38.7; FVD = 75.9; ENP = 15.7; EXP = −10.9.

distortion and the variation in astigmatism with conjugate change. As expected, a compensating change in center air space is required. The correction then holds stable between a magnification of 0 to -0.1. Distortion, which often is too large for lines to appear straight even under visual inspection, increases when focusing for closer objects.

The illumination is about equal to the $\cos^4 \Theta$ rule even with some vignetting, due to aberration of the entrance pupil and reduced obliquity of the off-axis rays in the image space.

There are many different design forms in popular use. The examples in Figs. 10 and 11 (Nakagawa, 1973; Glatzel, 1973), show the remarkable quality typical of such extreme lenses. The size of the 21-mm lens is particularly impressive.

D. TELEPHOTO LENSES

The telephoto lens has been reduced in length in recent years but not beyond that achieved previously for special applications. The size is usually determined by the overcorrected Petzval curvature and therefore relates directly to the field of view and, consequently, the focal length. For the 35-mm format, the relationships in the accompanying tabulation can be attained.

Focal length (mm)	Telephoto ratio
400	0.6
300	0.7
200	0.8
135	0.9
90	1.0

The telephoto ratio is defined as the front-vertex-to-focal-plane distance divided by the effective focal length.

The reduction in length has created two problems which have been solved by a variety of techniques. First, the problem of secondary color has been minimized through the use of artificial fluorite or by a judicious choice of ordinary and special optical glass. This latter method is, of course, more stable during temperature changes and less costly.

In Fig. 12, a very high-quality design of simple construction and ordinary glass is shown (Mandler and Schmidt, 1961).

The second problem is focusing. The aberration change, particularly in

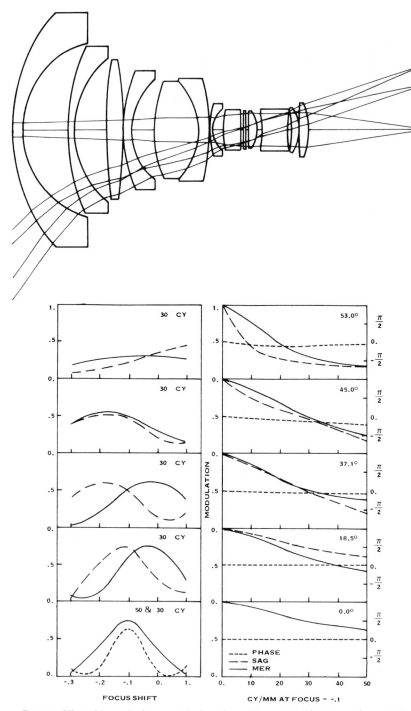

FIG. 11. Ultrawide-angle, inverse telephoto lens: 15 mm 1 : 3.5 for 35-mm format: $f/3.61$; 53.0° semifield; EFL = 15.4; BFL = 36.4; BRL = 93.7; FVD = 130.1; ENP = 33.9; EXP = − 16.1.

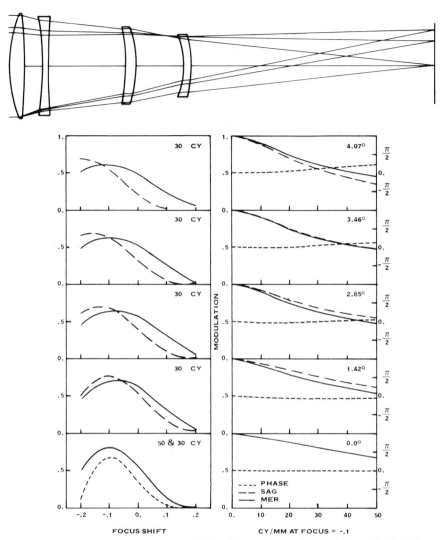

FIG. 12. Telephoto lens: 280 mm 1:4.8 for 35-mm format; $f/4.80$; 4.1° semifield; EFL = 278.9; BFL = 143.9; BRL = 107.4; FVD = 251.3; ENP = 144; EXP = −15.1.

aperture-dependent aberrations, is more pronounced for extreme tele-photo lenses. A rear corrector group is often employed to compensate during focusing either by moving or remaining fixed while the prime lens is moved. For very long-focal-length lenses, focusing for a magnification of -0.1 requires a large lens movement. With the use of a zoom lens design approach, internal elements can be moved a much smaller degree to achieve close focusing with aberration correction stability (Kreitzer, 1978) (Fig. 13).

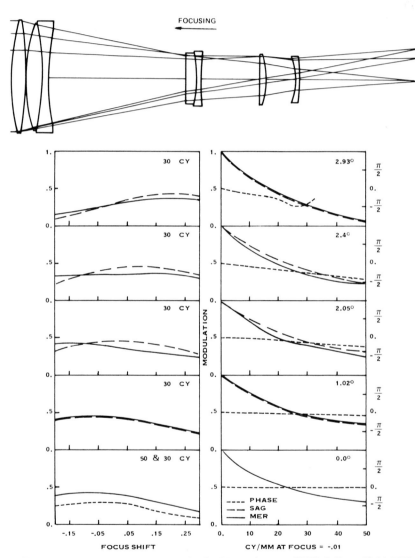

FIG. 13. Telephoto lens: 400 mm 1:5.6 for 35-mm format; $f/5.75$; 2.9° semifield; EFL = 389.9; BFL = 73.9; BRL = 182.1; FVD = 255.9; ENP = 211.8; EXP = -51.2.

E. Macro-Focusing Lenses

While most lenses now have a stable or corrected performance to a magnification of -0.1, focusing to -1.0 requires more than a simple extension of the focusing motion. With the use of a method similar to telephoto lenses with rear corrector groups, the aberration can be stabilized without any additional motions but with the addition of a fixed rear group (Betensky, 1974a). An example is shown in Fig. 14 for a magnification range of 0 to -0.5.

F. Zoom Lenses

Because of its inherent complexity, the zoom lens has considerable variation in form. While it might first appear that variations are decreasing as designs evolve toward certain standard types, as in the case of fixed-focal-length lenses, the opposite seems more likely. Accompanying the sophistication of design procedures are new requirements for compactness, low distortion, macrofocusing (magnification approximately -0.25 to -0.5), and greater range. Of course, a short discussion cannot begin to describe design trade-offs, weaknesses, or even the particular merits of any design form. The discussion here is intended to describe the design problem and show the direction of new solutions. The designs discussed are classified according to whether or not the iris diaphragm moves during zooming (focal length changing).

Fixed-iris-diaphragm design types are the most popular because of their low cost and high quality. Furthermore, with very little extra cost attributable to additional mechanical parts, these lenses can be designed to have a macrofocusing feature achieved by using the zooming groups with a separate set of motions (Watanabe and Betensky, 1972) (Fig. 15).

Although the macrofocusing feature is the source of a major constraint upon the Gaussian optics of the lens, there are several types of fixed-diaphragm zoom lenses being produced today. While some reach a true wide angle of $\pm 30°$, such as the lens in Fig. 16 (Betensky, 1974), most are of the long-focal-length variety, having zoom ranges in excess of 3:1. More typically, designs are limited to 2:1–3:1 ranges with an angular coverage of less than $\pm 20°$. These lenses have a positive front-focusing group followed by a moving negative zooming group and, depending upon the style, a positive or negative compensating group followed by a fixed rear group. While these lenses appear similar, many having a front-focusing group consisting of three elements, the aberration change upon focusing can vary considerably among design variations. Also, the movement of the entrance pupil position as a function of zooming has consider-

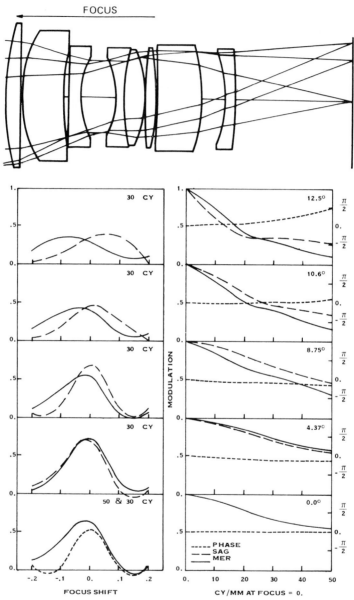

FIG. 14a. Macrofocusing lens: 90 mm 1:2.5 for 35-mm format; $f/2.50$; 12.5° semifield; EFL = 90; BFL = 40.5; BRL = 79.2; FVD = 119.7; ENP = 36.1; EXP = −38.3.

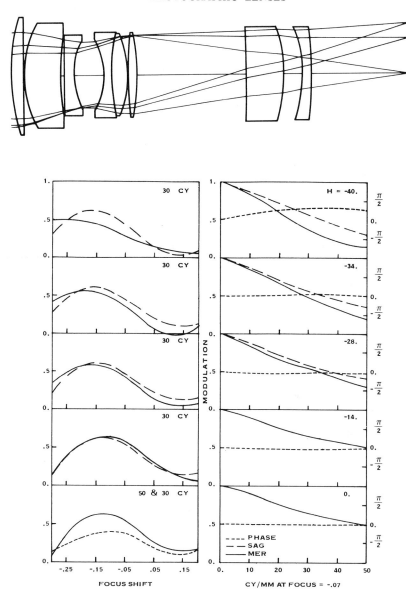

FIG. 14b. Macrofocusing lens: 90 mm 1 : 2.5 for 35-mm format: $f/3.75$; $H = -40.0$; $M = -0.5$; EFL $= 86.9$; OVL $= 404.4$; OBD $= -240.3$; IMD $= 40.5$; BRL $= 123.7$; FVD $= 164.1$; ENP $= 36.1$; EXP $= -76.6$.

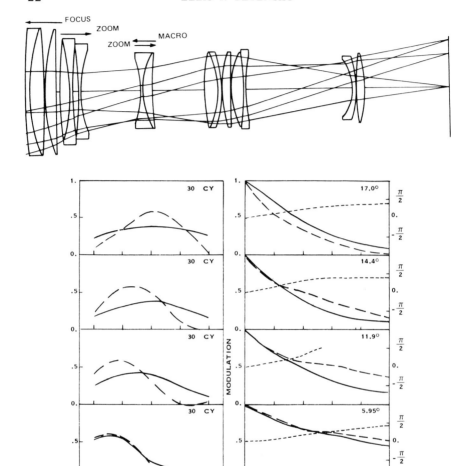

FIG. 15a. Zoom lens with macrofocusing: 70–210 mm 1:3.5 for 35-mm format; $f/3.80$; 17.0° semifield; EFL = 71.9; BFL = 39.9; BRL = 160.0; FVD = 199.9; ENP = 75.3; EXP −45.7.

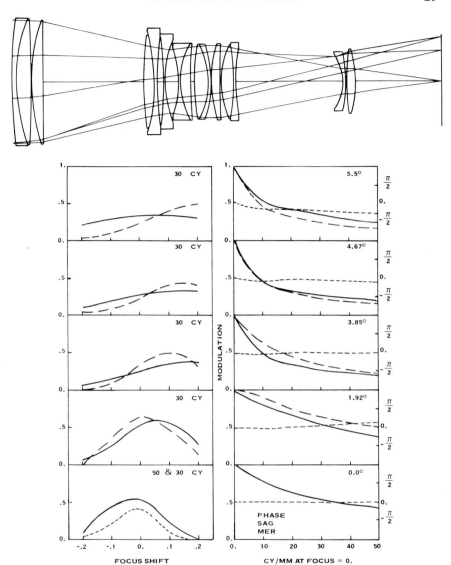

FIG. 15b. Zoom lens with macrofocusing: $f/3.80$; 5.5° semifield; EFL = 203.1; BFL = 40; BRL = 160.0; FVD = 200; ENP = 213.2; EXP = −45.7.

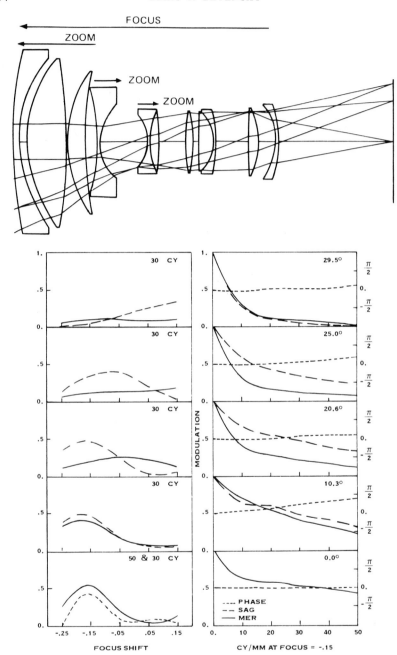

FIG. 16a. Zoom lens: 35–85 mm 1:2.8 for 35-mm format: $f/2.95$; 29.5° semifield; EFL = 36.0; BFL = 40.0; BRL = 93.7; FVD = 133.7; ENP = 43.6; EXP = −25.8.

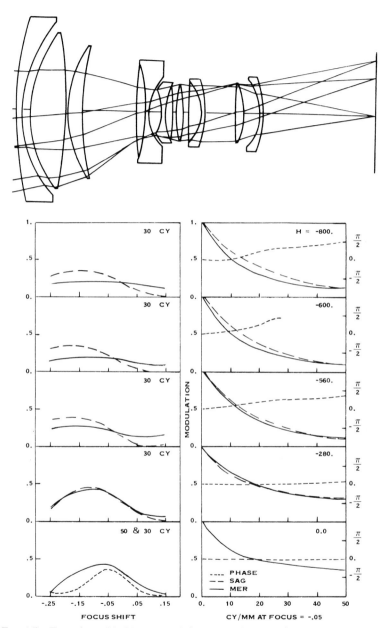

Fig. 16b. Zoom lens: 35–85 mm 1:2.8 for 35-mm format; $f/2.95$; $H = -800.00$; $M = -0.2500E$-01; EFL = 83.0; OVL = 3475.5; OBD = -3339.8; IMD = 42.1; BRL = 93.6; FVD = 135.7; ENP = 85.6; EXP = -25.8.

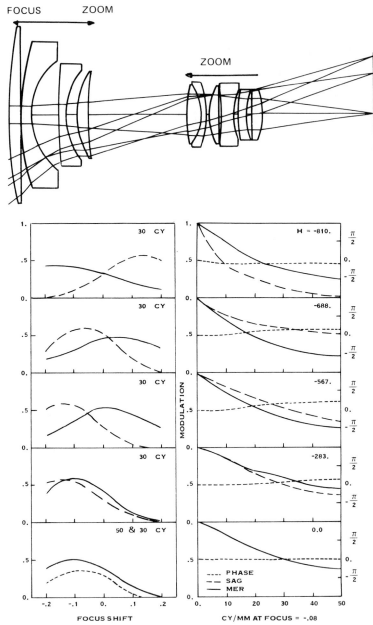

FIG. 17a. Zoom lens: 24–48 mm 1:3.8 for 35-mm format; $f/3.80$; $H = -810.00$; $M = -0.2500E\text{-}01$; EFL = 23.8; OVL = 1058.2; OBD = -929.8; IMD = 39.1; BRL = 89.3; FVD = 128.3; ENP = 30.1; EXP = -20.9.

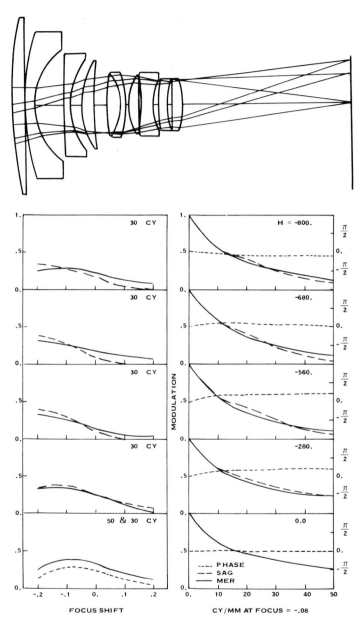

FIG. 17b. Zoom lens: 24–48 mm 1:3.8 for 35-mm format; $f/3.80$; $H = -800.00$; $M = -0.2500E\text{-}01$; EFL = 46.1; OVL = 1968.2; OBD = 1849.5; IMD = 58.5; BRL = 60.2; FVD = 118.7; ENP = 21.3; EXP = -20.9.

able variation, a characteristic system designers should determine for a particular lens.

Lenses having a fixed iris diaphragm cover less than ±30, because the lens diameter and/or length becomes excessive if the range is extended to wide-angle coverage. The astigmatism and distortion can be corrected but not with a compact overall configuration. When the iris diaphragm is allowed to move, the size problem is less objectionable. The design variations become very large and even the choice of positive or negative front group is not yet clear. With this freedom, high-quality zoom lenses of extremely compact size can be designed. The development limitation of these designs now lies mostly within the domain of the mechanical designer. For example, the difficulty of maintaining centration while zooming is significant.

With either a positive front group or a negative front group, a large zoom range can be covered, field angles in excess of ±40° zooming to less than ±10°. The choice of form depends upon size, cost, and focusing. A negative front group with a range of 2 : 1 is shown in Fig. 17 (Betensky and Lawson, 1974).

Since moving the entire lens for focusing requires a mechanism which varies travel as the square of the focal length, most zoom lenses are focused simply by moving the front group. For wide-angle lenses, this increases the diameter of the front group and thus is a very severe constraint upon the design. It is often suggested that all moving elements move as a function of both focal length and focus, but there are no lenses currently produced having these characteristics. There are, however, many attempts to improve front group focusing, as well as macrofocusing as discussed earlier.

G. WIDE-ANGLE, LARGE- AND MEDIUM-FORMAT LENSES

The wide-angle lenses for large-format cameras are not required to have a large back focal distance and thus are basically symmetric in form. This results in negligible distortion as compared to the inverse telephoto form. Also, the aberration correction is generally superior, particularly with regard to flatness of field. The relative lens size is greater, haveing a vertex-to-vertex length of typically one focal length, and rarely less than $0.8f$.

Since medium-format cameras are typically of the single lens reflex type, inverse telephoto objectives are used. The design problem is similar to that of the small 35-mm-format lenses, size being of significant concern. The lens in Fig. 18 (Glatzel, 1974b) demonstrates the quality level attainable.

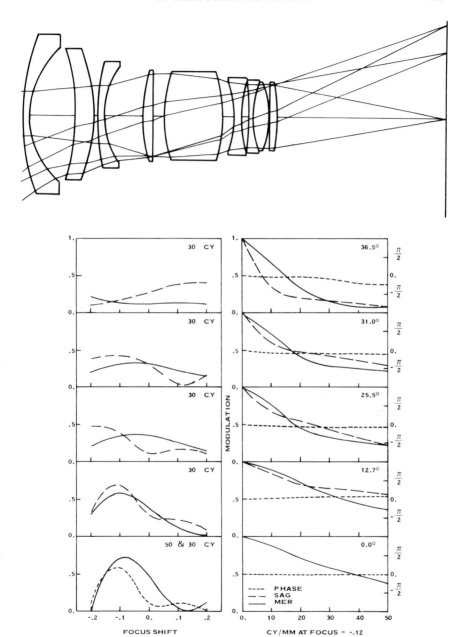

FIG. 18. Wide-angle lens: 50 mm 1:2.8 for medium format; $f/2.85$; 36.5° semifield; EFL = 51.8; BFL = 69.8; BRL = 106; FVD = 175.8; ENP = 39.6; EXP = -18.9.

ACKNOWLEDGMENTS

Selecting meaningful data for this presentation would have been impossible without the help of the following lens designers who sent the author their data and recommendations for further selection: E. Glatzel, Carl Zeiss, Oberkochen; K. Macher, Joseph Schneider Corporation, Kreuznack; W. Mandler, E. Leitz (Canda); S. Sakamoto, Olympus Optical Company, Tokyo; and Z. Wakimoto, Nippon Kogaku, Tokyo.

Their help is most appreciated. The author alone is responsible for the accuracy of the computations and for any errors or omissions which may have occurred.

REFERENCES

Behrens, K., and Glatzel, E. (1973). U.S. Patent 3,874,771 for Carl Zeiss Stiftung.

Betensky, E. (1974a). U.S. Patent 3,942,875 for Vivitar Corporation.

Betensky, E. (1974b). U.S. Patent 3,975,089 for Vivitar Corporation.

Betensky, E., and Lawson, L. (1974). Patent applied for, for Vivitar Corporation.

Crawford, C. (1978). Gaussian quadrature with tabular weight functions, presented at the Texas Conference of Mathematical Software.

Glatzel, E. (1973). U.S. Patent 3,864,026 for Carl Zeiss Stiftung.

Glatzel, E. (1974a). U.S. Patent 3,915,558 for Carl Zeiss Stiftung.

Glatzel, E. (1974b). U.S. Patent 3,958,864 for Carl Zeiss Stiftung.

Kreitzer, M. (1978). Patent applied for, for Vivitar Corporation.

Mandler, W., and Schmidt, H. (1961). D. B. Patent 1,096,057 for Ernst Leitz (Canada).

Mandler, W. (1975). Patent applied for, for Ernst Leitz (Canada).

Matsubara, M. (1973). U.S. Patent 3,830,559 for Olympus Optical Company.

Nakagawa, J. (1973). U.S. Patent 3,884,556 for Olympus Optical Company.

Watanabe, R., and Betensky, E. (1972). U.S. Patent 3,817,600 for Vivitar Corporation.

APPLIED OPTICS AND OPTICAL ENGINEERING, VOL. VIII

CHAPTER 2

Lens Mounting and Centering

ROBERT E. HOPKINS

Laboratory for Laser Energetics
College of Engineering and Applied Science
University of Rochester, Rochester, New York

I. INTRODUCTION

Most optical systems have rotational symmetry around the optical axis. This chapter describes several techniques used to mount lenses so that each lens element has its optical axis coaxial with the system axis. Attention is given to the design of the lens barrel, spacers, retainer rings, and lens edging. A cost-effective lens mounting depends on the use of proper measuring equipment, so comparisons among different methods of measurement are made. Since successful lens mounting also depends on consideration of tolerances as distributed between glass and metal parts, a section on tolerancing is also included.

31

The discussion may appear to place undue emphasis on precision lens mounting. Precision lens mounting requires attention to many details which may have minor effects in commercial lenses. These details should, however, be considered in any new design, for it is generally agreed that centering of the lens in the mount is one of the most important steps in lens fabrication.

II. BASIC PRINCIPLES IN LENS MOUNTING

A perfectly centered lens, such as the one shown in Fig. 1, has the centers of curvature of all the surfaces lying on a single axis. The spacer has matching curved surfaces and is the same shape as a lens with a hole drilled through its center. The overall thickness of the lens assembly is a minimum in this perfectly centered condition.

The mounting task is to put the lens in the centered condition and keep it there. The fact that the minimum overall lens thickness occurs when the lens is perfectly centered suggests that the lens will tend to center itself if assembled with the optical axis vertical. Slight vibration and tapping should encourage the assembly to settle into the centered position. Unfortunately, there are cases where self-centering does not occur at all, as illustrated in Fig. 2. This doublet will not self-center because surfaces 2 and 4 are completely concentric. The second element and spacer may rotate on surface 2 without changing the overall lens thickness. This is an unlikely case but, whenever two centers of curvature are close together, the self-centering condition is weak and not dependable. To ensure cen-

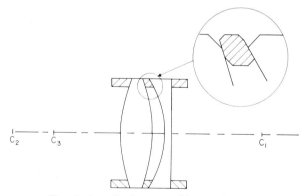

FIG. 1. A perfectly centered lens and spacer.

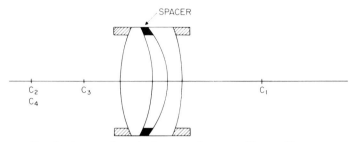

FIG. 2. A case where the lenses and spacer will not self-center.

tering, some other way has to be devised to locate the lenses in a centered position. The usual method is to use the edges of the lenses. Each lens and spacer is edged to a specific diameter. The cylindrical edge surface must be generated on an axis concentric with the optical axis. An assembly of perfectly centered lenses and spacers may be mounted in a true cylindrical barrel, and the assembly will be a centered lens system.

In practice lenses are never perfectly centered and edged to precise diameters. There must be clearance between the lenses and the barrel, and this creates the possibility of decentered elements. It is necessary therefore to specify tolerances. The following three are required:

(1) The inside diameter of the barrel and tolerance, i.d. $\pm \begin{smallmatrix} \text{tolerance} \\ 0 \end{smallmatrix}$,

(2) the outside diameter of the lens, o.d. $\pm \begin{smallmatrix} 0 \\ \text{tolerance} \end{smallmatrix}$, and

(3) the clearance between the o.d. of the lens and the i.d. of the barrel.

When the tolerances are specified, the maximum displacement of the optical axis is one-half the sum of the clearance plus the o.d. and i.d. tolerances. For example, a lens-mounting specification may read:

Barrel i.d. $= 50 \pm \begin{smallmatrix} 0.025 \\ 0 \end{smallmatrix}$ mm

Lens o.d. $= 49.975 \pm \begin{smallmatrix} 0 \\ 0.050 \end{smallmatrix}$ mm

Minimum clearance $= 0.025$ mm

There is a possibility that the lens could be displaced by $(0.025 + 0.050 + 0.025)/2 = \pm 0.050$ mm. The same tolerancing also applies to the spacers.

Section VII provides guidelines for tolerances, and the next few sections discuss some of the details of the design of barrels and spacers.

III. THE COMPONENTS IN A LENS MOUNT

A. The Barrel

The barrel i.d. should be that of a true cylinder. The wall thickness is important to help ensure roundness of the bore. If the cell warps after release from the turning operation, the straightness and roundness of the bore may be affected. If the cavity takes on an elliptical shape, the clearance will be determined by the minor axis. It is difficult to machine a long internal cylinder without some flare at the open end, so if the clearance is small this may cause problems.

Whenever possible, the barrel should be a straight i.d. barrel, for it is difficult to make the various i.d. cuts coaxial. If there is no alternative to having stepped i.d.'s, they should be bored from a single lathe setting. Many photographic lens barrels are made in two parts, a front and back, which screw into a shutter. This can be done if the tolerances are liberal but should not be done in a precision lens system.

Aluminum (2017) is a favorite material for lens barrels because of its machineability and light weight. Aluminum for the barrel for large precision lenses requiring small clearances is a questionable choice of material, for it tends to warp during machining. Lens barrels are usually made from bars or tubing. For some applications castings should be considered. An excellent aluminum for castings is Precedent 71-T5 (*Alloy Dig.*,1962).

Precision lens barrels are often made with brass (Series 360) or stainless steel (Cres 400) (Westort, 1977), but brass is used more often because it is more easily machined. For high precision the brass parts should be rough-machined and heat-treated before the final cuts are made.

Stainless steel of the Cres 400 series is excellent for lens barrels. With

TABLE I

EXPANSION COEFFICIENTS OF GLASS AND METAL
PARTS USED IN LENS MOUNTING[a]

Material	α (mm/mm C°)
BK7	71×10^{-7}
Stainless steel (304)	170×10^{-7}
Stainless steel (404)	86×10^{-7}
Brass (360)	205×10^{-7}
Aluminum (2017)	253×10^{-7}
Cast iron	120×10^{-7}

[a] From Richey (1974).

diamond turning, grinding or honing clearances as low as 0.01 mm can be achieved. This stainless steel takes an excellent black oxide coating.

The specified clearance has to take into consideration the difference in expansion between glass and metal. Table I lists the expansion coefficients for BK7 glass and metals used in lens mounts.

The barrel usually has to have a flange or reference surface perpendicular to the optical axis and located a prescribed distance from the lens focus.

B. THE SPACERS

The spacers should in principle have smooth, spherical curves ground into their faces in order to fit the lens surfaces accurately, but this is seldom done because of the costs involved. The next best spacer is one cut with cone surfaces which contact the spherical surfaces on a tangent plane. This type of spacer requires a different cone angle setting for each lens, which is expensive. It is also difficult to check the dimensions of a cone-shaped spacer in order to ensure the proper air space between the lenses. It is more practical to use a 45° bevel for all the spacers (see Fig. 3). The bevel is cut to ensure that the metal corners are definitely obtuse (135°). An obtuse corner is desirable, because the edge where the lens contacts the bevel is less likely to be damaged by a nick. Care must also be taken to make sure that the machining does not leave a burr on the corner the lens is to contact. The cut should be made toward the metal on such a edge. The designer should compute where the lens will contact the bevel and indicate this to the machinist.

A common lens design employs a convex–concave air space as shown in Fig. 4. This combination is used to balance high-order aberrations. The space between the elements and the centering of the two surfaces on the

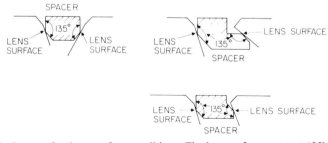

FIG. 3. Spacers for three surface conditions. The lens surfaces contact 135° corners on the spacers.

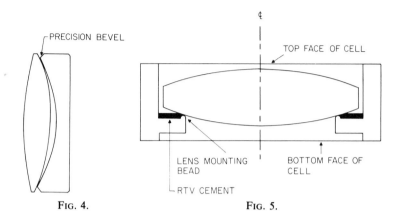

FIG. 4.   FIG. 5.

FIG. 4. A glass-to-glass, air-spaced lens.

FIG. 5. A cell design for cementing.

optical system axis are sensitive, and the air space is usually so thin that it is impractical to make a metal spacer. The solution to this problem is as follows:

Use three thin, shim stock spacers spaced by 120°—a delicate and expensive procedure.

Some designers specify glass-to-glass contact as shown in Fig. 4. A precision bevel on the lens is required in order to maintain the air space, and this is an expensive operation if the tolerance is small. An optical shop is not a good place to maintain precision dimensions, since most shops are not equipped to grind surfaces to precision dimensions. This type of lens mounting can be made cost-effective if there is sufficient production to warrant the proper tooling. It is not recommended if the bevel has to be ground by hand.

Some designers call for a bevel on the glass and a metal spacer to avoid glass-to-glass contact. This arrangement, however, still requires a precision bevel on the lens and means an extra spacer.

The lenses may be cemented in individual cells and spaced by the cells, as shown in Fig. 5. Mounting lenses in cells is discussed in Section III, D.

C. The Retainer Rings

Spacers are used to space the elements, but sooner or later a retainer must be inserted to hold the lenses in the barrel. Retainer rings are usually threaded. The lenses are inserted in the barrel with spacers, and the retainer ring screwed down until it just contacts the lens. The barrel is then tapped or vibrated to encourage the lenses to settle into the self-aligning (minimum-thickness) condition. Then the spacer is carefully tightened down on the lens. This procedure is widely used in lens mounting, but it is necessary to make sure that the resulting centering is adequate. The threaded retainer ring is an economical way to hold the lens stack in place, but it has several limitations:

(a)   The mating threads always have some play, which means that the retainer ring is slightly cocked as it is turned in. When it contacts a perfectly centered lens, it hits on just one side of the lens, causing the lens to tilt slightly.

(b)   The retainer ring is usually kept thin in order not to waste too much space. It then is difficult to keep it round and tilt is introduced into the lens it contacts.

(c)   If the retainer ring is tightened too much, it may warp the lenses.

(d)   The threads are often fine and easily crossed or stripped.

Retainers can be made with loose-fitting threads and a precision pilot diameter which fits accurately in the barrel. This type of retainer, however, is considerably more expensive to make and has to be installed and handled with care to avoid damage to the mating surfaces.

Some retainers are merely tight-fitting rings pressed in the barrel. They may be cemented or soldered, which involves a delicate operation not used often; however, with the new ultraviolet setting cements, cementing lenses in place is becoming more common.

D. THE LENS CELLS

Many precision lenses use individual cells for each lens. Figure 6 illustrates a lens mounted with the elements cell-mounted.

A typical cell is shown in detail in Fig. 5. The following items must be considered in this type of cell.

(a)   The o.d. has to be cut to precise diameter, for it to be inserted in a straight-bore sleeve or barrel. Since there is no self-centering in these cells, any difference between the cell o.d. and the barrel i.d. can result in a direct decentering of the lens.

(b)   The top and bottom faces of the cell must be cut perpendicular to the o.d. surface. An error in this cut will cause an error in spacing and can result in tilting of the element.

(c)   The lens mounting bead must be cut so that the surface of the bead contacting the lens is a circle in a plane which is also perpendicular

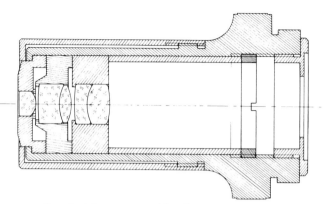

Fig. 6. A lens system with cell-mounted lenses.

to the optical axis. The bead should be cut as a portion of a sphere of known radius, so that the lens surface contacting it can be accurately located with respect to the bottom face. The mounting bead also has to be centered on the optical axis.

(d) The cell must not warp when released from the lathe chucking tool. Freedom from warping is dependent on the quality of the chucking procedure and the stability of the cell metal.

(e) The only surface not requiring precision machining is the inside surface of the cell, although consideration must be given to this surface. One of the major advantages of cell mounting is that the centering of the lens need not depend on the edge diameter of the lens. If the critical surfaces on the cell are made accurately, the lens can be centered in the cell without the edge contacting the cell. The lens can be cemented in place with a layer of cement as shown in Fig. 5. The inside of the cell must be cut to allow clearance from the lens and space to put the cement into the cell before the lens is positioned.

A precision cell requires excellent chucking of the bottom face and completion of all the precision cuts at one setup. These cuts are made on a precision lathe with a horizontal spindle. When all the critical surfaces are running smoothly, it is desirable to mount the lens in the cell and with a vacuum pull it against the mounting bead; then, by adjusting the lens, the wobble in its top surface can be eliminated. Once both surfaces run true, the cement has to be inserted and hardened. This is a difficult and time-consuming task to perform on a precision lathe, so usually the cell is removed and mounted on a special vertical spindle.

Some lens makers spin the lens into the cell before the cell is removed from the chuck. The cell has a spinning shoulder protruding above the lens. While the lens and spacer are spinning on their true center, the operator uses a spinning tool to bend the shoulder onto the lens. This action, if carried out properly, has a self-centering effect on the lens and results in a lens well centered in the cell. This method ties up a skilled operator and an expensive machine for a considerable time, so it is expensive and is seldom used.

When the lens is to be cemented into the cell, it is usually done on a vertical spindle. This means the cell has to be removed from the lathe and inserted into a vertical spindle equipped with a vacuum attachment. The vertical spindle should be as precise as the original lathe spindle. A mating cell to hold the lens cell is machined on the vertical spindle which is usually a plate of stable brass. Such a chuck is shown in Fig. 7. The lens cell sits snugly in the mounting chuck. The top edge of the cell and the o.d. should then be checked with an electronic indicator gauge to make sure the cell is seated properly. The vertical spindle should be rotated slowly enough to view the indicator movements but fast enough to ensure that the bearing is running true. Any movement of the indicator indicates one of the following conditions:

The cell is not seated properly. It should be removed, cleaned, and inspected for burrs.

The lens cell is not perfectly parallel.

The vertical spindle wobbles. It should be checked at various speeds to determine the optimum performance.

When the lens cell is seated properly, the lens mounting bead should be indicated. Any differential wobble with respect to the cell may be taken out with a tool mounted on the spindle table. A final touch-up cut also removes dents or scratches on the bead.

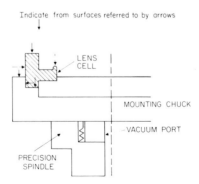

FIG. 7. A lens cell mounted on a precision spindle for centering and cementing. The arrows point to surfaces which should run true as checked with an indicator.

Once the mounting bead runs true, the lens is ready for cementing. There are several cements available for cementing optical components. For precision mounting room temperature vulcanizing (RTV) cements are commonly used. These cements never set up completely hard. They do, however, hold the lens against the cell without strain. The cement has elastic memory, so if the lens receives a severe shock, it may shift slightly but will return to its initial position. There are at least two excellent RTV cements for optical work: General Electric RTV-566 and Dow Corning RTV-Clear 732. These cements are stable over large temperature ranges and do not outgas under vacuum. They become tacky in approximately 40 min and set up in 4 hr, but take a week or more to become completely cured.

Some experience must be gained in designing cells for cementing. The cement cavity should not be too large, since a longer curing time would be required. The cell design should allow for contact of cement on the curved surface so that the curved surface of the lens is pulled down to the mounting bead. One should not cement around the edge of the lens. A small amount of cement should be put into the cell before the lens is inserted. Since the cement hardens slowly, it is necessary to provide another means to hold the lens in its centered position during the curing. This can be done with mechanical clips or by using heated beeswax cement applied at three points around the diameter of the lens. This technique enables one to remove the cell from the spindle before the RTV cement is hard. The lens and cell may then be left to harden without tying up the expensive spindle.

After the initial application is cured the cavity may be more completely filled with a second application of RTV cement.

It is feasible to use ultraviolet setting cements to cement lenses in cells, but they tend to set up much harder and may distort the lens.

Cementing lenses in cells is becoming an increasingly popular way to assemble lenses. It practically eliminates the need for centering the lenses in the optical shop and can be done in a clean room environment which is much more suitable for the precision measurements required. Cementing the lenses in cells also provides an excellent means of checking the cell-to-lens-surface dimension. The spaces can then be set accurately.

When the lenses are mounted in cells, they may be assembled in a barrel. Since the barrel and cell are both of metal, the clearance between them can be less than between a glass lens and barrel. If the fit is close, the lens cells must be inserted carefully. A few cells are stacked vertically as evenly as possible. The barrel is then lowered over them. The barrel should never be forced down, for if the cells become cocked there will be a problem. The cells and barrel can easily become damaged in the process of removing them. One should also disassemble the lens as few times as possible to avoid wear on the barrel and cells.

Tightening the retainer ring to hold the cells is also a delicate task. Even when the retainer ring is piloted, the retainer cell tends to turn the lens cells. This causes the lens to climb up inside the barrel and become decentered. The optimum retainer cell would not turn as it is tightened, but such a design is seldom used because of the increased cost.

IV. LENS CENTERING

A. BELL CENTERING (MECHANICAL)

The most widely used methods for mounting lenses depend on the lenses being centered and edged to a specified diameter. When the lenses are made in large production runs, they should be centered when the glass blanks are accurately edged prior to generating the curves. Unfortunately, this is not the case, for there are too many steps involved in the process. Extremely tight tolerances at every step add to the tooling cost. In practice, edging and centering are a final step in the lens-making process. The most common centering machines used to edge and center lenses are called bell centering machines. They employ a mechanical procedure using the principle of self-centering, in which the lens is squeezed between two bell-shaped tools (Fig. 8). The centering machine has two coaxial bearings, and the bell tools are mounted on these bearings. The bells edges must run true as they are rotated.

Bell centering machines are widely used in industry, and considerable

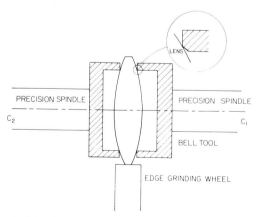

FIG. 8. Principle of mechanical centering on a bell centering machine.

literature and first-hand experience are available. The following points about bell centering should be noted.

There are several bearings in bell lens-centering machines. There are two bearings holding the bells, and the contact surface of the bells must run true with these bearings. The grinding wheel bearing is displaced but must be parallel to the bell bearings.

The grinding wheel must run true on its bearing and, if precision bevels are required, the wheel must have precision mating bevels.

The machine has features that hold the lens while the operator adjusts the lens into the centered position. It is then clamped tightly enough to ensure no movement during the grinding process. These adjustments require maintenance to achieve the mixture of requirements. If the clamping is too tight while the lens is being centered, the lens may not be centered and may damage the bell surface.

Some thought should be given to the cut on the bell tool. On a production run the bell should be cut with a curved surface, so that the lens contacts the surface tangentially. This ensures less wear on the tool and diminishes the chance of the surface receiving a nick which will decenter the lens. If a spherical surface is too expensive, the next best thing to use is a cone-shaped surface. The angle should be chosen to ensure contact away from the edges, which are easily nicked.

The bell tool should contact the lens outside the clear aperture, because some scratching is likely to occur during the centering of the lens.

Centering lenses on a bell centering machine depends on "squeezing out" the difference in edge thickness. Approximately 0.010 mm can be squeezed out. With some operator skill this can be reduced somewhat. An edge thickness difference of 0.010 mm introduces a beam deviation of the optical axis given by

$$\delta = 0.010(n - 1)/D \tag{1}$$

where D is the lens diameter in millimeters and n is the index of refraction of the glass (Smith, 1966). The precision of centering, however, depends on the curvature differences between the two surfaces of the lens. When the curvature differences are large, the lens is "centerable." A small decentering causes a large edge thickness difference. Manufacturers of bell-type machines provide curves indicating the centerability of lenses. These curves plot the equation

$$C = 0.5y/r$$

where y is the semiaperture of the lens, r the radius of the surface, and C the centerability. If C is less than 0.15 (corresponding to a contact angle difference of 17°), the lens is considered to be just centerable. If C is less than this, it is not centerable.

Bell centering is not adequate for some precision lenses, and in these cases it is necessary to resort to optical centering.

B. OPTICAL CENTERING

An optical centering machine differs from a production bell-type centering machine in that it uses only one bell. The principle is to mount the lens on a single bell and hold it in place with wax or a vacuum. The lens is then moved about the bell tool while the lens rotates. Centering is accomplished when the reflected images from both surfaces remain stationary as the lens rotates. The following features of this method should be noted:

(a) The bell tool.   The surface contacting the lens must run true and have no dents or scratches. The shape of the surface is also important.

(b) Holding the lens on the bell tool.   A vacuum appears to be an ideal approach if the spindle can be equipped with a vacuum attachment. A vacuum, however, has limited holding power, and it is necessary to have a reduced pressure difference while the operator is moving the lens about to center it. Once centered, the full atmospheric pressure difference can be used to hold the lens. Care must be taken during the glass-grinding phase to not move the lens. If a vacuum is used, light cuts must be taken on small lenses or they will move.

When a vacuum is used, it is preferable to employ a vertical spindle. The lens can then be moved into a center position before applying the vacuum. Once the lens is centered, the vacuum can then be used to hold the lens in place.

(c) Holding the lens with wax.   The traditional optical centering method is to use a horizontal spindle axis and to hold the lens on the bell tool with a hard wax that softens with heating. The brass bell tool is heated with a bunsen burner, and the wax is pressed against the bell edge until a thin layer of wax covers the surface that contacts the lens. The lens is then pressed against the bell tool. The wax is kept warm enough to hold the lens and allows the operator to move it about on the bell as it is rotated. By using a wooden stick the operator can move the lens into a centered position which is reached when both surfaces of the lens run true. This operation requires considerable practice and skill in order to maintain the proper heat on the brass tool so that the lens stays on but can be moved easily enough to achieve delicate centering. This kind of skill is becoming incompatible with our modern society, for it requires the full-time attention of an operator. It makes the operator completely dependent on this single specialized skill and vulnerable to the ups and downs of the economy.

V. TESTS FOR DETECTING DECENTERING

When optical centering is used, it is necessary to observe the decentering in order to remove it. There are three basic ways to do this.

A. DEVIATION OF A BEAM PASSING THROUGH THE LENS

A beam of light may be passed through the center of the lens, and any wobble will be observed in the image as the lens rotates. This requires a hollow, unobstructed bearing, an optical aid to view the transmitted image, and setup time for each change in lens type to be centered.

When the lens is not centered on the optical centering tool, it may be positioned as shown in Fig. 9. While the lens rotates, an indicator moves over the total indicated range (TIR). One can then consider that the first surface has been tilted by the angle TIR/D. This wedge will deviate an axial beam by $\delta = (n - 1)TIR/D$, where D is the clear-aperture lens diameter [Eq. (1)].

B. OBSERVING MOTION OF A REFLECTED IMAGE

One may observe the reflected images from the rotating lens. This method is more sensitive than the previous method for, if the surface is tilted by TIR/D, the central beam will be deviated by $2TIR/D$, which is

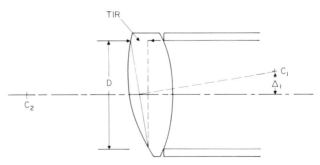

FIG. 9. Diagram showing a lens improperly mounted on a bell centering machine.

four times as large as by transmission. An experienced operator can detect as little as 6 sec of deviation. Increased precision can be achieved with optical aids. It is important to note that the reflected images from both surfaces must be observed, for it cannot be assumed that the lens is in contact with the bell when it is waxed on.

One problem with viewing the image motion with reflected light is that the images are located at a wide variety of distances and it is difficult to focus viewing optics for a range of positions. Figure 10 shows an optical system which can be used for optical centering. A point source is imaged by a low-power microscope at point A which is located fairly close to the surface being inspected for decentering. The reflected image is formed at B. As the light returns through the lens, it focuses at C, the eyepiece focal plane. Any surface tilt as the lens rotates causes motion at C, which is magnified by the eyepiece. The curvature of the surface being inspected influences where the eyepiece must be focused. If the return image cannot be reached with the provided focus range of the eyepiece, it can be brought within range by positioning the image A with respect to the surface being tested. For example, the reflected image B may be located in the objective. The eyepiece would then have to be moved all the way up to the objective, resulting in problems with the reflex plate. This can be remedied by moving A close to the surface under inspection. The sensitivity of the system can be increased by splitting the return beam as shown in Fig. 11. Since one beam reflects off a roof edge, there is an extra reflection. If the surface tilts as the lens rotates, the two images will rotate in two circles which contact at only one point. When a He–Ne laser is used as a source, the two return images form interference fringes. Any relative motion of the two images reflected from the rotating surface causes perceptible fringe movement. This method uses only a small portion of the surface to detect surface movement, but it is surprisingly sensitive when used with a prism and a laser source. It is a superior method for small lenses. For larger lenses ($D = {}> 100$ mm) an electronic or air gauge indicator is easier to use and more sensitive.

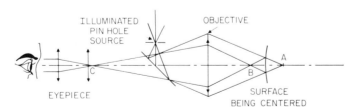

FIG. 10. Optical system used to observe decentering of a rotating surface.

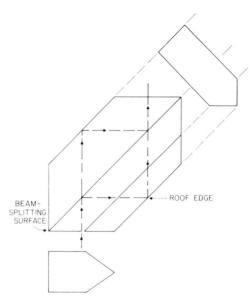

BEAM-
SPLITTING
SURFACE

ROOF EDGE

FIG. 11. Optical system used to double the centering sensitivity.

C. Mechanical Indicators

Centering can be detected by using mechanical or air gauge indicators. It is necessary to determine centering on both sides of the lens, and sometimes this is not easy to do when the lens is waxed onto the bell tool. For large lenses mechanical centering is as accurate as optical centering, but for small lenses (< 50 mm) optical centering is the preferred method.

VI. TESTING A LENS ASSEMBLY FOR CENTERING

After a lens is mounted, it should be checked for centration of the optical elements. The standard way to do this is to use a point source of light located near the optical axis of the lens. One then views the images reflected from the surfaces as shown in Fig. 12. When S_0 is located on the optical axis of a centered lens, all the reflected images (I_1, I_2, I_3, I_4) lie on the optical axis and appear so as the observer moves horizontally or vertically away from the optical axis. If the object source is displaced from the optical axis to O', each of the images will be displaced from the optical axis. The images I_1 and I_3 will move in the same direction as O, and the images I_2 and I_4 will move in the opposite direction.

When several lenses are mounted, the images cannot all line up in a

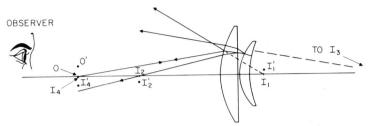

FIG. 12. In a centered system the reflected images appear on a straight line. When the object point O is displaced vertically from the optical axis, images I_1 and I_3 move off the axis in the same direction as O. The images I_2 and I_4 move in the opposite direction.

straight line if any of the centers of curvature are off-center. If the source of light is sufficiently intense, multiple reflections also occur and some of the images are even more sensitive to tilt. To enhance the sensitivity the reflected images may be viewed with a telescope, using a reticle. One problem is that the telescope has to be capable of focusing from $+\infty$ to $-\infty$ in order to inspect all the reflected images. This focusing requires moving lenses in the telescope, and the precision depends on the mechanical precision of the lens movements.

An added refinement is to use a laser as the point source; this provides a significant gain in that one obtains interference effects. With laser light each surface reflects a coherent wavefront. All the reflected wavefronts interfere with each other and cause a complex interference pattern. When the lens surfaces are perfectly aligned, all the reflected wavefronts interfere to form concentric interference fringes. By viewing this pattern with a moving lens telescope one can observe these interference patterns at any position in space. The viewing system does not affect the relative po-

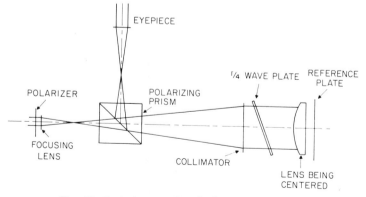

FIG. 13. Optical system for aligning optical systems.

sitions of the fringes, therefore, the absolute straightness of the moving telescope viewing optics is not critical. The image from any surface may be focused and will appear as a point source surrounded by a set of interference fringes. When the point is decentered with respect to the surrounding fringes, an error in centering has occurred. An optical system which could be used to make this observation is shown in Fig. 13. This principle can be used to statically center lenses and mount them without depending on the precision barrel (Hopkins, 1976).

VII. TOLERANCES FOR CENTERING

A. A Method for Estimating Tolerances

The effect of decentering can be appreciated by considering each lens in the system as made up of two plano lenses as shown in Fig. 14a. The lens shown is centered on an optical axis. An axial object pointing to the

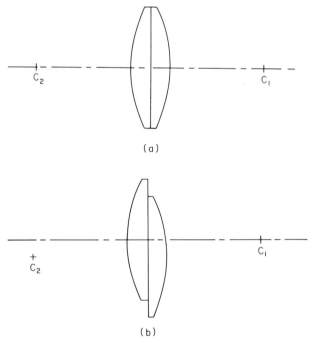

Fig. 14. A concept illustrating the effect of a decentered surface. (a) A centered lens may be considered as made up of two plan convex lenses, with the surface centers C_1 and C_2 on the system optical axis. (b) This diagram simulates the effect of C_2 being displaced from the optical axis.

left of the lens will be imaged to a point O′, and the image will appear rotationally symmetric around the optical axis. When the second surface is displaced by Δ as shown in Fig. 14b, the displacement introduces a wedge in the beam. The original optical axis is bent downward and, in addition, the symmetry of the image is destroyed. The upper ray is incident at a slightly larger angle than the lower ray, so it is bent more. The two rays then intersect below the ray through the center of the lens, resulting in coma. The expression for the optical path difference (OPD) caused by the displacement is

$$\text{OPD} = (n - 1) \left(\frac{y}{r} + \frac{1}{2} \frac{y^3}{r^3} + \cdots \right) \Delta \qquad (2)$$

The first term represents the tilt in the image, while the second term represents the coma. Tilt in the image plane can be corrected by tilting the receiving plane, but this introduces a slight amount of keystone distortion. Usually it is not noticeable, however, in the case of precision lenses used for metrology, it must be reduced to a small value.

Displacing a single surface introduces a tilt of the surface and a slight thickness shift. These two effects can be accounted for in estimating tolerances. It really is not necessary to investigate tilt and displacement in order to tolerance a lens. Some may argue that the above example is not valid for a real lens where the two surfaces are together in one piece of glass. However, when the single piece of glass is rotated around C_1, the second center of curvature C_2 rotates to the displaced position as in Fig. 15. The second surface is axially shifted by an extremely small amount.

Equation (2) may be extended to include shifts of both surfaces of the lens. The coma term of the optical path introduced by the lens displacement is

$$\text{OPD} = (n - 1) \left(\frac{y^3}{2r_1^3} \Delta_1 - \frac{y^3}{2r_2^3} \Delta_2 \right) \qquad (3)$$

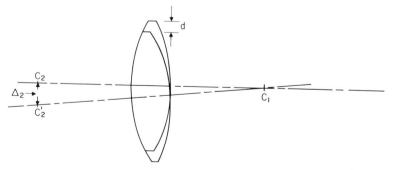

FIG. 15. Diagram showing that a tilted lens is equivalent to a decentering of the second surface.

A displacement of C_2 alone can occur only by rotation around C_1. When C_2 is displaced by Δ with rotation around C_1, the lens as a whole is displaced by (see Fig. 15)

$$d = [(r_1 - t)/(r_1 - r_2 - t)] \Delta_2 \tag{4}$$

where t is the lens thickness and r_1 and r_2 are the radii of curvature of the two surfaces. Rotation around C_2 to introduce a displacement of Δ_1 in C_1 results in a lens displacement of

$$d = -(r_2 + t)/(r_1 - r_2 - t) \tag{5}$$

The exact position of a lens in a lens mount is almost impossible to predict, so it is not significant to attempt to evaluate all these effects. The main purpose of the above concept is to enable one to pinpoint the sensitive areas and alert designers and manufacturers to consider the assigned tolerances carefully.*

B. Tolerancing a Telescope Objective

A telescope objective can serve as an example of tolerancing for centering. The objective specifications are shown in Table II.

The focal length of the objective is 10 cm, and it operates at $f/3.5$. The extreme marginal rays pass through the lens at $y = 1.4$ cm. If the tolerance is set at OPD = $\lambda/10$, then the tolerable individual displacements for the lens surfaces are given in Table III.

TABLE II

SPECIFICATIONS FOR AN $f/3.5$, 10-CM-FOCAL-LENGTH
TELESCOPE OBJECTIVE

$C = 1/r$ (cm^{-1})	t (cm)	n	V
		1	
0.16788			
	0.5	1.511	63.5
−0.24494			
	0.01	1	
−0.24379			
	0.30	1.649	33.8
−0.07386			

---

* Equation (2) is a third-order approximation. Modern lens design programs have features which can tabulate all the decentering and tilt tolerances. Equation (5) is provided merely to illustrate the considerations which go into determining tolerances and also shows how the sensitivity varies as $(r/y)^3$.

TABLE III

TOLERABLE SURFACE DISPLACEMENTS TO INTRODUCE $\lambda/10$ OPD

	Surface number			
	1	2	3	4
Δ	30.4λ	9.7λ	9.84λ	354λ

The most sensitive surface is the second surface. Since λ is approximately 0.5 μm, the displacement of C_2 must be no more than 0.005 mm. There are then the following cases that can occur in the first element:

(1) Displacement of lens; $d = 0.0020$ mm tolerance
(2) Rotation about C_1; $d = 0.003$ mm tolerance
(3) Rotation about C_2; $d = 0.006$ mm tolerance

The smallest tolerance is 0.002 mm. To meet this tolerance, the lens and cell specifications have to be written as follows:

Lens o.d. $= -0 + 0.001$ mm
Barrel i.d. $= -0.001 + 0$ mm
Clearance $= 0.002$ mm

This tolerance assumes that the lens is centered perfectly. When the lens is decentered, this figure must be subtracted from the mounting tolerance. A lens with an edge decentered by 0.002 mm would use up the entire mounting tolerance.

Most lens designers agree that the likely position for the lens is nested against the third concave surface. If it is assumed that the negative lens is mounted centered on the barrel mechanical axis, the positive lens will be twisted about C_2 and the tolerance will be 0.006 mm for the allowed displacement.

It is doubtful that anyone can really state what will happen when the lens is mounted with clearances. The tolerance to assign depends on many factors such as the capabilities of the optical shop and the machine shop, the assembly procedure, the cost of the lens, the number of lenses to be made, and a careful analysis of the initial statement that the lens should have less than a $\lambda/10$ OPD. It is easy to specify a tolerance without understanding the consequences. The above specification of $\lambda/10$ is difficult to achieve, and in any case, one assumes the lens will position itself. To meet this specification, cell mounting is definitely the preferred method.

Assigning tolerances is an extremely important but difficult task. It can only be approached wisely when the designer, the manufacturer, and the user clearly understand the important trade-offs. It can be expected that in practice experience will suggest alterations in the specifications. If each person involved in the trade-offs takes a "play it safe" approach, costs will soar. There are also few cases where a $\lambda/10$ tolerance can be justified.

C. Typical Tolerances

Table IV lists some of the wavefront (OPD) specifications found in various types of optics.

Lens centering tolerances are usually specified by the angular deviation of a beam passing through the mechanical center of the edged lens. In Fig. 9 a lens was shown centered on a bell chuck where the outside surface did not run true, and there was a measurable TIR. TIR/D is the wedge introduced in the lens if it is edged in this condition. The wedge causes a deviation of the optical axis equal to

$$\delta = (n - 1)\text{TIR}/D \qquad (6)$$

Table V lists a few guidelines for centering tolerances in commercial lenses in terms of δ.

From the geometry in Fig. 9 it can be seen that the center of curvature C_1 is displaced by

$$\Delta_1 = r_1\,\text{TIR}/D \qquad (7)$$

If the lens were inserted in a close-fitting i.d. cell, then it may assume this same orientation in the lens barrel. If so, the deviation of the optical axis is given by Eq. (6), and it follows that

$$\delta = (n - 1)\,\Delta_1/r_1 \qquad (8)$$

TABLE IV

TYPICAL SPECIFICATIONS FOR WAVEFRONT QUALITY

	Wavefront tolerances
Photographic lenses	$\pm 1\lambda$
Telescope objectives	$\lambda/4$
Microscope objectives	$\lambda/4$ to $\lambda/8$
Step-and-repeat lenses for photolithography	$\lambda/4$ to $\lambda/10$
Lenses used in laser applications	$\lambda/10$

TABLE V

TYPICAL SPECIFICATIONS FOR COMMERCIAL LENSES

Low-cost	$\pm\frac{1}{2}°$	Precision	±30 sec
Commercial	±6 min	Extra precision	$1-30$ sec

In the doublet example a tolerance for the displacement of C_2 was 0.003 mm. This tolerance would require the lens to be centered to within 8 sec and would be classified as an extra-precision tolerance.

VIII. FORMULAS FOR COMPUTING SURFACE SAG

In designing mounts for optical systems it is necessary to compute the thickness of the spacers in order to space the elements properly. This involves computing the sag of the optical surfaces. The problem is illustrated in Fig. 16. The following formulas are useful for computing the sag Z of a sphere.

$$Z = r(1 - \cos \Theta), \quad \text{where} \quad \sin \Theta = y/r \tag{9}$$

which can also be written:

$$Z = r[1 - \sqrt{1 - (y/r)^2}] \tag{10}$$

It is sometimes useful to expand the expression in a power series:

$$Z = (y^2/2r) + (y^4/8r^3) + (y^6/16r^5) + (5y^8/128r^7) + \cdots \tag{11}$$

For most calculations the first two terms of the series are adequate for mechanical calculations. With modern hand calculators Eq. (9) is probably the simplest formula to use.

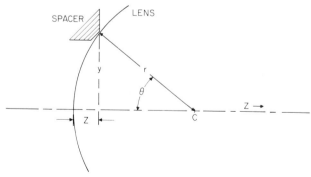

FIG. 16. The geometry involved in calculating the sag of a rotationally symmetric surface.

ACKNOWLEDGMENTS

Most of the useful information supplied in this chapter was obtained through my association with many people at Tropel, Inc. Mr. Joseph Kleiber was particularly helpful in discussions on lens mounting and tolerancing.

REFERENCES

Hopkins, R. E. (1976). *Opt. Eng.* **15**(5), 428.
Richey, C. A. (1974). Aerospace mounts for down to earth optics. *Mach. Des.* December 12, p. 121.
Smith, W. (1966). "Modern Optical Engineering," p. 420. McGraw-Hill, New York.
Westort, K. (1977). Design and fabrication of high performance relay lenses, Optical Society of America Workshop, Danbury, Connecticut.
(1962). *Alloy Dig.*, Aluminum alloy A1-113.

CHAPTER 3

Aspheric Surfaces

ROBERT R. SHANNON

Optical Sciences Center
University of Arizona, Tucson, Arizona

55

I. DEFINITION OF ASPHERIC SURFACES

A. GENERAL COMMENTS

An aspheric surface is, by simple definition, an optical surface that is not spherical in form. More than one radius of curvature is assigned to the surface, and the surface curvature generally varies across the aperture in a manner defined by an analytic formula, either explicit or implicit. In the most general case certain aspheric surfaces can be defined only through a table of numbers that defines the sag or shape of the surface locally at each coordinate.

It is well known that the focal length or power of a spherical surface is defined by a single parameter, the radius of curvature of the surface, along with of course the index of refraction of the media surrounding the surface. Aberrations inherent in the refraction or reflection of a bundle of rays indicate that the power parameter is a paraxial one only. Well-known but generally nonanalytic aberrations result from the passage of a ray bundle, and the nonparaxial surface power varies across the surface, except in very special cases. The use of a nonspherical surface permits a powerful and useful parameter, selective variation in the refractive or reflective power of the surface as a function of the coordinate location on the surface.

The most familiar application of an aspheric surface is the use of a reflecting parabola to correct the spherical aberration inherent in the use of a spherical reflecting surface to form the image of a point at infinity. The use of aspherics, however, has a wider range of applicability than just this simple example, as will be seen later.

In this chapter, the methods of defining useful general aspherics are discussed. The aberration-introducing and -correcting properties of aspherics are developed, and applications for aspherics demonstrated. Approaches to the design of systems with aspherics are described and, finally, methods of producing, testing, and specifying the purchase of aspherics are reviewed.

B. GENERAL ROTATIONALLY SYMMETRIC SURFACES

The most common general aspheric is that which can be described as a power-series departure from a simple base surface, such as a sphere. The base surface may well be a conic surface, at least in the design stages, but the aspheric departure generally is noted relative to some specific spherical reference surface.

A useful method of describing a spherical surface is in terms of its sag. Figure 1 shows a definition of the coordinate notation being used. By convention, the z axis coincides with the axis of rotational symmetry. The x- and y-coordinate axes form a right-handed set as shown. Again by convention, but a less generally observed one, the y–z plane is chosen as the meridional plane for the optical system when describing the off-axis imaging properties of the system. The z sag of the spherical surface is then given by the formula

$$z = \frac{c(x^2 + y^2)}{1 + \sqrt{1 - c^2(x^2 + y^2)}}$$

where c is the curvature, that is, the reciprocal of the radius of curvature, of the surface. Generally, x and y are limited to a circular region with a specified clear aperture, and z is much less than the maximum value of the allowed aperture.

This formula appears complicated but is actually the most convenient method for computing the nature of the surface for ray tracing, lens-mounting calculations, and measurement of lens thickness. For the case when the clear aperture is much less than the radius of curvature, the formula reduces, in close approximation, to the well-known parabolic sag formula for a parabola of focal length $f = r/2$:

$$z = \frac{x^2 + y^2}{2r}$$

although use of this approximation should be made with care.

To change the surface to a general rotationally symmetric aspheric, a power series in even terms of the aperture is added to the basic spherical surface. In order to carry this out with a minimum of confusing notation,

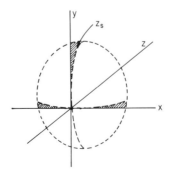

FIG. 1. Coordinate set used for describing aspheric surfaces.

define an aperture parameter

$$\rho = \sqrt{x^2 + y^2}$$

where ρ is required to be between zero and the value for the clear-aperture radius. Then the aspheric sag formula takes the form

$$z = \frac{c\rho^2}{1 + \sqrt{1 - c^2\rho^2}} + (ad)\rho^4 + (ae)\rho^6 + (af)\rho^8 + (ag)\rho^{10}$$

where the definition of the spherical and aspheric portions of the sag should be obvious. Figure 2 should help in determining the meaning of the formula. As a general practice, terms no higher than the tenth order in ρ are used in design.

The y–z section of the aspheric shown in Fig. 2 is somewhat of a simplification for the surface being described by the formula. The actual surface is a rotationally symmetric surface and has the local deviation described in Fig. 3. The shaded portion of this figure shows the deviation from the base spherical surface. The fact that the aspheric surface cannot be fit by a simple spherical surface of any radius should be clear from the contour lines shown on the figure.

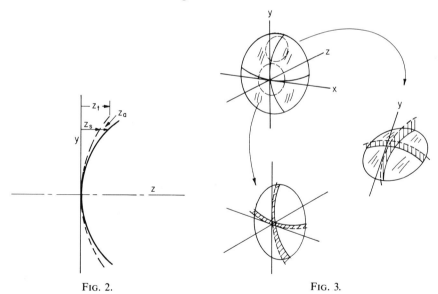

FIG. 2.   FIG. 3.

FIG. 2. Definition of the total surface sag z_t, which is the sum of the base sphere sag z_s and the aspheric sag z_a.

FIG. 3. Diagram indicating how the aspheric surface differs from the base spherical surface: – – –, reference sphere; ———, aspheric surface.

If it is desired to scale the chosen aspheric to a different set of dimensions or metric, a complicated method of scaling the coefficients of the polynomial is required. This occurs because the aperture height has units of length and the aspheric polymonial coefficients necessarily have units correcting the high algebraic power of length introduced in the computation. The coefficients thus scale reciprocally by one power lower than the order of the coefficient. Thus this formula appears to change drastically under a scaling of size. When in doubt, always check the scaled formula using a calculator.

C. Conic Surfaces

A special class of aspherics of great importance in design are conic sections with rotational symmetry. Each member of the class is generated by rotating a conic section about the axis of symmetry of the section. Some familiar members of the set are the paraboloids of revolution used to form aberration-free images of a single point at infinity, and the ellipsoidal sections used to form images of points at a finite distance. The conic sections of revolution form a continuum of aspherics that can be described in a closed form by the sag formula

$$z = \frac{c\rho^2}{1 + \sqrt{1 - (1 + \kappa)c^2\rho^2}}$$

where the value of κ is a conic constant defining the type of conic surface described by the formula.

The interpretation of the relation between κ and the type of rotationally symmetric conic the formula represents is given in Fig. 4. This figure shows how each member of the class of reflecting conics is obtained, but it is emphasized that the class of conics forms a continuous set. Each member of the set has an interpretation in terms of spherical aberration-free imaging characteristics for one particular set of conjugates. For example, ellipsoids form aberration-free images for a pair of real-image conjugates on the same side of the surface, hyperboloids for a pair of conjugates in which one is a virtual image or object on the back side of the surface, and so on. Changing the conjugates or changing the conic constant changes the amount of spherical aberration introduced. The judicious use of conic surfaces in a reflecting system is a powerful method of aberration control used extensively in optical design. Similar comments apply to the use of conics on refracting surfaces but, as shown in Fig. 5, the interpretation is not quite as simple because of the additional degree of freedom introduced by the choice of index of refraction. The

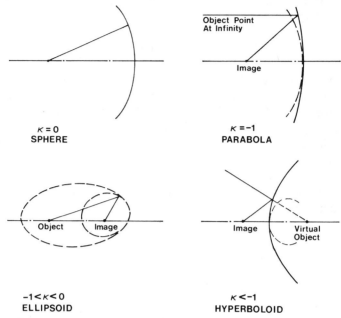

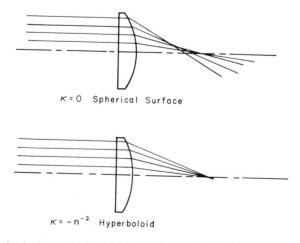

FIG. 4. Forms of reflecting conics producing perfect point imagery.

FIG. 5. A simple demonstration of the usefulness of a refracting aspheric surface.

aberration correction properties of conics are discussed in a general sense later in this chapter.

D. NEAREST-SPHERE DEFINITION

There are two characteristics of defining aspherics that need to be discussed. These are the definition of an aspheric based on the nearest or an arbitrary spherical surface, and the relation between the conic surface and the general polynomial aspheric first discussed.

For the latter point, it is necessary only to note that the difference between the base sphere and a conic can be found by expanding the conic sag formula into an infinite series and subtracting the terms resulting from a similar expansion of the base spherical sag formula. When this is done, the following formula results to eighth order:

$$z_s - z_c = \tfrac{1}{8}\kappa c^3 \rho^4 + \tfrac{1}{16}[(1 + \kappa)^2 - 1]c^5 \rho^6 + \tfrac{5}{128}[(1 + \kappa)^3 - 1]c^7 \rho^8$$

From the terms on the right side of the equation, it is evident that a conic can be expressed as a power-series form of aspheric to arbitrary accuracy. The conics are then seen as members of the set of general rotationally symmetric aspherics. The conic surfaces, by implication at this point, introduce certain added amounts of high-order aberration to the spherical base surface. The general aspheric adds a continuous-variation possibility to this set in which the ratios of the higher-order correction terms may be varied systematically.

The definition of the spherical base surface should be clear from the preceding formulas and figures. However, it is not necessarily the most useful surface to be used for describing the amount of asphericity developed from any spherical surface. A smaller aspheric departure can be obtained if a free choice of spherical radius is used, along with an added constant to make the new reference spherical surface touch the aspheric at some desired aperture. An interpretation of this concept is shown in Fig. 6. The effect of the reference spherical surface on the residual asphericity is shown for a particular parabola. As can be seen, there is a best choice for the reference surface if the amount of material to be removed from the surface during fabrication is to be minimized. The reference sphere used for this case is called the nearest sphere and is used in the initial generation of the spherical surface from which the aspheric could be produced. Note that the surface is rotationally symmetric, so that the actual choice of the nearest sphere must take into account the fact that there is much more area for a given zone at the edge of the aperture than for a zone near the center of the aperture of the surface. A similar interpretation applies to any rotationally symmetric aspheric.

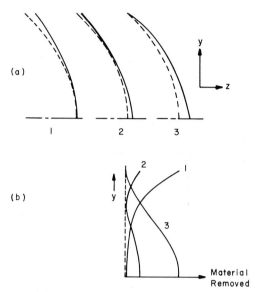

Fig. 6. An exaggerated drawing demonstrating the degree of fit of several base spheres (---) to a given conic aspheric surface (——): (a) The actual surface and the effective base spheres; (b) the resulting aspheric sag as a difference from the base sphere as the vertical dashed reference.

E. Asymmetric Aspherics

All the aspheric surfaces described above are rotationally symmetric. That is, they are derived by developing a formula for the sag of the surface from a vertex plane or from a specific base surface as a one-dimensional formula and then rotating the resulting value for the sag about an axis of symmetry. For example, a paraboloid is a parabola rotated in space about an axis through the vertex of the parabola. The requirement for axial or rotational symmetry is convenient for the majority of applications in optical systems but is not a requirement for general applications of such surfaces. In the most general case the surface can be defined as a two-dimensional polynomial or series form departing from a base plane or other low-order base surface. The discussion of such general surfaces is a very complicated problem that is not covered in this chapter because of the limited application in most areas of optical design. However, the concepts discussed regarding aberrations and fitting to closest base surfaces do apply to these more complicated two-dimensional surfaces. Usually such general surfaces have no axis of symmetry attached to the surface within the clear aperture. However, there are applications where the axis of symmetry is located outside the clear aperture of the aspheric and an apparent general two-dimensional aspheric surface is in reality an off-axis section of a rotationally symmetric surface. This has applications in unobscured-aperture reflective systems.

F. SPLINE FUNCTION-DEFINED ASPHERICS

Even a rotationally symmetric aspheric may be described by a one- or two-dimensional set of sag values for the purpose of fabrication. Such a table is useful for surface generation but is not essential to the surface description. There are useful aspheric forms that cannot be described by any closed formula. In such cases a tabular description of the surface is required even in the design stage. It is, however, desirable to handle such surfaces in a form where a smooth surface contour will result from the description. A concept using spline functions for this purpose has been developed.

Spline functions are mathematical functions which describe a surface locally but permit a smooth transition to be made from one locality on the surface to adjoining localities, in the manner of the familiar french curve used in drafting to smooth complex curves. The design then can be carried out on a computer using very complex ray-tracing methods. The complexity of such techniques makes the use of spline curve aspherics a very specialized and expensive proposition. This general method is available when needed but should be used with caution. The comments made above about the effects of two-dimensional surfaces apply as well here.

II. ABERRATIONS OF ASPHERICS

A. GENERAL COMMENTS

The principal purpose of aspherics in optical systems is to control the aberrations produced by the image-forming components of the system. The power of each surface is determined by the base curvature of the surface, but the use of a particular spherical surface in a specific location within the optical system brings with it a well-determined set of aberrations in the process of image formation. The addition of aspherics to specific spherical surfaces allows compensation for or control of the aberrations. The type of control achieved depends upon the location of the aspheric in the optical system. From the previous discussion of conic surfaces it is seen that a given aspheric surface can correct perfectly only for one point in the field for any given clear-aperture region on the aspheric. Therefore a solution to the imaging problem over a wide field of view is not necessarily obtained by introducing a number of aspherics. The influence of the aspherics on the entire design, and the resulting readjustment of base curves throughout the system, must be taken into account.

In general, aspheric surfaces of order n affect wave-front aberrations only of order n or higher. This means that ray aberrations of one order lower than the aspheric are not affected. The symmetry of the aberration is similar to the symmetry of the aspheric. For example, rotationally symmetric aspherics introduce all the aberrations commonly found in descriptions of symmetric systems. The amount of the mixture of off-axis aberrations, such as coma and astigmatism, depends upon where in the aperture of the aspheric the chief ray penetrates, or the location of the surface in the system relative to the stop or pupil location.

In the following discussion, third-order ray aberration contributions, ray-tracing methods, and some consideration of the high-order properties of aspherics are developed. Space does not permit an exhaustive description of the theory, but the principles are demonstrated.

B. RAY TRACING

Tracing of rays through aspheric surfaces is a somewhat more complicated operation than the tracing of rays through a given spherical surface. The formulas describing a general aspheric are usually not amenable to analytic solution for the intersection point of the ray and the surface. An exception applies to conic surfaces of rotation. An analytic solution exists for the intersection point of a ray on the surface, which is only slightly more complex than the formula for a spherical surface. Thus little penalty is found in computations which involve conic aspherics.

For a general aspheric defined by a polynomial, there is no analytic solution that can be used in determining the intersection of a ray with the surface. The technique used is to iterate a trial solution of the ray intercept and the surface. Usually a variation of Newton's method is used to construct a trial solution of the local normal to the surface and the intersection of the ray and the surface. The surface coordinate values obtained are then used to find a better approximation until the equations relating the ray and the surface are satisfied to within a given level of accuracy, usually determined by the word length of the computer being used in the calculation. This is a very accurate method of determining ray intersection, but it is time-consuming and potentially expensive. With present-day high-speed computers, the time required for ray tracing through aspherics is generally quite reasonable, but the additional time spent manipulating aspheric coefficients in the design process, as well as the longer ray-tracing time, need to be considered in the decision to use general rotationally symmetric aspherics.

Tracing of rays through nonrotationally symmetric aspherics is even more complex because more surface coefficients are involved and consequently more computation time for the ray intersection. The extreme is the general irregular surface, which can only be represented as a set of surface deviations given in a list along with the appropriate surface coordinates for each point. The surface is fit locally by a spline function. The ray intersection operation then becomes very complicated and time-consuming. Even with present-day fast computers, the difference in cost and time for designing with spline-fit surfaces is significant. Thus, serious consideration must be given before using such surfaces in a design. There are cases in which a very general irregular surface is required for a successful design, but it is often found that the irregular surface can be sufficiently closely represented by a simpler aspheric, with an attendant improvement in design time and cost.

C. LOW-ORDER ABERRATION COEFFICIENTS

In Section I, the definition of an aspheric as a higher-order departure in sag from a spherical base surface was discussed. The principal contribution to the aberration of an optical system is the spherical aberration produced by this high-order departure. This aberration can be introduced in an amount proportional to the high-order aspheric depth to compensate for aberration produced by the base surface or elsewhere in the system. Since the amount of this aberration is in effect tunable by the choice of the depth of the aspheric departure, this provides an extremely powerful method for controlling the correction of any optical system. Often, as in catadioptric systems, it is the sole means of compensating for or balancing aberrations.

The spherical aberration contribution does not appear as spherical aberration alone, except when the aspheric is being traversed centrally by the ray bundle, that is, when the aspheric is located in or adjacent to the aperture stop or pupil of the lens system. When the aspheric surface is attached to a surface some distance from the pupil, the aspheric is traversed by the bundle of image-forming rays in a nonsymmetric manner, and the spherical aberration introduced by the aspheric appears as spherical aberration plus coma, astigmatism, and distortion. The basic Petzval field curvature conditions in the system are not altered by introduction of the aspheric.

The above discussion shows that the entire aberration balance of an optical system can be favorably altered by the judicious addition of aspheric deformations to various spherical (or plane) base surfaces. An understanding of the process by which the introduction of aberration occurs can be had by referring to the following figures and equations. In this discussion we consider only the so-called third-order aberrations; the higher-order contributions are discussed later. Third-order ray aberrations are produced by slope errors in a wavefront with fourth-order optical path errors. The introduction of aberrations can most readily be demonstrated by considering the wavefront error. Figure 7 shows graphically the introduction of a fourth-order wavefront error by a fourth-order aspheric deviation on a plane base surface. The fourth-order deviation in the surface is seen to be replicated by a fourth-order wavefront error with a optical path difference (OPD) given by

$$OPD = (n' - n)\alpha\rho^4$$

The conversion of this wavefront aberration into a spherical aberration contribution coefficient as conventionally used in optical design is given by

$$B = - B(n' - n)\alpha\rho^3$$

where B is the spherical aberration contribution coefficient for the specified surface. This is converted to an amount of transverse third-order

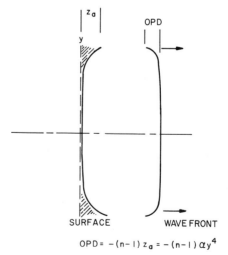

$$OPD = -(n-1)z_a = -(n-1)\alpha y^4$$

Fig. 7. Development of the spherical aberration contribution resulting from the fourth-order aspheric figuring on a refracting surface.

spherical aberration blur of radius given by SA3 in the formula

$$SA3 = -B/2n'u'$$

where u' is the final paraxial ray slope angle. Another interpretation of the amount of aberration is the depth of the wavefront error measured in wavelengths, as given by

$$W = -By/8\lambda$$

where W is the wavefront aberration, in this case being the amount of spherical aberration introduced by the aspheric surface of depth $(ad)\rho^4$ giving the spherical aberration blur radius SA3.

When the surface is traversed off the axis by the bundle of rays of semi-diameter given by y, with the center at the chief ray intersection height $\bar{y}$, the resulting wavefront aberration is a portion of the spherically aberrated wavefront and contains asymmetric contributions to the aberration given by

$$B = B, \qquad F = B\bar{y}/y, \qquad C = B(\bar{y}/y)^2, \qquad E = B(\bar{y}/y)^3$$

where y is the paraxial marginal ray height, $\bar{y}$ the paraxial chief ray height, B the spherical aberration contribution, F the coma contribution, C the astigmatism contribution, and E the distortion contribution where the interpretation of the amount of aberration in the image plane is in agreement with conventions normally used in lens design.

D. HIGH-ORDER ABERRATION CONTRIBUTIONS

The effect of aspherics on high-order wavefront aberrations is similar to the effect on the fourth-order aberrations described in Section II,C. One can readily imagine that the sixth-, eighth-, and higher-order aspherics introduce spherical aberrations of the sixth, eighth, and higher order. This is substantially true, except that the introduction of third-order aberrations is not dependent on the obliquity of the wavefront on the surface, but only on the paraxial properties describing the passage of a bundle of rays through the surface. The higher-order aberrations are dependent on the intrinsic and transferred aberrations of the next lower order and thus can be described only approximately by the rules described in Section II,C. In addition, the fourth-order aspheric error generally introduces all orders of aberration, in diminishing amounts for the same obliquity reason described above. It is, however, generally true that the major contribution to the wavefront aberration is of the same order as the aspheric coefficient describing the surface asphericity.

A conic surface, as will be shown, contributes aberrations of all orders because of the asphericity, but with a specific mixture of the amounts of the various orders of aberration. This concept will be used shortly to describe the utility of conics as basic aspherics in optical design, as well as the virtues of conics as readily tested surfaces.

E. CHROMATIC ERRORS

The index of refraction of the material on each side of the aspheric surface enters into the calculation of the wavefront aberration. Therefore, the normal variation in the index of refraction with wavelength introduces a variation in the amount of aberration introduced with the color of the image-forming light. This introduces a variation with wavelength of the amount of spherical aberration introduced by the aspheric and thus contributes spherochromatism to the image. In normal design practice, this must be balanced by the introduction of a selected amount of low-order chromatic aberration to reduce the net image blur. The same is true for the oblique aberrations that may be introduced by use of the aspheric surface.

To obtain an approximate idea of the amount of variation in aberration with wavelength that will be introduced, the special, but most generally used, case of an aspheric on a glass–air interface may be considered. A simple consideration shows that the amount of spherochromatism is given by

$$\delta B = B/V$$

where V is the reciprocal dispersion, or V number of the glass used. For most crown glasses, V is the order of 60 over the visual spectral range (C to F). Then, as a rule of thumb, an aspheric on glass that introduces 10 wavelengths of spherical aberration introduces $\frac{1}{6}$ wavelength of spherochromatism. Expressed another way, the amount of spherochromatism is negligible and may be ignored if the spherical aberration introduced is less than about 15 wavelengths. Up to about 60 wavelengths, the introduction of a compensating second-order, or spherical, term to the aspheric can add enough compensating longitudinal chromatic aberration to balance the spherochromatism to below the diffraction limit. Above this level, spherochromatism can become the limiting aberration in the system unless some compensating spherochromatism exists elsewhere in the lens.

A concept of an achromatic doublet aspheric plate consisting of crown and flint glass components is sometimes used, extending the range of allowable asphericity by a factor of about 40. Such doublet corrector plates are required in very high-numerical-aperture systems. A penalty is paid, however, in that excessive compensating aspheric power is required, analogous to the excess power required in an achromatic doublet.

F. REFLECTIVE ASPHERICS

A reflective aspheric surface is one of the most important surfaces in optics. It can be thought of as a spherical base surface which provides the optical convergent or divergent power necessary to form an image in a desired location or with a specified magnification, but with high-order aspheric deformation to compensate for the spherical aberration introduced by the spherical surface to a specified degree. The application in a single-surface imaging system is obvious; the spherical radius of the surface is chosen to place the image in a desired location, and the asphericity is added to correct the spherical aberration in the image. If more than one reflecting surface is to be used, the amount of asphericity on each surface can be varied to produce the desired compensation for both spherical aberration and coma. Three mirrors permit compensation for spherical, coma, and astigmatic aberrations. The imaging and the aberration contribution are, of course, independent of the color of light and are thus achromatic. There are some limitations on the amount of correction that can be achieved, because of the required geometry of a useful reflective imaging system, but the principles involved are extremely useful.

III. APPLICATIONS AND DESIGN

The vast majority of lenses and optical systems are based on the use of spherical image-forming surfaces. This is quite understandable because of the complications inherent in the manufacture of aspheric surfaces. In this section a discussion is given of the applications where aspherics are either essential or provide a significant advantage over the use of only spherical surfaces. It is likely that the development of improved manufacturing techniques will expand the list of useful applications.

A. CONIC SURFACES

The application of conic aspheric surfaces is one of the classical developments in geometrical optics. The principal advantage of such surfaces is that they provide aberration-free imaging between one pair of conjugate image points, with the magnification and location of these aplanatic image points determined by the base curvature and conic constant of the surface. The imagery between other pairs of conjugate points and across the field of view of the component are not, however, free of aberration. The action of a conic aspheric surface on the wavefront can be described in the classical manner or using the concepts described in Sections I and II. The latter approach is used here, since it is more readily extended to the general use of asperics and is especially useful in describing the concept of null testing of aspheric surfaces.

The action of reflecting and refracting surfaces is similar, but the extent of the useful detail in describing the use of each surface differs. An example of the use of a conic aspheric refractive surface is shown in Fig. 5. Here a plano-convex lens is used to form an image of a point at infinity. If a spherical surface is used to form the image, the image is spherically aberrated. If the convex refracting surface is made hyperbolic, the imagery is perfect for the one set of conjugate points. The conic surface is described by the choice of a conic constant equal to minus the square of the index of refraction of the surface. A similar relation exists for any other pair of conjugate points, a different conic being required. The higher-order aberrations may not be corrected for other conjugates because of the effect of the front surface of the lens on the aberration when the incoming ray bundle is refracted at the front surface. Thus, in order to retain perfect imagery for one point for a different conjugate, it is necessary to use a different radius of curvature on the front surface. A similar condition applies if the lens is "bent" to form a double convex lens; the curved front surface viewing the object at infinity will introduce some aberration and change the conjugate relations the rear surface sees. A different aspheric is required, which may not be exactly a conic, to remove the aberration completely. The conic surface is, however, a close fit to the required surface in this case.

Conics in reflection are used for aberration-free imagery in cases where a large aperture or freedom from wavelength dependence of the imagery is desired. The subject of conics of reflection is dealt with in more generality, both because of the importance of the subject and because of the fact that it is tractable to useful analysis. To carry out the discussion, it is important to note the equivalence between conics and asperics using a spherical base but with a polynomial description of the aspheric portion of the surface. This can be accomplished by expanding the conic sag expression into the power-series form:

$$z_c - z_s = \tfrac{1}{8}\kappa c^3 y^4 + \tfrac{1}{16}(\kappa + 2)\kappa c^5 y^6 + \tfrac{5}{128}(3 + 3\kappa + \kappa^2)\kappa c^7 y^8$$

Here it can be seen that the aberration contributions of various orders appear in approximate proportion to the size of the coefficient of the particular order. The third-order spherical aberration contribution is given by the sum of the surface contributions due to the base spherical power and the aspheric contribution due to the fourth-order sag coefficient. A formula relating the net spherical aberration to the conjugate positions is

$$B_{tot} = (2y^4c^3)\left[\kappa + \left(1 - \frac{r}{l}\right)^2\right] = (2y^4c^3)\left[\kappa + \left(\frac{1-m}{1+m}\right)^2\right]$$

Here m is the magnification of the surface and l/r is the relative object distance with respect to the radius of the surface. There is an obvious choice of conic constant for each choice of conjugate. In the case where the object is a point source with no transferred aberration, the correction for spherical aberration is exact to all orders. The relation is very close to exact for relayed objects. Figure 4 shows a few of the possible imaging relations which are free of spherical aberration.

Two conic surface systems are of particular importance (Fig. 8). These systems are free of spherical aberration for an object at infinity. Changing the conic constant from that which provides freedom from spherical aber-

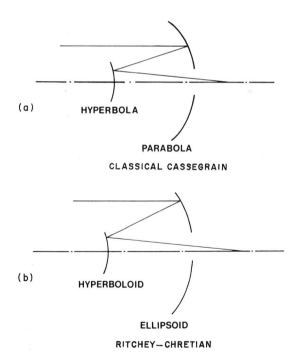

(a)   HYPERBOLA

PARABOLA

CLASSICAL CASSEGRAIN

(b)   HYPERBOLOID

ELLIPSOID

RITCHEY—CHRETIAN

FIG. 8. A simplified diagram of the two most generally used forms of a two-mirror aspheric reflecting system.

ration for both the intermediate and final images separately permits a solution in which the coma can be corrected as well. These are called aplanatic systems and are more difficult to fabricate, since the individual surfaces are not testable independently unless special arrangements are made.

B. SCHMIDT SYSTEMS

One of the most important concepts in the design of systems using aspherics is the Schmidt principle. This principle consists of using a refractive aspheric plate located at the center of curvature of a reflecting spherical surface to provide effective "parabolization" of the spherical surface. The plate acts as the stop for the surface, which exhibits no coma or astigmatism because of the location of the stop at the center of curvature. Symmetry confirms this, as the surface is seen from the same aspect for all field angles. Symmetry also indicates that the image is located on a curved spherical surface, with a radius of curvature equal to the distance to the stop, which is of course the focal length or half the radius of the surface. Proper choice of the fourth-order figuring of the Schmidt plate eliminates the spherical aberration of that order. The refractive nature of the plate indicates that there will be a variation in the amount of correcting spherical aberration introduced at different wavelengths, or spherochromatism. This variation is compensated for by introducing a second-order, or paraxial, curvature to the plate to balance the fourth-order curvature. A balancing amount of axial chromatic aberration is thus introduced, usually by placing a minimum on the shape of the aspheric surface at the 0.707 zone of the aspheric. This yields the common form of the Schmidt

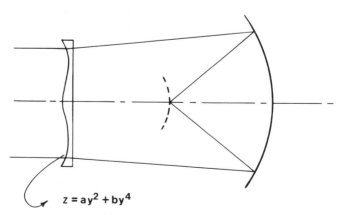

$$z = ay^2 + by^4$$

FIG. 9. Diagram of a Schmidt-type system with a refracting aspheric plate located at the center of curvature of a spherical mirror.

plate as a chromatically balanced aspheric plate. An example of a basic Schmidt system is shown in Fig. 9.

The Schmidt system is not perfect beyond the fourth order, however. In fact, it is usually necessary to add higher-order terms to correct the high-order spherical aberration for f numbers less than 4. In addition, the obliquity of the entering bundle of rays from an infinite object shows oblique spherical aberration, a fifth-order aberration. It is this latter aberration which limits the useful field of view of a simple Schmidt system.

The design and fabrication of Schmidt systems is a subject of its own and has been dealt with in several references. Therefore no time is spent on it in this chapter.

C. Refractive Imaging Systems

Aspherics have not been used extensively in normal imaging systems because of the economics of production. In fact, for camera lenses, recent efforts to use aspherics have generally led to a redesign of the product using additional spherical surfaces rather than a single aspheric surface. As technology develops, it is likely that the advantages of a smaller number of aspheric surfaces will be encountered in practice. The use of a single aspheric in a lens, located on a surface adjacent to the stop, is an extremely powerful approach in the design of lenses with high numerical apertures. A pair of aspherics, located symmetrically some distance from the pupil, can be used to control the oblique spherical aberration which is the field-limiting aberration for fast wide-angle lenses.

The increased use of optical plastics has already encouraged use of aspherics in camera lenses. Since such lens elements are produced by injection molding, the economics of producing a nonspherical surface are much more attractive. The low index of refraction and small selection of materials available in plastics also make the use of aspherics attractive.

D. Catadioptric Systems

An area where aspherics have come into prominence is that of very high-numerical-aperture (low-f-number) lenses. These lenses find use in many applications, principally low-light-level electrooptical devices and night vision systems. A compound corrector plate is used with a spherical or close-to-spherical primary mirror to obtain lenses with diameters commonly about 100 mm at $f/0.9$. The tolerances for the aspheric are somewhat relaxed, because these lenses are commonly used with image tubes or detectors with relatively poor linear resolution.

Infrared systems also employ high-speed catadioptrics. Since the transmission of many materials is poor in the infrared, refractive elements must be as thin as possible. Aspherics offer the only possibility of aberration control in some types of systems.

E. ILLUMINATING SYSTEMS

The control of illumination in devices ranging from condensers for motion picture projectors to searchlights and street lamps is a major area for the use of aspherics. The tolerances are generally relaxed, because no image is being formed. Slope tolerances on the order of minutes of arc can generally be tolerated. Usually, a high numerical aperture is desired at the end of the system collecting the light from the source. Both the above conditions lead to aspherics, generally of the conic form; although general spline-defined aspherics are appropriate for some specialized illumination conditions. Refractive condenser lenses for slide and other projectors generally use one or more elements with a conic surface in order to obtain reasonable aberration control with a minimum number of elements. Since such lenses are close to the source, heat from the projector lamp is a severe problem. Condenser elements can be molded of glass into an aspheric form and often do not require any additional processing.

IV. DESIGN METHODS USING ASPHERICS

A. SYSTEM LAYOUT

The power distribution and the basic aberration content of the lens system are determined during the initial layout process. Generally it is known from experience that aspherics will be required for the design. A nominal basic aspheric, usually to the fourth order, is selected for the starting point. The design is then submitted to a computer, and correction of the low-order aberrations called for. The use of an aspheric permits a different approach to the design to be carried out. Low-order aberrations can be corrected exactly, rather than balanced against the high-order residuals.

After the initial steps, the high-order aspheric terms can be introduced and used for correction of the high-order aberrations. At this stage, it is appropriate to carry out the correction using exact rays. Usually, the correction is made by correcting the lateral intercept error of each of the rays, or sometimes by correcting the optical path along the rays. The direction of the solution on a computer is usually determined by the correction, or by minimization of a merit function which consists of the sum of the squares of all the ray aberrations, weighted by amounts determined by the designer. The choice of the rays to be used by the designer is of critical importance in correcting systems with aspherics. Enough rays must be selected to provide errors to be corrected by use of the aspheric coefficients. The placement of the rays in the aperture must be carefully chosen, since the effect of the aspheric as defined by a polynomial coefficient depends on the height in the aperture. An erroneous choice of distribution of weights among the ray errors in the merit function, or in the distribution of rays in the aperture, can lead to aspherics with large coefficients of alternating sign, rather than the minimum aspheric required to meet the imaging requirements.

B. ABERRATION DEFINITION (DESIGN MERIT FUNCTION)

The aberration definition used in the design is critically important to the success of the design. Generally the choice of the aberrations to be corrected is determined by the goals of the design. A narrow-angle system, such as a laser beam expander or collimator, may require only correction of the spherical aberration. The comments in Section IV,A then apply to the choice of rays in the aperture to be corrected. For most cases, a finite field of view is required. The principal correction is for spherical aberration, coma, and astigmatism of various orders. As a general rule, it is not possible to correct for more aberrations than there are aspheric coefficients or parameters available for correction.

The use of a single aspheric, usually in the pupil, corrects one aberration. Spherical aberration is the choice for correction in this case. However, use of the aspheric frees the remainder of the lens parameters to find another region of solution which might contain spherical aberration for the base spherical system but has improved imagery for, say, coma. In this respect use of the single aspheric can be said to permit the correction of off-axis aberrations, although the actual correction is due to a rebalancing of the aberrations in the entire lens system. Similar logic, and consideration of the effect of an aspheric located remotely from the pupil, lead to the obvious conclusion that two separated aspherics can be used to correct for two aberrations.

Several techniques have been described in the literature for the calculation of shapes of aspherics satisfying the above two conditions. As a practical matter, the best approach for use in lens design is to add the aspherics to the basic automatic correction system and include the aspheric coefficients as variables in the correction of the entire set of aberrations being worked on.

C. Aspheric Shape Variation

The use of aspheric coefficients as variables in lens design has been shown to have many advantages, but there are also some pitfalls to be avoided. Modern lens design programs have been developed to the extent that some of the divergences that formerly occurred when aspheric coefficients were mixed with curvatures or other variables have been eliminated. The problem arises from the fact that the size of each coefficient scales with a high power of the scale of the system. The effect of a change in the aspheric coefficient has an effect on the aberration change that may differ significantly from the change in a base curvature, for example. The matrix to be solved so that the lens design can proceed is then dominated by the aspherics, and extremely strong aspherics may be obtained as the program attempts to make all the corrections with the aspheric terms rather than distributing the correction changes throughout the variable set.

The approach for controlling this is to weight the aberrations and the variables carefully, when possible, to produce a more uniform correction matrix. Some modern programs will do this automatically if so instructed. A second approach is to choose the ray set to be used in constructing the aberrations such that the aspherics are used to correct high-order aberrations essentially uniformly. For instance, some designers have found that selection of the ray heights in the aperture as the nodes of an nth-order Chebyshev polynomial produces a more uniform fit of the aspheric coefficients to the minimum required aspheric depth. A designer who explores the problems of design with aspherics will likely develop a series of such approaches that will fit the type of problem to be solved.

An important question to ask when a complex high-order aspheric is indicated in the design is how close the aspheric is to the best fitting conic. The ratio of high-order coefficients that match a conic is such that simpler fabrication, alignment, and testing occur than for a general, strong, high-order aspheric. In many cases, significant departure from the conic form is required, but the question should always be investigated.

The design of aspherics using spline-fit forms has already been noted to be a difficult, expensive task. These comments apply equally well to this type of aspheric, except that the absence of rotational symmetry may force the use of such surfaces.

V. FABRICATION OF ASPHERICS

A. MACHINING

An obvious method for producing aspherics, or at least the first step in production, is precision machining the surface. The accuracy required in producing a surface is generally on the order of a tenth of a wavelength to a quarter of a wavelength, or 0.00005 mm to as loose as 0.0002 mm, for some refractive aspherics. This is at the limit of accuracy for most machine tools. The accuracy requirements are limited both by tool placement and by accuracy of tracking or smoothness of the surface. In this last respect, the type and shape of the tool are of significant importance. Generally diamond tools are used in order to obtain a clean cut and to limit the amount of wear, which affects the ability of the tool placement drive to produce a repeatable location. The use of diamonds is, of course, required for glass materials, either in the form of a single diamond tool point or as a diamond-impregnated lap or wheel. The use of diamonds for cutting all types of metal or plastic materials is becoming common practice.

Traditional aspheric machining processes produce surfaces that are rough or of various levels of fine-ground nature. A rapidly rotating diamond-impregnated wheel is used, producing a surface sufficiently smooth so that polishing with a flexible lap can proceed directly. When the greatest smoothness is required, some additional loose abrasive grinding may be required prior to polishing. For some applications, a direct diamond-fined surface may be acceptable.

A new surface generation technology called single-point diamond turning has emerged as a major new process for fabricating a wide variety of optical surfaces. The technique uses a precision two-coordinate drive to move a single-point diamond tool across the surface of a rotating workpiece. The diamond removes material from the surface in a planing action that produces a finished surface almost of lap-polished character. The accuracy is limited by the characteristics of precision machinery to about 0.0001 mm. The process has produced surfaces of even greater smoothness for large, flat components and may yet develop into a major production capability for aspherics used in the visible wavelength region. The principal application at this time is for surfaces used in infrared systems. A principal motivation is the fact that such surfaces, at least on metals, show excellent resistance to damage from high-intensity laser radiation. A second advantage for general use is that the process has the potential capability of being the most economical method of aspheric production, with predictions for the future of a reduction in the cost of optical components of one order of magnitude. To date, the principal results have been that many metals can be worked in this manner, with surface smoothness of a level applicable for 5- to 10-μm wavelengths or longer. Whether this process can be applied to visible wavelength systems without a required polishing stage is still to be demonstrated. As yet, this process is applicable to only a few metals and special infrared materials, but the technique has great potential.

B. Grinding

The grinding of an optical surface refers to the generation of a rough base surface by means other than machining with a diamond tool driven by a precise mechanism. Tools of steel or cast iron are used in a lapping technique with a conventional low-precision optical machine. The form of the tool stroke across the surface, and the resulting time–pressure–velocity profile of the tool with a loose abrasive slurry between the tool and the workpiece control the material removal on the surface. This method is less direct than the machining process but uses standard types of machinery and is much less capital-intensive in cost. It is also less accurate in producing the desired surface after one pass through the process. There are techniques, such as plug grinding, in which a rigid tool with the desired profile is rotated over the workpiece with no random motion to average the wear across the surface, which can be used repeatedly in a production environment. Most grinding techniques depend on the ability of the operator, or a programmed controller, to move the lap over the surface in a quasi-random track that will result in the desired surface. This method has the most promise for large, high-precision aspherics.

The result of any grinding operation is that a series of abrasives of successively finer grit size are used to produce a finer grind on the surface at each stage. The final stage usually employs a 3-μm grit, which leaves a surface roughness of about 3–4 μm and a crack layer or subsurface damage layer of about 10–12 μm. This crack layer or subsurface must be removed in the polishing stage without seriously altering any figure shape produced in the grinding stage.

C. Polishing

The surface of any useful aspheric must be polished. The principal problem in attaining a polished surface is retaining the basic accuracy of the machined or ground surface as the polishing takes place. The main difficulty is that the curvature of the surface varies over the aperture of the surface, thus the polishing tool must necessarily be resilient or flexible. Maintainance of an accurate surface through control of the lap position is therefore not possible. Generally, the approach is to control the lap motion or the lap pressure to average the effects of the polishing action, and to disturb the shape of the surface as little as possible. Testing of the surface shape is needed to determine the extent to which the process has converged to the desired aspheric figure, with an iteration of the process until success is achieved within a stated tolerance. The polishing of aspherics is still closer to being an art than a defined practice, and it is at this stage that the craftsman finds his niche.

Some attempts have been made to formalize the process into a computer-directed method. They have been successful in a few cases and probably will continue to improve in technique until the manufacture of aspherics is as economical as that of spherical surfaces, at least for precision optical applications. The most empirical area of the process lies in conversion of the ground surface to the initial polished form. This procedure is nonlinear, especially since the aspheric polishing process requires small laps, leaving various portions of the surface in different stages of completion at different times.

The polishing of spherical surfaces is generally better understood in most shops, since large laps can be used to average the effect over the surface. Thus an alternate approach to the fabrication of relatively weak aspherics can be made through the polishing of a base sphere and subsequent polishing of the aspheric form. Since no conversion of the ground surface to a polished state is going on, the process is very close to linear and is amenable to computer-directed control. The time involved in surface fabrication is much longer with this approach, but the elapsed time for production may be much less because of the better rate of convergence of the process to the final surface.

D. FIGURING

After a polished surface is achieved, the shape of the surface must be made to lie within a stated tolerance. This almost inevitably leads to localized polishing or surface correction, usually the most expensive and time-consuming portion of the work for a precision aspheric. All accumulated fabrication errors must be removed, avoiding the introduction of zonal errors on the surface as a result of using a small lap.

The figuring process is controlled through the use of knowledge that the removal of material by a polishing lap is proportional to the integrated pressure, velocity, and area product for each portion of the surface. A mathematical or empirical model of the process can be developed on this basis and provides excellent results for portions of the surface over which the lap is in uniform contact with the surface. If the lap is sufficiently large so that the spring constant of the flexible lap does not permit uniform contact, the wear function so defined will not be static but will be changing dynamically and related to the local surface error. Near the edge of the surface, the lap will necessarily overhang the edge of the workpiece, and a similar effect will result. These problems can be dealt with by empirically modifying the model for portions of the surface where the aspheric slope is great and for regions near the edge of the element. In most cases, aspherics still are figured using an empirical "cut-and-try" method based on the optician's experience.

Any of the parameters of pressure, area, or velocity can be used in defining the process. Sometimes the area variable can be used successfully by trimming the lap with a razor blade to modify the wear function. At other times the parameters are held as constant as possible, and a complex path for the tool over the surface is used to provide material removal designed to produce the desired surface but leave a minimum of zonal artifacts on the surface.

VI. TESTING OF ASPHERICS

A. System Tests

A common way of testing the aspheric surfaces in a lens system is to test the overall system against a particular source, such as a distant star source for a telescope or a resolution chart at a finite distance for a camera lens. Determination of the level of image quality produced by the system is then expressed in the units desired by the system engineer. For example, a telescope may be evaluated by the percentage of incident light collected or imaged within a particular diameter, usually measured in arc seconds. Sometimes the reference is with respect to the diffraction limit of performance, such as the Strehl ratio which can be determined from photometry of the source image. Values of the resolution obtained under specific imaging conditions and with a particular detector can likewise be used to ascertain whether the system performs adequately.

Such tests detail the acceptability of the system but do not provide useful indications of the corrections to be made in figuring the aspheric within the system. Overall quality tests are useful as an acceptance test (or perhaps as a rejection test in production) but are of little use in improving the surface figure or alignment. Such tests are related more to the surface tests to be described in Section VI,B. Measurements of the wavefront or the ray displacement are made and referred to a specific surface in the system by careful subtraction of the contributions from other surfaces in the system.

B. SURFACE TESTS

Testing the surface shape of a sphere is simple in most cases. A spherical test glass is fitted to the surface, and the difference from the test glass is determined by counting the fringes appearing between the surfaces. When a test glass is not appropriate, an interferometer can be used, or one of the many geometrical ray tests, such as a knife-edge test. An interpretation in terms of the deviation of the surface from the ideal spherical form is readily obtained. Surface testing of an aspheric is somewhat more complicated because the surface does not have a natural null characteristic when examined at the center of curvature. The surface deviation in such cases must be measured, and then the design values of the deviation from the reference sphere subtracted from the measurements in order to determine the surface corrections necessary to obtain the desired aspheric. Determination of the most appropriate reference sphere must also be made so that the results will have meaning. A change in the focus or reference position of the measurement location will change the apparent shape of the measured surface, but the actual surface will not, of course, change.

As an example of the type of surface testing that may be carried out, consider the case of a reflecting parabolic surface. As noted above, the difference between a spherical surface and a parabola with the same base radius can be described, at least in the leading terms, by a fourth-order surface departure. Since this fourth-order departure introduces third-order spherical aberration, a test at the center of curvature of the surface will indicate the presence of spherical aberration of a particular amount. The deviation of the surface from the desired parabolic form is determined by comparing the apparent spherical aberration to the desired amount for the particular surface. Some opticians use their judgment of the appearance of the spherical aberration as a determination that a parabola has been fabricated. A quantitative evaluation of the surface shape is found by measuring the zonal behavior of the rays reflected from the surface and fitting the measurement to a third-order form. Alternatively, the measured ray deviations may be integrated to obtain the surface error, as compared to that for a base sphere. In the latter case, the origin of the coordinates for the integration is arbitrary, and correction for any of the base sphere effects shown in Fig. 6 is required. Subtraction of the known design value of the deviation of the aspheric from the chosen base sphere is required to determine the exact surface error to be corrected. Figure 10 illustrates this type of test, presuming that the ray deviation is measured with a knife edge in the classical Foucault manner. A similar observation applies to a zonal test in which the focus position along the axis of each of the surface

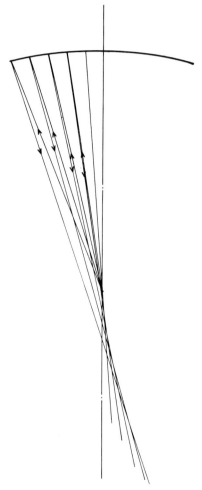

FIG. 10. Development of the caustic of rays formed at the center of curvature of an aspheric under test.

zones is recorded. This test may be used to monitor a surface in the various stages of polishing in the aspheric profile.

Tests that may be used in this manner are the Foucault knife-edge test, the Hartmann test, and any of several interferometric tests. Often the difference to be subtracted from the base reference for the desired aspheric is sufficiently large that a computer must be used to determine the correction to be made. The last point is generally true if a measurement is desired for the entire two-dimensional surface of the aspheric.

C. NULL TESTING

A convenient method of removing the design difference from the base sphere from the test results is the use of additional optics in the test setup such that the result of the test when the desired aspheric is attained shows no error indicated on the surface. The simplest type of null test is selection of the proper set of conjugates such that a perfect image is obtained only when the aspheric surface in the test setup has the right form. For a parabola, this means placing one conjugate at infinity, perhaps through the use of a full-aperture reflecting flat. Light observed at the focus would then show no surface error only when the surface being tested was a parabola. Similar methods apply for hyperboloids, as shown in Fig. 11. Accurate testing of the surface requires that the test setup by designed so that no walk-off errors appear between the location of a residual error on the surface and the apparent location of the error on the surface. While such tests are null in the limit of a perfect aspheric being tested, they may produce mapping errors in the early stages of fabrication when significant

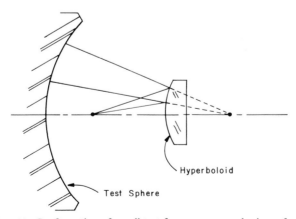

Hyperboloid

Test Sphere

FIG. 11. Configuration of a null test for a convex aspheric surface.

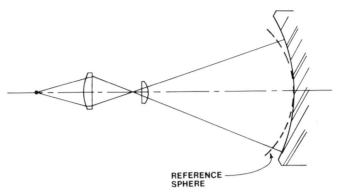

REFERENCE
SPHERE

FIG. 12. Configuration of a null test for a concave aspheric surface.

surface errors are present. Decisions regarding the figuring corrections to be made on the surface may well be very faulty, leading to divergence of the correction process. Usually, these errors can be avoided if the surface errors are measured with respect to a set of aperture coordinates imaged onto the surface being tested.

Figure 12 shows another form of null test in which a special refractive lens is designed to produce a wavefront for testing at the center of curvature of the surface; this particular wavefront is aberrated by the null lens such that a perfect return from the surface is obtained only when the aspheric matches the desired aberration. The null lens is designed such that the proper imaging of the pupil of the test system onto the surface being tested takes place.

VII. SPECIFICATION OF ASPHERICS

A. ASPHERIC SURFACE DESCRIPTION

A designed aspheric may fall into one of several classes as described in the Section I. The amount of information that must be conveyed to the shop for fabrication is clearly determined by the class of aspheric. For example, a parabola is entirely specified by stating the diameter and focal length. Most shops can then devise the proper test and tooling procedure required for fabrication. Any other conic requires a statement of the conic constant, or eccentricity, of the surface. This is only partially useful for the shop; usually a statement of the type of conic is required, along with the conjugates for which a null test is possible. For hyperbolic surfaces which do not have a useful null arrangement, it is usually necessary for the designer to provide useful test information suitable for production, even if this requires the design of a null lens or other auxiliary optics.

A general rotationally symmetric aspheric requires more information, since it is generally not obvious what test arrangement is possible, or even what the depth of the desired aspheric may be. The designer must specify the base curvature, a conic constant, if appropriate, and the coefficients of the defining polynomial. A table of representative aspheric surface sags and total surface sags is appropriate for the drawing of the element. A detailed surface sag table may be required if a numerically controlled generating machine is to be used. Generally these data are specific for a particular machine, and no general rules can be given for such information. As in the case of conics, a suggested radius of the surface such that minimum material removal from the suggested reference radius occurs is useful to the shop. For very steep aspherics, some values for local slope of the aspheric surface are desirable. In most cases the lens designer ends up being responsible for defining the aspheric test procedure.

B. TOLERANCE SPECIFICATION

Tolerancing an aspheric surface is sometimes a very difficult task. The designer must determine the amount of optical path error acceptable for the entire system and then assign some budgeted amount to the aspheric surface. The tolerance for the accuracy of the base sphere for the surface can generally be stated separately, as the aberrations introduced may vary slowly with a base curvature change. The amount of surface asphericity, on the other hand, will result in a direct amount of high-order aberration being introduced into the wavefront. The usual specification is to determine the allowable peak or rms error allowable as a deviation from the specified aspheric surface. The shop then has to devise a test suitable for obtaining this measured error.

Some additional considerations exist for aspherics, as compared to spheres. Since smaller laps are usually required, fine zonal errors may readily occur on the surface and lead to large slope errors of short distance on the surface. These errors may be of small magnitude with respect to the desired optical path error but can lead to scattered light that is apparent in the image, especially in low-numerical-aperture systems. Thus some statement of allowable slope errors is useful in determining the process of fabrication to be used.

APPLIED OPTICS AND OPTICAL ENGINEERING, VOL. VIII

CHAPTER 4

Automated Lens Design

WILLIAM G. PECK

Applications Division, Genesee Computer Center, Incorporated
Rochester, New York

I. INTRODUCTION

The generally accepted purpose of lens design is to determine values for the constructional parameters of a lens system such that the system will satisfy a given set of requirements. The classical procedure for solving this problem has been described by Kingslake (1965). It may be briefly summarized as follows:

(1) *Judiciously* select or devise several possible initial designs capable of refinement to a satisfactory solution.

(2) Determine the performance of the initial designs and select the most promising as a starting lens.

(3) Alter the construction of the selected design in a fashion expected to improve the performance.

(4) Determine the performance of the altered design.

If upon completion of all four steps performance is considered satisfactory, the design is completed. If performance has improved but is not yet satisfactory, repeat steps (3) and (4). If performance is unimproved and no other alteration is expected to improve the design, return to step (1).

Prior to the introduction of electronic computers, the repetition of steps (3) and (4) often required many months of tedious calculations by the designer and a staff of assistants. Since it was not feasible to predict performance reliably by computational methods, it was not uncommon to manufacture samples of intermediate designs. In addition to checking design performance, this procedure yielded information concerning possible improvements. It also further increased time and costs.

Despite the substantial time and expense involved, it was quite possible for a design effort to fail. This failure could result from an inadequate starting design or from the inability of the lens designer to find an appropriate route to the solution. Such a defeat could be devastating both to management and to the design teams. The risk of course could have been reduced by investing in alternative design approaches based on different starting lenses. Disregarding the added expense, this approach had the unfortunate tendency of dividing the design teams between winners and losers, according to the final design selected. Yet, where alternative parallel design efforts were not pursued, there existed the possibility of disastrous delays. Again, the psychological difficulty involved in discarding many months of work without convincing proof of the impossibility of further performance improvement was enormous. Indeed, most designs of reasonable complexity could have been improved significantly beyond the point at which they were either considered completed or abandoned. Unfortunately, the computational power required for such refinement was beyond the resources of the times.

The present state of the art is dramatically different. Since World War II an evolutionary change has taken place (Feder, 1963), brought about by the enormous increases in speed and storage capacity of the automatic

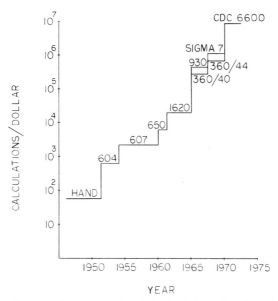

FIG. 1. Decrease in the cost of optical calculations since World War II.

computing equipment commonly available to the lens designer. The concomitant decrease in the cost of optical calculations is shown in Fig. 1. The horizontal axis, representing time, is linear; the vertical axis, representing the number of calculations obtained for a given cost, is logarithmic. The figure shows that the number of optical calculations obtained per dollar has increased approximately 10-fold every 4 years. This growth in computing resources has reduced the time required to carry out the traditional lens design process from a span of many months to a period of a few weeks or days, or even to just a few hours.

A more important, if less obvious, effect has been the addition of three new capabilities to the lens designer's repertoire:

(1) The ability to determine optimum designs for a given application from a given starting point,

(2) the ability to predict reliably the performance of a tentative design without building a sample, and

(3) the ability to generate timely alternative designs based on system trade-offs and alternative starting points.

The first of these capabilities represents the ultimate realization of the major function of the classical lens design procedure. It is accomplished by the process of automated lens design. Automated lens design, together with the ability to predict performance, naturally leads to the generation of alternative designs. This in turn allows an objective evaluation of various compromises which might be made among such factors as optical performance, illumination, size, weight, and relative cost of production. With the capabilities resulting from modern computing equipment and automated lens design programs, designers routinely meet specifications that would have been considered naive and foolish 20 years ago.

The following sections describing automated lens design show that it is now both practical and economical to produce a number of different finished lens designs for the sole purpose of exploring the trade-offs available among the optical, mechanical, and electronic subsystems. First let us examine the assumptions upon which automated lens design is based. Then we examine the actual process in detail.

II. THE ASSUMPTIONS

Automated lens design programs are based on several general assumptions. A successful design effort requires that the conditions implied by these assumptions be satisfied.

A. SUITABLE INITIAL LENS

The classical lens design process requires an initial lens layout capable of refinement to the desired performance characteristics. Automated lens design programs may be thought of as mimicking the iterative refinement process of the classical designer. For this reason automatic programs also need a starting lens capable of refinement to the desired characteristics. For example, currently available automatic programs are neither capable of recognizing the need for an additional element nor of inserting it without intervention on the part of the lens designer. Of greater importance, however, is the inability of an automatic program to guarantee a global solution to the refinement problem, a shortcoming shared by the manual process. The difficulty is relatively easy to understand.

The basic design procedure is to seek out progressively improved designs based on small test changes in the design parameters. When the procedure is successful, the test changes indicate that any further change will cause performance to deteriorate. However, the fact that such a point has been reached is no guarantee that the optimal solution to the original problem has been found. It may be that, by making changes that cause deterioration in performance, new possibilities leading to substantial improvement will be found. It is a bit like driving in the country. A better road north may be found by traveling a few miles in some other direction. For lens design it is usually more practical to reexamine the potential of the initial design and start the refining process in a new direction or to choose a different initial design. These are the kinds of decisions which depend upon the experience and judgment of the lens designer. A major advantage of automatic improvement procedures is that, although a suitable starting lens is still required, unsuitable lenses are recognized in days (sometimes on the first computer run), thereby allowing the lens designer the freedom to explore the possibilities of many designs.

B. SUFFICIENT VARIABLE PARAMETERS

The use of automated lens design programs is based on the assumption that a sufficient number of construction parameters are available for alteration during the design process to effect the desired change in performance. The maximum number of variables that can be used is naturally dependent upon the number of elements in the initial lens. However, the number and nature of constraints placed upon the manipulation of the elements are frequently more important in determining the number of parameters available. For example, the small number of parameters available with a single lens element (singlet) were sufficient to create the landscape lens (Fig. 2) used in innumerable early box cameras. However, as simple a constraint as requiring the use of the lens itself as the aperture stop of the system completely destroys the concept of the landscape lens.

The comparative miracles brought about by automatic design programs have led to greater and greater expectations on the part of project engineers. Consequently, a veritable plethora of such constraints has arisen. The following are typical examples of this problem.

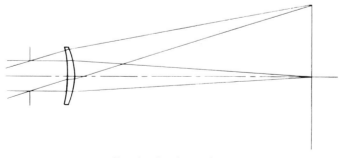

FIG. 2.   Landscape lens.

1. *Glass Must Be Selected from the List of Glass on Hand*

Inevitably missing from such a list is the glass type which would have made the design effort trivial. An accommodating designer may spend several extra weeks on the design in order to comply with this request. Indeed, the time and cost involved in working around such a constraint can exceed that involved in obtaining the appropriate glass in the first place.

2. *The Distance from the Front Vertex to the Image Plane Must Be 6 in. or Less.*

Clearly, when all the thicknesses and air spaces except one have been chosen, the remaining one is constrained to be less than a certain amount and may be effectively lost as a variable. In fact, such a constraint often prevents several variables from reaching their optimum values. This is particularly apt to occur if the constraint has been judged "reasonable" on the basis of a thin-lens first-order layout.

3. *The Distance from the Front Vertex to the Back Vertex Must Be 2 in. or Less*

The comments on the previous example apply here as well. In addition, constraints of this type lead to an implied limitation on the number of elements that may be used.

4. *The Diameter of the Lens Elements Must Not Exceed 2.75 in.*

This apparently straightforward constraint actually limits both the front-vertex-to-back-vertex distance and aperture stop location in a single blow. The limits which result are dependent on the focal length, field of view, f number, and vignetting specifications.

5. *The Focal Length Must Be 9 in., and the Lens Must Be Designed to Provide Unit Magnification for an Object-to-Image Distance of 36 in.*

This combined specification of magnification, focal length, and object-to-image distance is based on a particularly common misconception that real lens systems, composed of a number of elements, behave like a single thin lens. Normally, the desired magnification and object-to-image distance are known. The focal length should be allowed to vary in order to achieve an optimum distribution of power among the various elements.

As these examples have shown, satisfying the assumption of a sufficient number of variable parameters requires:

(1) An adequate number of elements,

(2) freedom to vary the constructional parameters of the elements over an appropriate range, and

(3) freedom from restrictions which indirectly reduce the range of variation or number of elements to inadequate levels.

C. COMPUTABLE CRITERION OF IMPROVEMENT

Automatic design improvement is achieved by a process of differential correction. Small test changes are made in the variable parameters, and the resultant changes in performance are stored in the computer. A variety of mathematical techniques may then be applied to determine new values for the variable parameters that result in better performance of the lens system.

Implicit in this description is the assumption that an unambiguous definition of "better performance" is available for use in the computer program. Such a definition must, in a meaningful way, take into account the vast array of performance requirements, mechanical constraints, and cost limitations that may be imposed by the specifications. Furthermore, it must have the capability of reducing all this information to a single number which can then be recognized as indicating improvement or deterioration.

In practice this definition, which is commonly called a merit function, is a mathematical function constructed by using information supplied by the designer. This information includes instructions for sampling the various defects that may be present and further instructions for determining the relative importance of the individual defects. The resultant merit function approximates the extent to which the system approaches the desired specifications.

Proper definition of the merit function is critical to the economical and successful use of the automatic optimization procedure. If the sampling of defects is inadequate or poorly chosen, the merit function may give little indication of design performance. On the other hand, the computational time required for each step in the improvement process is roughly proportional to the number of defects that must be calculated. Irrelevant or inconsistent defects may obscure or block the optimum route of the refinement process. Statements have been made to the effect that modern lens designers design merit functions instead of lenses.

D. Predictable Change in Performance with Change in Variables

Currently available computer programs for design optimization are based on the assumption that changes in the merit function, brought about by changes in individual design parameters, are predictable over a limited but significant range of variation for each parameter.

Most of these programs are based on two further assumptions:

(1) Changes in the individual defect items, from which the merit function is constructed, are proportional to the parameter changes within the same limited range of variation.

(2) The individual proportionalities are maintained when a simultaneous set of changes is applied to a number of variable parameters.

The basic assumption of predictability is required for virtually all refinement procedures except random search. The two additional assumptions on proportionality allow the programmer to utilize a number of powerful optimization techniques developed for application to linear problems.

III. THE INITIAL LENS

A. Basic Types of Systems

Lens systems may be classified in many different ways (Cook, 1965; Kingslake, 1965) depending upon constructional characteristics and intended application.

1. *Classification by Construction*

A lens system having all the following properties might be said to constitute a routine or normal design problem from the standpoint of construction:

(a) There is rotational symmetry of all the lens elements about a given line (known as the optical axis).

(b) The object and image lie in fixed planes perpendicular to the optical axis.

(c) All surfaces are spherical.

(d) All surfaces are refracting.

(e) None of the lens elements are movable.

(f) There is only one channel or mode of operation.

(g)   The image lies at a finite distance from the lens.

A departure from these routine conditions is apt to constitute a "special" (code word for expensive) design project.

Since the preponderance of existing designs have these properties, they are commonly designated in more detail by the general layout and shape of the elements. Names are usually based on the inventor, use, or a manufacturer's trade name. Examples are double Gauss, Cooke triplet, landscape lens, rapid rectilinear, Tessar, and gunsight triplet.

Lenses which do not meet the construction conditions of a routine lens are normally referred to according to the abnormal characteristic. Such characteristics include:

reflecting	toroidal
catadioptric	eccentric pupil
afocal	front-focusing
aspheric	interfocusing
tilted component	scanning
variable conjugate	projected reticle
multiple channel	bang-bang
flip-flop	obstructed pupil
zoom	Fresnel
off-axis	

A wide variety of exotic conditions such as these are handled by automatic optimization programs. For example, an apparently simple, bright-framed viewfinder might be classified as an obstructed-pupil, projected-reticle, triple-channel, aspheric, afocal, catadioptric, inverted Gallilean. For many reasons computer use charges are normally substantially less when such nonstandard construction is avoided.

2. *Classification by Use*

Both the entire optical system and various subsystems may be classified by use or application. Thus, systems may be referred to as visual, photographic, projection, reproduction, microscopic, telescopic, detector, condenser, or illumination. Subsystems may be referred to by such names as objective, field lens, relay, eyepiece, primary, secondary, variator, and compensator. It should be said that there are computer programs which can routinely handle all the above systems and subsystems.

It should also be pointed out that, although lenses may be similar in construction, they may be required to have substantially different performance depending on their application. For this reason classification by use is normally more meaningful than classification by other traits.

B. Choosing a Starting Lens

An initial lens may be chosen by any of three methods:

(a)   Dictates of external circumstances,
(b)   reference to prior art (previous designs) and designer's experience, and
(c)   application of theory.

1. *Dictates of External Circumstances*

There are any number of valid reasons for management or other outside influences to dictate a starting lens. One very common reason is that the project is actually a continuation of a previously completed project. For example, glass has been ordered for a finished design; meanwhile, the performance criteria have been changed. It is then obviously more efficient to use the completed lens design as the new starting lens than to start over completely. Another external circumstance might be the need to circumvent a patent with a design that is identical from a marketing viewpoint. Still another problem might be the need to improve an existing product with a minimum of retooling. In this case, if the original design was accomplished by manual methods, a day of the designer's time and between $5 and $25 of computer time may achieve wonders.

Under circumstances such as have been described, the actual selection and evaluation of a starting lens have already taken place. The project then becomes solely one of refinement, which can be handled totally by computer.

2. *Reference to Prior Art*

Under the classical design procedure, the starting design is selected with great care from voluminous detailed files of prior art. Time and care in the selection of a well-corrected initial design from such files could result in substantial savings in the iterative refinement process. With automated lens design it normally takes less time to refine a crude prototype chosen for a routine project than to carry out a meticulous selection procedure. It is not unusual for a designer, relying on experience, to select a basic form such as a triplet, double Gauss, or Tessar and then to proceed by sketching the form and estimating the values of the construction parameters from the sketch. Of course, it is advisable to exercise greater care with special systems in order to reduce computer charges and to be certain that the initial system is feasible.

3. *Application of Theory*

Occasionally, the designer encounters projects for which there is no adequate prior art. For example, a problem may arise from an innovative systems concept, an unusual zoom lens requirement, or a routine application with unusual constraints. A number of analytic methods exist which assist the designer in determining a first-order or Gaussian layout, that is, the locations and focal lengths of the various subsystems involved. Many automatic design programs have sufficient flexibility to solve, or aid in the solving of, such problems. In effect, these methods allow the designer to obtain the proper image size, location, and illumination for a given object without requiring unrealistic subsystems. This may be viewed as the crudest useful approximation to a performance criterion.

The next approximation to performance, based on Seidel (third-order) aberration theory, provides a method for arriving at a set of closed solutions for certain construction parameters (as contrasted to the iterative procedures of the automatic refinement methods). Each solution provides initial values for the glasses and the radii of curvature of the surfaces of the individual lens elements. Of great utility is the fact that a multiplicity of solutions may be found. All the resulting lens systems are similar in performance to the degree of approximation provided by the Seidel theory and correspond to performance levels preselected by the designer. Because the theory only approximates performance, thorough analysis of the resulting performance using more nearly exact theories normally reveals substantial differences among the solutions.

There is a computer program available for calculation of these solutions. It is of greatest use with complex systems such as zoom lenses.

C. TYPICAL COMPUTER INPUT

1. *Data Required*

After selection of the initial lens system the designer prepares data describing the lens for processing by the computer, if this was not already done during the selection process. The minimum data for a routine project include specifications of:

(a) Construction parameters (radii of curvature, air space and element thickness, glasses),
(b) field of view,
(c) axial aperture or speed, and
(d) aperture stop or pupil location.

Some of these specifications may be indirect. For example, an algorithm may be incorporated into the computer program allowing the designer to specify that the value of the last radius of curvature be calculated in such a manner that the entire lens system will have a given focal length. This indirect specification of a parameter by a constraint is commonly called a solve.

If the design project is not routine, additional information must be provided regarding the unusual characteristics. Examples of such additional specifications include:

(a)   Unusual surface characteristics,
(b)   spaces which change for zooming or focusing, and
(c)   the surfaces used in the various channels (configurations) of a multichannel system.

All the data are prepared in a format suitable for input to the computer and for processing by automatic programs for evaluation and refinement of performance.

The detailed requirements for preparing input data differ widely among the various programs available. At one extreme programs have been written such that the value of each possible parameter must appear in a specific set of columns on a specific line or card of input and must be encoded according to precise notational form. Numeric or alphabetic codes, at specified input locations, may be used to indicate solves or unusual characteristics of the system. Input data of this type are often referred to as rigid- or fixed-format input. Commonly, one or two cards or lines of input are required for each lens surface.

At the other extreme is free-form input. Normally appearing on the free-form card or line of input are an alphabetic mnemonic specifying the particular data item(s) present and one or more numeric values separated

```
RN* 1   RUN00 F/16 LANDSCAPE LENS EFL=1
NST 4  2      3  5  1  3  2  3 -1           3
NTU
C 1
X 1      .166429
C 2    -2.
X 2   1 .022          1.50013      61.42
C 3    -3.942
X 3   1 1.01467
C 4
X 4
DEL
CRT                                   -10.
APM -1 .03125        .28735     10.      .65629      .48613
```

FIG. 3. Typical fixed-format input for the landscape lens in Fig. 2.

```
LENS
RUN00 F/16 LANDSCAPE LENS EFL=10
PARAX      -.3+8  .3125
TH         1+8
AIR
ASTOP
TH         1.664
AIR
CV         -.2
TH         .22
GLASSK11
CV         -.3942
PARHS
AIR
AIR
```

FIG. 4. Typical free-form input for the landscape lens in Fig. 2.

by commas, spaces, or other convenient delimiters. From one to three lines of input are required for each lens surface, and additional lines are used for solves or unusual characteristics. An almost universal convention for free-form input is that the specification of the glass or material following a particular surface is the last line of input for that surface.

Each type of input has advantages and disadvantages, related primarily to the nature of the computational facilities. Fixed-format input is well suited to an environment where the designer enters the data on preprinted forms which are then processed by a clerical staff specializing in computer entry. Free-format entry is a virtual necessity when the designer personally enters the data into an interactive computer which processes each line as it is entered. Figure 3 shows a typical fixed-format input, and Fig. 4 a typical free-form input, for the landscape lens in Fig. 2.

IV. VARIABLE PARAMETERS

A. TYPES OF VARIABLE PARAMETERS

Any of the construction parameters of a lens is potentially a variable parameter for the purpose of optimizing or refining the performance of the lens. It is probable that there is some range of variation for each parameter such that a change in performance is predictable as a function of a change in parameter. Thus radius of curvature, thickness, index, and dispersion (variation in index with wavelength) may be regarded as potentially variable, along with any special construction parameters such as aspheric deformations, focusing motions, zoom motions, tilts, and decentrations.

B. Constraints on Variations

Any constraints on the variation in construction parameters may reduce the effective number of variables and prevent attainment of performance goals. Yet certain constraints on the variation in construction parameters are necessary. For example, air spaces and thicknesses must have positive values. Lens surfaces cannot intersect or "cross." The index and dispersion of a glass must be consistent with real materials.

In addition to these mandatory constraints, additional constraints may be imposed by cost, weight, and size requirements. Some of the optional constraints commonly handled by automated lens design programs include:

(a) Distance between two given surfaces,
(b) diameter of a lens element,
(c) location of pupils,
(d) equality or other linkage of two parameters,
(e) range of variation in a parameter,
(f) sensitivity to construction defects,
(g) focal length of an element or group of elements, and
(h) concentric surfaces.

In theory at least, these constraints may be regarded as optional. An optimum overall system design requires that the trade-offs, both among these optional constraints and between these optional constraints and the mechanical and electronic subsystems, be studied in detail. It may be necessary to complete a number of designs, based on different values of the optional constraints, in order to obtain adequate information on the trade-offs available.

There are three common methods of satisfying constraints in the course of an automatic refinement procedure. In order of the degree of control exercised, they are direct solution or barrier, overriding consideration, and inclusion in the performance criterion.

A direct solution (or solve) involves the assignment of one or more parameters for the sole purpose of maintaining the constraint. In the barrier, or "brick wall" approach, the solve consists of temporarily "freezing" a variable parameter attempting to violate its constraints. In both cases, these parameters are clearly lost as optimization variables.

Control by overriding consideration is normally carried out by employing the method of Lagrange multipliers. All the variable parameters are assigned in an attempt to satisfy the constraints in a fashion consistent with maximum improvement (or minimum deterioration) of the performance criterion.

C. TYPICAL COMPUTER INPUT

A fixed-form input program typically requires one or more lines of input containing a list of minimum values for variable parameters and additional lines with a list of maximum values. Specification of the variable parameters may be accomplished by any one of a number of coding schemes. Solves are generally indicated by a code included with the surface data.

When free-form input is employed, several options may be available for specifying variables and constraints on variation. Direct solves are indicated by including a specific line with the surface data. For instance, a line such as

$$\text{PICKUPCV} \quad 5$$

indicates that the curvature (the reciprocal of the radius of curvature) of the current surface is to be held equal to the curvature of surface 5. A line such as

$$\text{VCV} \quad 5 \quad -.01 \quad .01$$

might be used to indicate that curvature 5 can be varied between values of -0.01 and $+0.01$. The line

$$\text{ATMIN} \quad 2.5$$

indicates that the minimum axial thickness following the current surface is to be 2.5.

V. THE PERFORMANCE CRITERION

As stated before, the performance criterion in automated lens design is commonly called the merit function. One of the most difficult tasks in modern optical design is the establishment of an appropriate merit function for automatic optimization. Two major difficulties are involved. First, image quality must be approximated in a manner that allows economical computation. Second, the overall merit function must be constructed in such a manner that a subtle balance can be achieved among image quality for various points in the field of view, the mandatory constraints, and the optional constraints. A merit function composed of a weighted sum of the squares of individual defect items has properties useful in tackling both these difficulties. There is a sound basis in optimization theory for a number of other criteria, such as minimizing the maximum defect or minimizing the sum of the absolute values of the defects.

A. Gaussian Properties and Physical Constraints

A constraint by direct solution or the barrier method is not contained within the merit function. A constraint by overriding consideration, if accomplished through the merit function, is generally removed from the display of the merit function. Gaussian properties and physical constraints are most frequently handled by one of these two methods. However, direct solves may create undesirable "cross-talk" during optimization by concealing changes in defects with respect to changes in variables. The brute-force characteristics of overriding considerations and direct solves may result in the entire optimization process going off in a new direction when a new constraint is added or an existing constraint changed. This may be undesirable, particularly when the design has approached completion. In addition, trade-off studies may be facilitated by considering these defect items on the same basis as the image defects.

B. Image Characteristics

Image quality is the raison d'être of automated lens design. Even a crude estimate of image quality represents an enormous number of calculations. Yet, one subtle distinction among many, such as an additional 10% depth of focus at the extreme field of view, is often the deciding factor in the selection of a design for manufacture. In essence, image quality requirements vary greatly for different applications. As a result, it is virtually impossible to define a merit function which is useful as an optimization criterion and simultaneously provides an unequivocal means of ranking image quality. There are, however, theoretical relationships among certain computationally practical criteria, such as wavefront variance, rms spot size, and the commonly accepted modulation transfer function.

In practice the designer learns from experience that certain types of merit functions are the most effective for certain types of performance specifications. In the course of the design process the merit function is modified, on the basis of the results being obtained, in order that the idiosyncrasies of the specific project may be accounted for in a more accurate manner.

C. Manufacturing Considerations

There are a number of costs associated with the actual manufacture of an optical system. Modern automated design programs present an opportunity to effect substantial savings on at least three of these costs:

(a) Direct charges in coin of the realm,
(b) elapsed time until completion, and
(c) performance degradation due to manufacturing error.

While there are clear and strong relationships among these costs, the need and methods for reducing them may be more apparent when they are considered separately. Many manufacturing considerations are subject to control through physical constraints. A number of them, such as the ratio of the lens diameter to the radius of curvature, are well known to designers and received close scrutiny even with classical design methods.

A distinct advantage of automated design methods, however, is found in the capability of inexpensively computing and controlling manufacturing sensitivity. Substantial improvement in the *finished product* may be achieved by allowing a slight degradation of the *idealized computer model* in order to achieve less degradation in the manufacturing process. Needless to say, this reduced sensitivity may result in substantial reductions in time and charges as well.

D. AUTOMATIC CRITERIA VERSUS FLEXIBILITY

As noted previously, there are theoretical relationships allowing the development of merit functions which are a useful measure of image quality. In general, however, the defect items and the relative emphasis or weighting of the individual defect items for such merit functions must be developed by complex calculations. For this reason some automatic design programs use a preprogrammed merit function, based on such calculations, to represent the image quality.

However, design problems occasionally arise with unusual requirements that cannot be conveniently stated within the framework of the automatic merit function. Also, it can prove difficult for even an experienced designer to guide the optimization process along desired lines using an automatic merit function. Yet, if an automatic criterion is not used, it may be necessary for the designer to define as many as 100 or more defect items.

In practice a compromise between these situations may be obtained with either automatic or nonautomatic merit function programs. The designer normally has the freedom to choose the sample points, in the aperture or field of view, used in constructing an automatic merit function. Also, provision is generally made to assign different relative weightings across the field of view. Optional access to the weightings of individual defects may be provided. In the alternative case, where the designer defines the merit function, the definition may be saved and revised many times, with minor adjustments, for design projects with similar requirements.

It is the author's conclusion, based on substantial observation, that a novice designer is inevitably more productive when using a predefined merit function. Further, more experienced designers are eventually more efficient and productive when using such a program for routine projects.

E. Typical Computer Input

The characteristics of computer input are substantially different for programs using an automatic merit function, as opposed to those using a flexible merit function.

1. Automatic Merit Function

Three basic types of input are commonly provided for in a program using a predefined merit function:

(a) Sampling of the field of view,
(b) relative weighting of the sampled field points, and
(c) sampling across the aperture for each field of view.

With multimode systems additional input may be used to assign relative weights to the different modes of operation.

Fixed-form input typically consists of one line each describing the sampling and weighting of field points and an additional two lines per field point describing aperture sampling.

Free-form input may have a variety of arrangements. Generally, however, one line is input describing each field point to be sampled, followed by additional lines describing the individual aperture samples at the given field point.

2. Flexible Merit Function

The input data for a flexible merit function are logically divisible into two sections. The first section is similar to the input for a predefined merit function, except that only the sampling specification, not the weightings, is given. The second section of input data describes, for each sample point, the individual defects to be calculated and the weights on these defects. This section may become quite lengthy. The very flexibility of this approach militates against fixed-form input, unless a complete catalog of all possible defects is used in conjunction with some form of check-off procedure. Free-form input normally consists of one line for each defect to be included, in addition to the sampling specifications.

Regardless of whether the merit function is fixed or flexible, sampling specifications are normally given as fractions of the full field and aperture, with either fixed- or free-form input. This allows the same input to be used for a wide range of projects.

VI. AUTOMATIC IMPROVEMENT ALTERNATIVES

Although a number of optimization methods for optical design have been reported, only two basic methods have as yet received widespread currency in commercially available automatic design programs. The first of these methods, known as damped least squares, was popularized by C. G. Wynne. The second method, known as orthonormal optimization, was introduced by David Grey. Each method has a sound mathematical basis which helps to explain its characteristic behavior. If one assumes strictly proportional variation in the individual defect items with respect to the variable parameters, and that there are no constraints, the two methods yield theoretically equivalent results; although from the standpoint of numerical analysis, the orthonormal method may be expected to yield more precise numerical results. In actual practice, however, the results may be strikingly dissimilar. This occurs primarily because of an interaction between inaccuracies in the predicted changes and the detailed steps taken to produce a solution. Grossly oversimplified numerical examples are used to illustrate the fundamental concepts involved. It should be stressed that the actual mathematics of the processes involved are extremely complex and that the examples have been intentionally constructed so that the bulk of the operations normally carried out may be ignored.

A. INEXPENSIVE INITIAL IMPROVEMENT

As will be seen, the damped least squares method is well suited to obtaining rapid initial improvement of a crude prototype. However, certain circumstances may make it difficult to attain an optimum solution. As an example, consider the data displayed in Table I. Assume that they were obtained in the following manner.

An initial lens was evaluated, and the amount of defect 1 present was calculated to be 100 units. The amount of defect 2 was calculated to be 1 unit. A test change of 1 unit was made in variable (parameter) A and the defects were recalculated and found to be 90 units and 1 unit, respectively. Variable A was reset to the initial value, and variable B was changed by 1 unit. In this case the values of the defects were found to be 150 and 0.9 units, respectively.

TABLE I

STARTING SYSTEM: CALCULATED RATES OF CHANGE
AND RESIDUAL DEFECTS

Defect item	Variables		Residual defect
	A	B	
1	−10.0	50.0	100.0
2	0.0	−0.1	1.0

The goal is to reduce both defects to zero. The criterion by which progress is measured (the merit function) is the sum of the squares of the defects. By inspection the solution is to change variable A by 60 units and variable B by 10 units, using the assumption of proportionality. If there is perfect proportionality, the merit function will be reduced from 10,001 to 0.

Assume, however, that because of nonlinearities and numerical difficulties the values of 50 and −0.1 obtained are only approximations to true values of 52 and −0.09 for the rates of change of defects 1 and 2 with respect to variable B, as shown in Table II. Then the change of 60 units in variable A and 10 units in variable B results in a system with 20 units of defect 1 and 0.1 unit of defect 2. The new merit function will be 400.01, which is still substantially better than the 10,001 of the starting system. This solution is considered acceptable, and further improvement is sought using the information in Table III.

In this particular circumstance it is clear that, by repeating the process outlined above a sufficient number of times, any desired accuracy of the solution can be obtained. Convergence can be hastened by recalculating the rates of change if the inaccuracies are due to nonlinearity (failure of the assumption of proportionality), as opposed to numerical difficulties. The process appears to be rapid and efficient.

Now consider a starting system as represented in Table IV. Both the calculated rates of change and the true rates of change are assumed to be

TABLE II

STARTING SYSTEM: TRUE RATES OF CHANGE
AND RESIDUAL DEFECTS

Defect item	Variables		Residual defect
	A	B	
1	−10.0	52.0	100.0
2	0.0	−0.09	1.0

TABLE III

IMPROVED SYSTEM: CALCULATED RATES OF CHANGE
AND RESIDUAL DEFECTS

Defect item	Variables		Residual defect
	A	B	
1	-10.0	50.0	20.0
2	0.0	-0.1	0.1

the same as in the previous example. Note that the merit function, at 101, is substantially better than the merit function for the previous starting system.

The solution by inspection for this system is to change variable A by 51 units and variable B by 10 units. Because the true rates of change with respect to variable B are those given in Table II, application of the indicated changes results in a system with defect 1 equal to 20 units and defect 2 equal to 0.1 unit. Since the merit function, at 400.01, is almost four times the merit function of the starting system, the solution must clearly be rejected. Almost any reasonable merit function leads to the same conclusion. It is instructive to note, however, that the rejected solution is identical to the system in Table III which yields readily to a solution. Any starting system with a defect 2 of 1 unit and the indicated rates of change results in the same solution.

The damped least squares method is specifically intended to permit optimization to proceed in the face of such obstacles by means of the following artifice. Find a solution such that the merit function plus the weighted sum of squares of the variable changes is a minimum. If the merit function of the proposed solution is worse than the merit function of the starting point, the weighting on the sum of squares of the variable changes is increased. In the case under consideration, a typical resultant solution is to change variable A by $+0.33$ unit and variable B by -0.067

TABLE IV

ALTERNATE STARTING SYSTEM: CALCULATED RATES OF
CHANGE AND RESIDUAL DEFECTS

Defect item	Variables		Residual defect
	A	B	
1	-10.0	50.0	10.0
2	0.0	-0.1	1.0

unit. Applying these changes results in a system with defect 1 approximately 3.33 units and defect 2 approximately 1.006 units. The resulting merit function of 12.2 is clearly superior to the starting value of 101. Control of defect 1 is straightforward. However, control of defect 2 is deteriorating. In practice it is most difficult to regain control, as the error rate for defect 1 is greater than the correction rate for defect 2. Under the relatively favorable assumption that the difficulty is due to second-order effects on the proportionality of defect 1 with respect to variable B, it may be possible to correct as much as 20% of defect 2 in a single change. Additional correction will require recomputation of the rates of change. Eventual attainment of the solution will require a very large number of computations in comparison to the number required to reach a solution from the starting point in Table I. More work is required to reach a solution from the better-performing starting point.

B. Ultimate Refinement of Solution

It is clear from the examples in Section VI,A that, in certain circumstances, the damped least squares method is not well suited for obtaining a truly optimum solution of a lens design problem. However, these same examples may be used to show that the orthonormal optimization method of David Grey is well suited to the ultimate refinement of a system. Again, as in the description of damped least squares, many steps in the method have been omitted for the sake of clarity in presenting certain fundamental concepts.

Again, consider the problem displayed in Tables I and II. The solution proceeds as follows. First, using *only* variable A, minimize the merit function. Clearly a change of 10 units in variable A is indicated. The predicted values of the defects after such a change are defect 1 equal to 0.0 and defect 2 equal to 1.0. Application of the indicated change yields the expected results, and the solution is accepted. The status of the new system is displayed in Table V. Next, construct a new variable, variable B_1, using both

TABLE V

Improved System: Calculated Rates of Change
and Residual Defects

Defect item	Variables		Residual defect
	A	B	
1	−10.0	50.0	0
2	0.0	−0.1	1

TABLE VI

IMPROVED SYSTEM—NEW VARIABLE: CALCULATED RATES
OF CHANGE AND RESIDUAL DEFECTS

Defect item	Variables		Residual defect
	A	B_1	
1	−10.0	0.0	0
2	0.0	−0.1	1

variable B and variable A. The construction of variable B_1 is to be such that variation in B_1 is not expected to "undo" or destroy any of the accomplishments of variable A. Since variable A obtained an optimum value for defect 1 and did not affect defect 2, the construction must be such that the expected rate of change in defect 1, with respect to variable B_1, is zero. This is accomplished if 1 unit of change in variable B_1 is defined as consisting of a 5-unit change in variable A and a 1-unit change in variable B. The variables available for optimization are now B_1 and A, as displayed in Table VI. A change of 10 units in variable B_1 (50 units of A and 10 units of B) is clearly indicated. This change is applied; and the now familiar values of defects 1 and 2 are obtained, since the true rates of change are those given in Table VII. Defect 2 has been substantially reduced, as was predicted, and the new solution is provisionally accepted. It is noted that defect 2 changed by −0.9 unit instead of the expected −1.0 unit. This results in the current calculated rate of change in defect 2 with respect to variable B_1, −0.1, being replaced with the value −0.09. Attention is now turned to defect 1, which has a residual of 20 units instead of zero, as expected. A 2-unit change in variable A restores defect 1 to zero, as expected. Calculation of the merit function shows that the system is improved, and the solution is accepted. The 2-unit change in variable A corresponds to 0.2 unit of change in variable A for each unit of change made in variable B_1. Variable B_2 is therefore defined such that a 1-unit change in B_2 corresponds to a 5.2-unit change in variable A and a 1-unit change in

TABLE VII

IMPROVED SYSTEM—NEW VARIABLE: TRUE RATES
OF CHANGE AND RESIDUAL DEFECTS

Defect item	Variables		Residual defect
	A	B_1	
1	−10.0	2.0	0
2	0.0	−0.09	1

TABLE VIII

SECOND IMPROVEMENT AND VARIABLE CHANGE: CALCULATED
AND TRUE RATES OF CHANGE AND RESIDUAL DEFECTS

Defect item	Variables		Residual defect
	A	B_2	
1	−10.0	0.0	0
2	0.0	−0.09	0.1

variable B. The rates of change in defects for B_2 are the same as the recalculated rates just obtained for B_1. Table VIII shows the new status of the problem. As the calculated rates of change are now equal to the real rates of change, an immediate solution can be found by reoptimizing with respect to variable B_2.

Now consider the example shown in Table IV, which gave such difficulties with the method of damped least squares. Applying 1 unit of change in variable A leads immediately to a system with the status shown in Table V. The orthonormal optimization process will lead to a solution in a fashion identical to that outlined in the previous example.

In practice a substantial number of system evaluations are required in the variable-by-variable approach called for above. This results in an increase in computer costs relative to damped least squares, but the orthonormal approach is noted for finding solutions that the damped least squares method ignores.

C. COMPUTER ALTERNATIVES

The reader is warned that much of the material in this section may well be obsolete before it is published, because of the rapid decrease in cost and increase in performance evident in the area of microprocessor-based computers. Nonetheless, some of the material may prove to be of utility.

Perhaps the single most distinguishing characteristic of computing equipment from the lens designer's point of view is the degree of access available, assuming an adequate optimization program is available. If the designer does not have the option of specifying that output from an automatic design run is to be returned within a short time after submission of the run, typically 1 hour or less, apprehension that the project will not be completed on a timely basis may arise. The resultant atmosphere of fear, and concomitant hostility among designers competing for the same computing resources, can quickly become intolerable if the organization routinely attempts to produce designs for complex systems based on an optimistic time table. Historically, upon the introduction of automated lens design programs within a design group, these problems soon became so severe that improvement of computer "turnaround time" was established as a top-priority project. The approaches taken to achieve this goal are, in essence, a chronicle of the tremendous changes that have occurred in computer technology.

In the early 1960s most lens design departments could not justify the expense of a computer capable of carrying out automated design procedures. An apparently reasonable alternative was to share the expense of a suitable computer with other departments within the institution. An agreement would be arranged whereby the lens design group received access to the computer for stated periods of time. Unfortunately, unilateral abrogation of such agreements, based on the priority of turning out a payroll or a management report, was not uncommon. In addition, the physical separation between the designers and the computer facility might be so great as to require a messenger service for the transmission of input and output between the two sites. Private consultants at least had the option of traveling to a computer site where they had arranged to purchase machine time on an "as used" basis. The advent of less expensive computers, economically justified for a staff of perhaps 5 to 10 designers, was indeed welcome.

It did not take long to realize that turnaround time for a group of 10 designers, each of whom submitted a new 20-min-long computer run close on the heels of receiving the output from a previous job, was still half a day. To fill the idle hours the practice of having each designer work on several projects at the same time was introduced. In an inexorable application of the laws of physics, turnaround time rose to a day. Fortunately, arithmetic processing speeds rose rapidly; and the situation was alleviated to a degree. However, inexpensive, hard-copy output was obtained with mechanical devices, and a practical limit on the order of 5–10 min print time was rapidly reached, regardless of whether the arithmetic processor was located on-site or at a remote facility. At this point interactive processing was becoming widely available.

Certain advantages of interactive systems were soon apparent. In-house systems capable of supporting programs with comparable capabilities represented a fixed investment of $5000 or more per month. Batch access to comparable programs at remote facilities required a terminal and associated equipment with a fixed cost in excess of $1000 per month. Terminals suitable for access to interactive programs could be obtained for less than $100 per month. The prospect of returning intimate control of the optimization process to the designer was overwhelmingly appealing (at least to the designers). Instant access to the computer from any location with a telephone was ensured. There was also utility in the concept that there would be directly visible evidence that a designer was working diligently to complete his or her current project. The disadvantages were not so apparent.

Extended trials soon revealed two major disadvantages. First, the charges for computer arithmetic, using interactive commercial services, were typically 3–10 times the charges for the same work performed by a remote batch or an on-site system. Second, charges ranging from $10 to $20/h, for connection of the terminal, could easily exceed the total cost of performing the work with a batch system. These disadvantages could have been overlooked if the increased control of the designer had led to a more efficient and effective design process. There is a reason why this did not generally occur.

In Section VI,A it was shown how one poor estimate of rate of change in a defect, with respect to change in a variable parameter, could bring an automatic design process to a virtual standstill. Circumventing such difficulties is a major function of the lens designer. In a system with dozens of variables and hundreds of defects, a considerable amount of information must be obtained if this is to be accomplished in an orderly fashion. Typical interactive terminals are very slow, requiring from 2 to 10 h to print the same information that would be printed in 10 min by a batch system. Thus the instant-access, intimate contact of an interactive system paradoxically provides the designer with an order of magnitude less information than would be obtained from a batch system. Again it appears that computer technology will come to the rescue.

At current price levels (May 1977)* all the computer hardware components required to construct a computer system capable of running the most exotic design programs available, in an interactive mode and with high-speed, hard-copy output, can be purchased for less than $25,000. Following a shakeout of suppliers, it can be confidently expected that integrated systems, complete with an operating system and compilers, will be available for less than $10,000 by 1980. There is an even chance that the cost will be less than $5000; but the cost of the optical design software will almost certainly exceed the hardware cost in either case. Most existing programs will need extensive modifications to utilize fully such facilities. In fact, it is likely that such systems will lead to the first change in basic concepts of the automated design process since the introduction

---

* *Note added in proof:* These estimates, made at the time of the initial draft, may be of interest. Actual pricing in early 1980 is approximately $15,000 if purchased from a systems house which assembles the system using components from several different manufacturers. Manufacturers' prices for similar systems range upward from $20,000.

of automatic optimization. This will result from the unique combination of interactive access and high-volume output.

D. COMPUTER OUTPUT

As mentioned in Section VI,C, the output available from an automatic design program can have a substantial impact on the designer's ability to perform in an effective, efficient manner. The minimum output at the end of each optimization cycle normally consists of a new value for the merit function, a listing of existing or attempted violations of constraints, and the damping factor for damped least squares optimization.

Optionally available output for each cycle may include such items as current status of individual defects and variable parameters, degree of predictability of changes, effect of boundary constraints on defects, and improvement obtained with each variable for orthonormal optimization.

Additional output, available at the end of the optimization run, typically includes evaluation by classical methods and optical transfer function, graphic representations of the system and performance, and a representation of the new system suitable for resubmission to the optimization program.

It appears reasonable that an effective interactive program should present information regarding the predictability of individual defects which cause poor predictability of the merit function during optimization, particularly in programs employing a flexible merit function.

In any event, the output actually selected provides the basis from which additional automatic optimization runs are prepared.

VII. ITERATIVE USE OF AUTOMATED LENS DESIGN

Most lens design projects involve a number of automatic design runs on the computer. Inadequate performance may be attained on the initial run, specifications may be modified, or there may be indications that a more suitable solution is attainable.

A. DIFFERENT GROUP OF VARIABLES

There are numerous reasons for adding or deleting potentially variable parameters on the variable list for continued automatic optimization. A variable may become ineffective as a result of boundary constraints, while another variable, previously unused, is capable of producing the desired effect. The optimization may proceed to a point where it is appropriate to reinsert such a variable. A variable leading to poor predictability of the merit function may be temporarily deleted. Solves may be used to convert curvature variables to new variables called bending and power, or vice versa.

B. Restatement of Performance Criterion

There are three fundamental reasons for restatement of the performance criterion or merit function: (a) change in specifications, (b) failure of the merit function to represent the specifications, and (c) circumventing a defect which hampers optimization. The need in each case is obvious.

C. Alternate Improvement Procedure

In working with crude designs the damped least squares method tends to offer three advantages over orthonormal optimization: (a) lower cost, (b) less thought required, and (c) more facile control of boundary conditions. However, orthonormal optimization is clearly superior in the capacity to obtain ultimate refinement of the solution. Thus it has become a frequent practice to initiate a project with damped least squares optimization and conclude it with orthonormal optimization. It should be noted that this can be accomplished in a single computer run.

D. Select a New Starting Lens

A new starting lens may be selected as a result of changes in specifications, inadequate results from optimization, or optimization results that are far better than required. Where inadequate results have been obtained, either of two approaches may be taken. The first is to change the parameters in an effort to circumvent optimization difficulties. The second is to select a new lens type, normally more complex.

If optimization results are far superior to specifications, a redesign may be attempted with a less expensive lens type or with cost-saving constraints on the variable parameters.

VIII. SUMMARY

Automated lens design is a very complex process. For a given problem results may be almost instantaneous. For another a seemingly endless stream of changes in starting lenses, performance criteria, variable parameters, and optimization programs may be required. Future improvements may be expected in optimization procedures, the definition of merit functions, and computer hardware. Yet there exists a mathematical basis for the conjecture that no optimization routine can be developed which ensures a global solution to a given lens design problem. Thus, lens design per se remains an art.

REFERENCES

Cook, G. H. (1965). *In* "Applied Optics and Optical Engineering," Vol. III (R. Kingslake, ed.), p. 98. Academic Press, New York.

Feder, D. P. (1963). Automatic optical design, *Appl. Optics* **2**, 1233–1238.

Kingslake, R. (1965). *In* "Applied Optics and Optical Engineering," Vol. III (R. Kingslake, ed.), p. 1. Academic Press, New York.

APPLIED OPTICS AND OPTICAL ENGINEERING, VOL. VIII

CHAPTER 5

Radiometry

WILLIAM L. WOLFE

Optical Sciences Center
University of Arizona, Tucson, Arizona

Radiometry is the science of the measurment of radiation. Such measurements are usually made only in the optical part of the spectrum, which encompasses the ultraviolet, visible, and infrared—between 0.1 μm and 1 mm. At shorter wavelengths, where the photons are highly energetic, counting techniques are used; at longer wavelengths the radiation is often coherent, and heterodyne methods are applied. The domain of radiometry has been the optical region of the spectrum, and techniques have usually been limited to the heating of a thermal detector and the design and analysis of instruments based on geometric optics. Although these techniques are still valid and useful, new methods are gradually taking their rightful and necessary place alongside the old. These include considerations of partial coherence in both space and time, heterodyne measurements, and generalized flux concepts.

Radiometry is considered by some to be the calculation of flux values in optical instruments of various types. This is a sensible viewpoint involving the use of such concepts as throughput and the optical invariant as far as they apply. Also included are some of the less well-known approaches to vignetting, cosine-fourth falloff, and the irradiation of detectors by optical elements.

This chapter is of course flavored by my experiences in and preferences for certain portions of the field, although I have tried to make it reasonably comprehensive and balanced. The examples have been chosen to illustrate the principles discussed and to illustrate a few special points.

There is a rich but diffuse literature on radiometry. Some workers study measurement techniques for their own sake, while others include a discussion of their radiometric measurement technique as part of the study of meteorology, astronomy, biology, or some other discipline. The engineering field of heat transfer is also a fertile area in which to search for information on radiative transfer. Temperature-measurement publications usually include information about pyrometry of various sorts.

For brevity's sake, several of the important survey articles are cited here rather than the many articles that made the original contributions. Jon Geist (1976) describes radiometry as he perceives it from the early reports of Bouguer and Lambert in about 1750. Fred Nicodemus has provided two valuable literature reviews. The first of these (Nicodemus, 1969) is an Optical Resource Letter which gives about 30 references culled from over 1000. The second is a collection of reprints published by the American Association of Physics Teachers. This latter publication deals mainly with questions of normalization and units and has good tables and information on conversions among radiometric, photometric, and photon fluxes. Two fields of radiometry not covered by review articles are the design of cavity-type blackbody simulators and the influence of partial coherence on measurements and transfer. These subjects are both beyond the scope of this chapter.

I. NOMENCLATURE, SYMBOLS, AND UNITS

A. Introduction

For some reason the field of radiometry is plagued with problems involving the naming of certain physical quantities. This seems to come in part from the fact that the science of the measurement of electromagnetic radiation is useful and necessary in a variety of disciplines. Astronomers deal with power, intensity (usually power per unit area), and specific intensity (power per unit area and solid angle). Illumination engineers deal with the power that causes a visual sensation (lumens), the area–solid angle distribution of lumens (luminance), and the distribution of lumens (illuminance). Biologists are sometimes interested in quantities that cause photosynthesis or similar reactions in biologically active radiation. Here quantities such as fluence arise.

All these concepts usually separate into spectral distributions, spectrally effective quantities, and geometric and time distributions. The first category is discussed at some length in Section III, while the geometric properties are covered in this section.

B. *Système International* Units and Symbols

Système International (SI) units and symbols are used throughout this chapter. *Commission Internationale de l'Éclairage* (CIE) symbols and nomenclature with a few additions and modifications are incorporated as described in this section. Table I gives the symbol, definition, and units for each quantity. In general the symbols are given in groups of three for the number of photons, energy, and visible radiation.

The main radiometric symbols and terms of the CIE system are Q, Φ, M, E, and I for energy, flux, emitted flux concentration, received flux concentration, and intensity, respectively. The subscripts q, e, and v are used to indicate whether these are, respectively, photon, energy, or visible quantities. In this chapter the subscript e is omitted. Any radiometric quantity without a subscript is an energy quantity (or a quantity for which no specificity is necessary). Several remarks should be made about this

TABLE I
SYMBOLS, NOMENCLATURE, AND UNITS

	Symbols	Equation	Units
Quantity			
Number of photons	N	—	—
Energy	Q_e, U	—	J
Luminous energy	Q_v,	$\int Q_v(\lambda)V(\lambda)\,d\lambda$	ℓm s
Flux			
Photon rate, flux	Φ_q, $\dot{N}$	$\partial N/\partial t$	s^{-1}
Power, flux	Φ_e	$\partial U/\partial t$	W
Luminous flux	Φ_v	$\partial Q_v/\partial t$	ℓm
Density			
Photon density	n	$\partial N/\partial V$	m^{-3}
Radiant energy density	u	$\partial U/\partial V$	J m^{-3}
Visible energy density	—	$\partial Q_v/\partial V$	ℓm s m^{-3}
Concentration			
Photon flux concentration	ϕ_q	$\partial\dot{N}/\partial A \cos\theta$	s^{-1} m^{-2}
Photon exitance	M_q	$\partial\dot{N}/\partial A \cos\theta$	s^{-1} m^{-2}
Photon incidance	E_q	$\partial\dot{N}/\partial A \cos\theta$	s^{-1} m^{-2}
Flux concentration	ϕ_e	$\partial\Phi_e/\partial A \cos\theta$	W m^{-2}
Radiant exitance	M_e	$\partial\Phi_e/\partial A \cos\theta$	W m^{-2}
Radiant incidance	E_e	$\partial\Phi_e/\partial A \cos\theta$	W m^{-2}
Luminous flux concentration	ϕ_v	$\partial\Phi_v/\partial A \cos\theta$	ℓx
Luminous exitance	M_v	$\partial\Phi_v/\partial A \cos\theta$	ℓx
Illuminance	E_v	$\partial\Phi_v/\partial A \cos\theta$	ℓx
Intensity			
Photon intensity	I_q	$\partial\dot{N}/\partial\Omega$	s^{-1} sr^{-1}
Radiant intensity	I_e	$\partial\Phi_e/\partial\Omega$	W sr^{-1}
Luminous intensity	I_v	$\partial\Phi_v/\partial\Omega$	ℓm sr^{-1}, cd
Sterance, specific intensity			
Photon sterance	L_q	$\partial\dot{N}/\cos\theta\ \partial A\ \partial\Omega$	s^{-1} m^{-2} sr^{-1}
Radiance	L_e	$\partial\Phi_e/\cos\theta\ \partial A\ \partial\Omega$	W m^{-2} sr^{-1}
Luminance	L_v	$\partial\Phi_v/\cos\theta\ \partial A\ \partial\Omega$	ℓm m^{-2} sr^{-1}

list of quantities. The first is that Q_q is ambiguous. Is it the energy of a photon or is it the total number of photons? Here N is used for the latter quantity. The second is that the SI system provides no symbol or name for flux concentration independent of whether it is emitted or received. I use ϕ for this quantity, although Φ_A is another logical designation. I also use the word "concentration" rather than "density" for a distribution with respect to area; I reserve "density" for a volume distribution. Last, I prefer the symbol u rather than w for energy density, and U rather than Q for energy.

It should be clear from the table that the visible, photon, and energy quantities are all closely related. In fact, other quantities related to photosynthetic response, lead-sulfide detector response, etc., are just different spectral distributions. Jones (1963) made use of this underlying generality when he introduced the concept of phluometry. The following is my version of this treatment (with my spelling).

C. FLUOMETRY CONCEPTS

The beginning point is a quantity of something, which is designated Q. This symbol can stand for a quantity of photons or of energy or of some "effective" version of these.* One can then define a variety of useful derivations of this quantity:

$$\partial Q/\partial t = \text{flux or rate} \quad \text{(flux)}$$
$$\partial Q/\partial V = \text{volume density} \quad \text{(density)}$$
$$\partial^2 Q/\partial V\, \partial t = \text{flux of volume density}$$
$$\partial^2 Q/\partial t\, \partial A = \text{flux areal density or flux concentration} \quad \text{(incidance and exitance or areance)}$$
$$\partial^2 Q/\partial t\, \partial A\, \partial\Omega \cos\theta = \text{flux areal solid-angle density or flux solid-angle concentration} \quad \text{(sterance)}$$
$$\partial^2 Q/\partial t\, \partial\Omega = \text{flux solid-angle concentration (intensity or pointance)}$$

The phrases describing these derivatives are long and cumbersome, so the words in parentheses have been used as shorthand instead. The words "exitance," "incidance," "sterance," and "intensity" were coined by Jones, and "sterance," "aerance," and "pointance" by Nicodemus. These last two somehow do not convey the right meaning to me, but they are economical and clear.

Just as luminous quantities can be defined in terms of the response of

---

* The concept of "effectiveness" is discussed in Section III.

the eye, photon quantities can be defined in terms of the response of a detector which responds uniformly to photons of all energies. The reverse is also true. One can start with the number of photons N and derive all the geometric distributions accordingly. Then the energy relations can be obtained by considering them normalizations (see Section III) with the response function hc/λ.

II. PROPERTIES OF BLACKBODY RADIATION

A. Introduction

The ideal source and ideal receiver is a so-called blackbody (Kirchhoff, 1860). It is an idealization that has never quite been achieved, a body that in thermal equilibrium emits all the radiation possible and absorbs all the radiation incident upon it. More realistic bodies can be described in terms of how they relate to blackbodies in spectral distribution of radiation, in angular distribution, and in total radiation. So in this section I describe the radiation properties of a blackbody.

B. Spectral Distributions

The flux density of a blackbody is distributed with respect to wavelength λ, frequency ν (in hertz), wave number $\tilde{\nu}$ (in reciprocal centimeters), or the normalized dimensionless frequency x ($= hc/k\lambda T$). Each of these different functions can be written as $D(e^x - 1)^{-1}$, where D is the density-of-states function that depends upon both the independent and dependent variables. Table II gives expressions for D for the many different variations of the function. Table III gives the relationships among the independent spectral variables. The maximum for each of these functions can be found be setting the derivatives with respect to the spectral variable equal to zero and solving the resultant transcendental equation

$$xe^x/(e^x - 1) = |m|$$

where m is the exponent of the spectral variable. Table IV contains solutions to this equation.

The relationships among these variables can be found by a physical interpretation of the chain rule of differentiation. A change in the spectral radiant exitance is the same no matter what the spectral variable. Thus

$$dM = \frac{\partial M}{\partial \lambda} \, d\lambda = \frac{\partial M}{\partial \nu} \, d\nu = \frac{\partial M}{\partial x} \, dx = \cdots$$
$$= M_\lambda \, d\lambda = M_\nu \, d\nu = M_x \, dx = \cdots$$

TABLE II

Density of States Function D in the Planck Function[a]

Function	D					
	k	$\bar\nu$	ν	$x = h\nu/kT$	λ	ω
$n_\nu = N_\nu/v$	k^2/π^2	$8\pi\bar\nu^2$	$8\pi\nu^2/c^3$	$8\pi(\kappa T/ch)^3 x^2$	$8\pi\lambda^{-5}$	$\omega^2/\pi c^3$
$u_\nu = h\nu n_\nu$	$chk^3/2\pi^2$	$8\pi ch\bar\nu^3$	$8\pi h\nu^3/c^3$	$8\pi ch(\kappa T/ch)^4 x^3$	$8\pi ch\lambda^{-5}$	$h\omega^3/2\pi^3 c^3$
$M_{q\nu} = \bar{c}n_\nu/4$	$ck^2/4\pi^2$	$2\pi c\bar\nu^2$	$2\pi\nu^2/c^2$	$2\pi c(\kappa T/ch)^3 x^2$	$2\pi c\lambda^{-4}$	$\omega^2/4\pi^2 c^2$
$M_\nu = u_\nu c/4$	$c^2 hk^3/8\pi^3$	$2\pi c^2 h\bar\nu^3$	$2\pi h\nu^3/c^2$	$2\pi c^2 h(\kappa T/ch)^4 x^3$	$2\pi c^2 h\lambda^{-5} = c_1\lambda^{-5}$	$h\omega^3/8\pi^3 c^2$
$L_{q\nu} = M_{n\nu}/\pi$	$ck^2/4\pi^3$	$2c\bar\nu^2$	$2\nu^2/c^2$	$2c(\kappa T/ch)^3 x^2$	$2c\lambda^{-4}$	$\omega^2/8\pi^3 c^2$
$L_\nu = M_\nu/\pi$	$c^2 hk^3/8\pi^4$	$2c^2 h\bar\nu^3$	$2h\nu^3/c^2$	$2ch(\kappa T/ch)^4 x^3$	$2c^2 h\lambda^{-5}$	$h\omega^3/8\pi^4 c^2$

[a] The independent variable is given as y. The functions are number density n_ν, energy density u_ν, photon exitance $M_{q\nu}$, radiant exitance M_ν, photon sterance, $L_{q\nu}$, and radiance, L_ν. Each entry is D, the multiplier of $(e^x - 1)^{-1}$ in the Planck expression.

TABLE III

RELATIONSHIPS AMONG SPECTRAL VARIABLES

$$\nu = c\bar{\nu} = (2\pi)^{-1}\omega = \left(\frac{c}{2\pi}\right) k = \left(\frac{kT}{h}\right) x = c\lambda^{-1}$$

$$\lambda = c\nu^{-1} = \bar{\nu}^{-1} = (2\pi c)\omega^{-1} = 2\pi k^{-1} = \left(\frac{hc}{kT}\right) x$$

$$d\nu = c\, d\bar{\nu} = (2\pi)^{-1}\, d\omega = \left(\frac{c}{2\pi}\right) dk = \left(\frac{kT}{hc}\right) dx = c\lambda^{-2}\, d\lambda$$

$$d\lambda = -c\frac{d\nu}{\nu^2} = -2\pi c\frac{d\omega}{\omega^2} = -2\pi\frac{dk}{k^2} = -\left(\frac{hc}{kT}\right)\frac{dx}{x^2}$$

$$\frac{d\lambda}{\lambda} = -\frac{d\bar{\nu}}{\bar{\nu}} = -\frac{d\omega}{\omega} = -\frac{dk}{k} = -\frac{dx}{x} = -\frac{d\nu}{\nu}$$

Therefore one can always write

$$M_x\, dx = M_y\, dy$$

where x and y are any two (independent) spectral variables. In addition, when converting from a frequency variable to a wavelength variable the simple relationship

$$M_\nu/\nu = M_\lambda/\lambda$$

is often useful.

Blackbody radiation varies with temperature, and the change in it with respect to temperature for a single wavelength can be written

$$\frac{\partial B(\lambda)}{\partial T} = \frac{xe^x}{T(e^x - 1)} B(\lambda) = f(x,T)B(\lambda)$$

where $B(\lambda)$ represents any spectral radiometric quantity.

TABLE IV

MAXIMA OF THE DEPENDENT AND INDEPENDENT VARIABLES
FOR DIFFERENT ISOTHERMAL PLANCK SPECTRAL DISTRIBUTIONS

Function				
Dependent variable	Independent variable	m	x_{max}	R_{max}
Photons	$\bar{\nu}$	2	1.59362426	0.6476
Power	ν	3	2.82143937	1.4214
Photons	λ	4	3.92069039	4.7796
Power	λ	5	4.96511423	21.2036
Power contrast	λ	6	5.96940917	115.9359

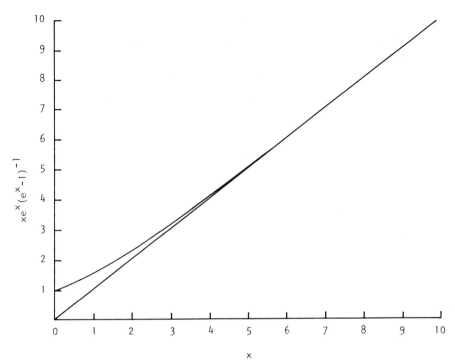

FIG. 1. The functions $xe^x(e^x - 1)^{-1}$ and x as a function of x.

Figure 1 shows the functions $xe^x(e^x - 1)^{-1}$. The maxima of all these contrast functions can be found from the equation

$$\frac{xe^x}{e^x - 1} = \frac{m + 1 + x}{2} \approx m + 1$$

Table IV lists values of maxima for different distributions. Table V lists values of maxima for the contrast functions.

C. INTEGRALS OF BLACKBODY FUNCTIONS

It makes no difference of course which frequency variable is used, since the integral from 0 to ∞ of photon flux is always the same. The integral of radiant flux is also independent of the variable of integration. Thus

$$\int_0^\infty M_\lambda \, d\lambda = \sigma T^4; \qquad \int_0^\infty M_{q\lambda} \, d\lambda = \sigma_q T^3$$

The infinite integrals of the rates of change in flux quantities can be written

TABLE V
CONTRAST MAXIMA FUNCTIONS

Function				
Dependent variable	Independent variable	m	x_{md}	x_{md}/x_{max}
$\partial M_{qv}/\partial T$	ν	2	2.575678910	1.5
$\partial M_{v}/\partial T$	ν	3	3.830016097	1.3
$\partial M_{q\lambda}/\partial T$	λ	4	4.928119359	1.25
$\partial M_{\lambda}/\partial T$	λ	5	5.969409172	1.20

$$\int_0^\infty \frac{\partial M_\lambda}{\partial T} \, d\lambda = \frac{\partial}{\partial T}(\sigma T^4) = 4\sigma_q T^3$$

$$\int_0^\infty \frac{\partial M_{q\lambda}}{\partial T} \, d\lambda = \frac{\partial}{\partial T}(\sigma_q T^3) = 3\sigma_q T^2$$

The finite integrals can be found by expanding $(e^x - 1)^{-1}$ or $(e^x - 2 + e^{-x})^{-1}$ in a series. The results of these expansions are given below. The details are in the Appendix.

$$\int_0^\lambda L_\lambda \, d\lambda = 2c^2h \left(\frac{T}{c_2}\right)^4 \sum_{m=1}^\infty m^{-4}e^{-mx}[(mx)^3 + 3(mx)^2 + 6mx + 6]$$

$$\int_0^\lambda L_{q\lambda} \, d\lambda = 2c \left(\frac{T}{c_2}\right)^3 \sum_{m=1}^\infty m^{-3} e^{-mx}[(mx)^2 + 2mx + 2]$$

$$\int_0^\lambda \frac{\partial L_\lambda}{\partial T} \, d\lambda = \frac{2c^2h}{T} \left(\frac{T}{c_2}\right)^4 \sum_{m=1}^\infty m^{-4}e^{-mx}[(mx)^4 + 4(mx)^3 + 12(mx)^2 + 24mx + 24]$$

$$\int_0^\partial \frac{\partial L_{q\lambda}}{\partial T} \, d\lambda = \left(\frac{2c}{T}\right)\left(\frac{T}{c_2}\right)^3 \sum m^{-3}e^{-mx}[(mx)^3 + 3(mx)^2 + 6mx + 6]$$

Table VI gives values for these integrals, and Figs. 2–5 give normalized curves.

D. APPROXIMATIONS FOR THE PLANCK FUNCTION

In the region where $x \gg 1$ the Planck function reduces to $c_1\lambda^{-5}e^{-x}$ (where c_1 is the first radiation constant, equal to $2\pi c^2h$), and in the region where $x \ll 1$ it becomes $c_1c_2T\lambda^{-4}$ (where c_2 is the second radiation constant, equal to hc/k_B). The first approximation is called Wien's law; the second is the Rayleigh–Jeans expression. The relative errors are, respec-

TABLE VI

VALUES OF THE TOTAL INTEGRALS OF
SOME PLANCK FUNCTIONS

Distribution	Variable	Equation	Value
Photon density	n	$\int_0^\infty n_\nu \, dy$	$\dfrac{4}{c} M_q$
Energy density	u	$\int_0^\infty u_\nu \, dy$	$\dfrac{4}{c} M$
Photon exitance	M_q	$\int_0^\infty M_{q\nu} \, dy$	$\dfrac{2\pi k^3 T^3}{c^2 h^3} (2.4041) = 1.5202 \times 10^{11} T^3$
			$= \dfrac{\sigma T^4}{2.75 kT}$
Radiant exitance	M	$\int_0^\infty M_\nu \, dy$	$\dfrac{\pi^5 k^4}{45 c^2 h^3} T^4 = \sigma T^4$
Photon sterance	L_q	$\int_0^\infty L_{q\nu} \, dy$	$\dfrac{M_q}{\pi}$
Radiance	L	$\int_0^\infty L_\nu \, dy$	$\dfrac{M}{\pi}$
Contrast	$\dfrac{1}{R}\dfrac{\partial R}{\partial T} dT$	$\dfrac{dT}{T} \int_0^\infty \dfrac{x e^x}{e^x - 1} dx$	$\zeta(2) \dfrac{dT}{T} = \dfrac{\pi^2}{6}\dfrac{dT}{T}$

tively, $-e^{-x}$ and $1 - (e^x - 1)x^{-1}$. These relative errors are shown in Fig. 6.

E. GEOMETRIC DISTRIBUTIONS AND RELATIONS

Classical theory based largely on thermodynamic arguments shows that the energy density everywhere inside a true blackbody cavity is the same. Because this is so, the radiation pattern from the blackbody aperture is such that the radiance is the same in every direction. Thus, one can say that an isothermal blackbody is isoenergetic and isoradiant. More recent theories show that the radiance is constant in its angular distribution and that there is a degree of coherence in blackbody radiance. The coherence length in the aperture is on the order of the median wavelength of the distribution and for most problems in classical radiometry is insignificant.

The geometric radiation properties can be derived by the following argument. The number of photons per unit volume in a narrow spectral interval can be written n_λ. In a time Δt, which is assumed to be very short, there exists a volume $c \, \Delta t$ long and $dA \cos \theta$ in cross section. We can imagine that each of these tubes at $t = 0$ terminates with its left end as

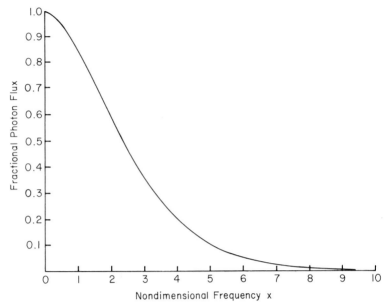

FIG. 2. Fractional photon flux in the spectral band from $x = \infty$ to x.

shown in Fig. 7 in the blackbody and its right end at the aperture; at time $t = \Delta t$ the left end is at the aperture and the right end is outside the blackbody cavity. Therefore in time Δt an array of tubes full of photons will have left the cavity. The general expression for the number of photons in each tube during a small spectral interval is

$$n_\lambda \, dA \cos \theta \, c\Delta t$$

The radiant exitance of photons is the total number of photons radiated per unit time into a hemisphere per unit area of the source. Since the photons can travel in any direction after they leave the aperture, one integrates over the hemisphere the quantity given by

$$n_\lambda \, dA \cos \theta \, c/4\pi$$

Therefore

$$M_q = N/A = \int n_\lambda \, dA \cos \theta \, c \, d\Omega/4\pi \int dA$$

$$= cn_\lambda \int \frac{\cos \theta \, d\Omega}{4\pi}$$

$$= cn_\lambda \int_0^{\pi/2} \int_0^{2\pi} \frac{\cos \theta \sin \theta \, d\theta \, d\phi}{4\pi}$$

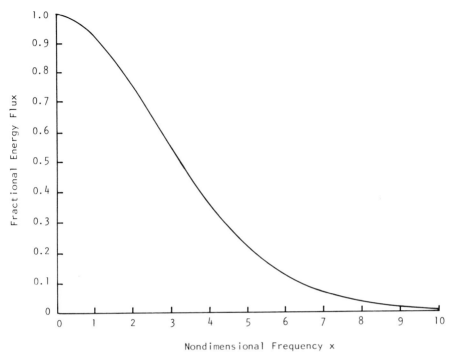

FIG. 3. Fractional energy flux in the spectral band from $x = \infty$ to x.

The result for a hemisphere is

$$M_{q\lambda} = cn_\lambda / 4$$

F. BLACKBODY RADIATION IN MEDIA OF ANY REFRACTIVE INDEX

One formulation of Planck's law derives the density of states in k space. The number of modes with radian wavenumber between k and dk is determined. In this case the number of photons per unit volume in the interval between k and dk can be written

$$n_k \, dk = k^2 \pi^{-2} (e^x - 1)^{-1} \, dk$$

The term k^2 / π^2 is the density of states, and $(e^x - 1)^{-1}$ is the average occupancy per mode. Therefore the expression is the average number of photons per unit volume in the radian wave-number interval dk:

$$n_\lambda \, d\lambda = n_k \, dk = 4\lambda^{-2} (e^x - 1) \, dk$$

A similar result can be obtained by considering the basic equation for radiance as a function of wavelength:

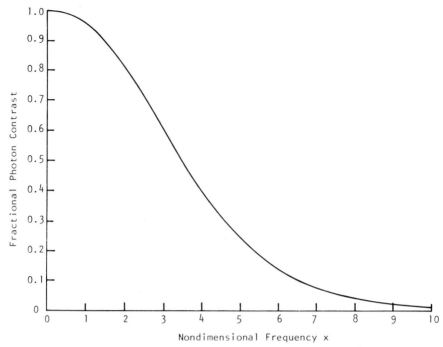

FIG. 4. Fractional photon contrast in the spectral band from $x = \infty$ to x.

$$L_\lambda \, d\lambda = 2c^2 h\lambda^{-5}(e^x - 1)^{-1} \, d\lambda$$

If a subscript 0 indicates the value of a quantity in a vacuum, then

$$L_\lambda \, d\lambda = 2c^2 n^{-2} h(n/\lambda_0)^5 (e^x - 1)^{-1} n^{-1} \, d\lambda_0$$

Therefore

$$L_\lambda \, d\lambda = n^2 L_{\lambda_0} \, d\lambda_0$$

It is tempting to think of increasing the radiance of a blackbody by filling the chamber with a high-index material. This is a losing proposition, as we shall see later, unless it is done for other, technical reasons. The volume density is proportional to the cube of the index. The flux is proportional to the square of the index.

III. RADIATION PROPERTIES OF REAL BODIES

A. INTRODUCTION

Any object which radiates by equilibrium processes (rather than the stimulated emission of lasers or luminescence, for instance) radiates less

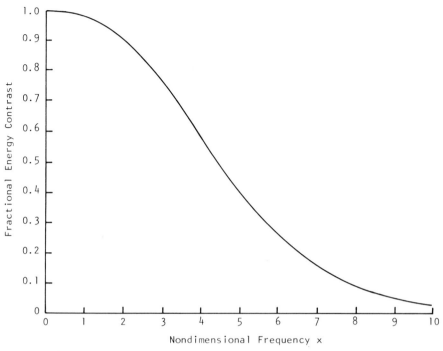

FIG. 5. Fractional energy contrast in the spectral band from $x = \infty$ to x.

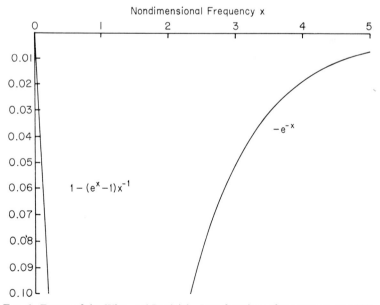

FIG. 6. Errors of the Wien and Rayleigh–Jean functions (for errors up to 10%).

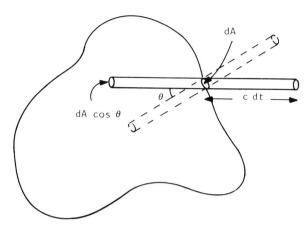

FIG. 7. Geometry of flux leaving a cavity.

flux than a blackbody at the same temperature. "Emissivity" is the word used to describe the efficiency of radiation of any given body. In general, emissivity depends upon wavelength, temperature, polarization, and angle. Care must be taken to describe and understand the type of emissivity under consideration. Hemispherical emissivity is the ratio of the flux per unit area (exitance) radiated into a hemisphere to that of a blackbody. The quantity can be written for photon flux or radiant flux as $\epsilon_{q,\lambda,\mathrm{hem}}$ or $\epsilon_{e,\lambda,\mathrm{hem}}$. For quasi-monochromatic radiation these quantities are identical.

B. Total Hemispherical Emissivity

Total hemispherical emissivity is the ratio of flux per unit area (exitance) of a given body radiated in the entire spectrum to the same quantity radiated by a blackbody:

$$\epsilon_{e,\mathrm{tot,hem}} \equiv \int \epsilon'_{\lambda,\mathrm{hem}}\, 2\pi c^2 h\, \lambda^{-5}(e^x - 1)^{-1}\, d\lambda/\sigma T^4$$

$$\epsilon_{q,\mathrm{tot,hem}} \equiv \int \epsilon_{\lambda,\mathrm{hem}}\, 2\pi c^2 h (ch)^{-1}\lambda^{-4}(e^x - 1)^{-1}\, d\lambda/\sigma_q T^4$$

These two emissivities are not equal but are related as follows:

$$\epsilon_{e,\mathrm{tot,hem}}/\epsilon_{q,\mathrm{tot,hem}} = \sigma_q ch \int \epsilon_{\lambda,\mathrm{hem}}\lambda^{-1}M^{\mathrm{BB}}_{q\lambda}\, d\lambda/\sigma T \int \epsilon_{\lambda,\mathrm{hem}}M^{\mathrm{BB}}_{q\lambda}\, d\lambda$$

$$= \sigma_q ch \int \epsilon_{\lambda,\mathrm{hem}}\, M^{\mathrm{BB}}_{\lambda}\, d\lambda/\sigma T \int \epsilon_{\lambda,\mathrm{hem}}\, \lambda M^{\mathrm{BB}}_{\lambda}\, d\lambda$$

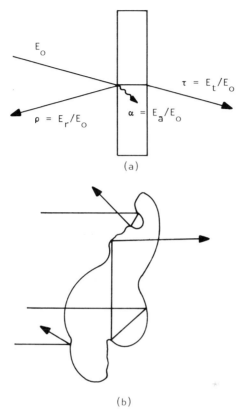

FIG. 8. Reflection, transmission, and absorption of radiation.

For simple bodies, like the plane-parallel plate shown in Fig. 8a, the radiation is reflected only in the left half-plane and all transmitted light goes into the right half-plane. More complicated structures like that shown in Fig. 8b need more careful interpretation, but the theory applies to them as well. If the incident, reflected, absorbed, and transmitted fluxes are, respectively, Φ_0, Φ_r, Φ_a, and Φ_t, then by energy conservation one can write

$$\Phi_0 = \Phi_r + \Phi_a + \Phi_t, \qquad 1 = \rho + \alpha + \tau$$

where $\rho = \Phi_r/\Phi_0$, $\alpha = \Phi_a/\Phi_0$, and $\tau = \Phi_t/\Phi_0$. Each of these quantities is the ratio of two total fluxes, the total incident flux and the total absorbed, reflected, or transmitted flux. Perhaps the simplest example of the use of this relationship is the application to a plane wave incident normally on a plane-parallel plate. The reflectivity can be calculated for each surface using Fresnel's formulas and taking note of all the interreflections. The remainder must be the flux absorbed.

The total power law certainly applies when radiation of all wavelengths is included, but does it apply generally for quasi-monochromatic radiation? The answer seems to be *no*, as the following counterexample shows.

If a high-energy laser irradiates an almost transparent plane-parallel plate, a portion of the radiation ρ will be reflected as described above. A portion τ will be transmitted, and a portion α will be absorbed. The absorbed radiation will increase the temperature of the body and thereby increase the radiation emitted in both the right- and left-hand planes. The measured reflectance and transmittance is larger than it should be. This problem can of course be avoided by making the measurement immediately upon illumination. In equilibrium the reradiated radiation need not be considered absorbed. In this sense the law is upheld. Conservation of energy is not violated but indicates that interpretations and applications must be made with care.

Do nonlinear processes obey the monochromatic power law? In second-harmonic generation the input light is absorbed and reradiated in part at a different frequency. In active media, a metastable state has been established. Strictly, the system is not in equilibrium. A portion of the power of the metastable state is seen in the transmitted beam.

For conditions for which the monochromatic power law is true one can derive Kirchhoff's law for total power and for monochromatic power:

$$\alpha = \epsilon, \qquad \alpha(\lambda) = \epsilon(\lambda)$$

These equations apply for the hemispherical quantities. Do they apply for the directional quantities? A little thought generates a positive answer, and the same is true for the individual polarization components. These relationships between photons and energy apply also to other geometric forms of emissivity.

C. DIRECTIONAL EMISSIVITY

Directional emissivity can be defined in terms of the ratio of radiances:

$$\epsilon(\theta,\phi) = L(\theta,\phi)/L^{B}(\theta,\phi) = L(\theta,\phi)/L^{B}$$

The second equation emphasizes the fact that a blackbody is Lambertian and a Lambertian radiator has a radiance that is independent of angle. This concept of directional emissivity is important in the design of cavity simulators and in many radiation problems for which uniformity of radiance is an insufficient approximation. Emissivity is a property that depends upon wavelength, angle of emission, polarization, the temperature and composition of a body, and its surface.

D. REFLECTIVITIES

The concept of reflectivity is different from that of emissivity in that the incident beam must be specified as well as the reflected beam. One can then imagine situations in which either the incident or the emergent beam is collimated, is a convergent or divergent cone, or is hemispherical. Judd (1967) considered convergent and divergent beams to be equivalent and defined nine different types of reflectivity.

A more general description is the bidirectional reflectance distribution function (BRDF) introduced by Nicodemus (1970). This description of reflectivity is defined as the ratio of reflected radiance to incident irradiance:

$$\rho(\theta_i,\phi_i;\theta_r,\phi_r) = L(\theta_r,\phi_r)/E(\theta_i,\phi_i) \qquad sr^{-1}$$

The usefulness of the function can be described by considerating several examples. The first and simplest to describe is isotropic hemispherical illumination of a Lambertian surface reflected into a hemisphere. The irradiance is the uniform radiance integrated over the hemisphere onto the projected area:

$$E = L \int_0^{\pi/2} \int_0^{2\pi} \cos\theta \sin\theta \, d\theta \, d\phi = \pi L$$

The irradiance is not a function of the angle of incidence. Because the surface is Lambertian, it is easy to show that the reflected radiance is some constant ρ_0 times $E\pi^{-1}$. Therefore $\rho = \rho_0$. This, according to Judd's scheme is called hemisphere–hemisphere reflectivity. If a collimated beam is incident upon a Lambertian reflector and collected by a hemisphere, then ρ is ρ_0/π. The collimated beam comes from a single direction, so that its radiance can be described as $L_0\delta(\theta - \theta_i, \phi - \phi_i)$.

Those interested in remote sensing of the earth from satellites, for instance, often need to calculate on their instruments the irradiance arising from reflected sunlight. The BRDF accounts for this nicely if it is known. Assume that the sun is at the (zenith) angular position (θ_s,ϕ_s) and the radiometer (or multispectral sensor) is at a position indicated by (θ_r,ϕ_r). The solar spectral irradiance $E_s(\lambda)$ is a quantity that has been measured over the years and is known to an accuracy of about 5–10% (depending upon spectral region and altitude). If the BRDF of the earth in the region of interest is called $\rho(\theta_s,\phi_s;\theta_r,\phi_r)$, then the reflected radiance is given by

$$L(\lambda) = \rho(\theta_s, \phi_s; \theta_r, \phi_r) E(\lambda; \theta_s, \phi_s)$$

The instrument irradiance is

$$E_r(\lambda) = L(\lambda; \theta_s, \phi_s; \theta_r, \phi_r)\Omega = \rho(\theta_s, \phi_s; \theta_r, \phi_r) E(\lambda; \theta_s, \phi_s)\Omega$$

where Ω is the solid angle subtended by the radiometer at a point on the earth. Other applications for such measurements include scattering in optical systems using lasers and particularly those used to measure very low flux levels in the presence of much higher levels near the field of view.

There are several useful ways to describe the BRDF. Nicodemus has suggested the use of a delta function, representing the specular component and a constant. Harvey (1976) has suggested that the BRDF be described logarithmically with the angular arguments given as the difference between the direction cosine of the specular reflection direction and the direction cosine at which the measured value is taken. His studies, a discussion of which is beyond the scope of this chapter, indicate that for many circumstances BRDF values are invariant with respect to shifts of incident angles in this coordinate system.

IV. RADIATIVE TRANSFER IN NONABSORBING, NONSCATTERING MEDIA

A. INTRODUCTION

The main application of this simplified theory is to calculate the flux on the detector of a measuring instrument. One often assumes that first-order optics applies, that solid angles can be approximated by areas divided by the square of distances, and that there is no loss (or gain) by scattering or absorption in the intervening medium.

B. FUNDAMENTALS

Radiance is the distribution of flux with respect to projected area and solid angle:

$$L = \partial^2 \Phi / \cos \theta \, \partial\Omega \, \partial A$$

One can then write the differential element of flux as

$$d\Phi = L \, dA \, d\Omega \cos \theta$$

The differential solid angle $d\Omega$ can be written

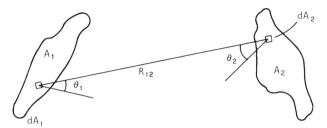

FIG. 9. The geometry of radiative transfer.

$$d\Omega = dA \cos \theta / r^2$$

These can be made more specific by reference to Fig. 9. Then the differential element of flux radiated from surface 1 to surface 2 can be written

$$d\Phi_{12} = L_1 \, dA_1 \cos \theta_1 \, dA_2 \cos \theta_2 (r_{12})^{-2}$$

Similarly the differential element of flux radiated from surface 2 to surface 1 can be written

$$d\Phi_{21} = L_2 \, dA_2 \cos \theta_2 \, dA_1 \cos \theta_1 (r_{12})^{-2}$$

These can each be written in terms of the projected solid angle, which I designate $d\Omega'$:

$$d\Phi_{12} = L_2 \, dA_1 \, d\Omega'_{12}, \qquad d\Phi_{21} = L_2 \, dA_2 \, d\Omega'_{21}$$

For Lambertian surfaces the radiance is constant and given by $M\pi^{-1}$. Therefore one can write

$$d\Phi_{12} = M_1 \, dA_1 \, G_{dA_1 - dA_2}, \qquad d\Phi_{21} = M_2 \, dA_2 G_{dA_2 - dA_1}$$

The factor G is defined for differential transfer by these equations and is called variously angle, shape, geometric configuration, or view factor. Sparrow (1970) provides an extensive discussion of the properties of G and a catalog of G values for different geometries. The interchange involving finite surfaces can be calculated by realizing that the view factor represents the fraction received by the second surface of the total radiation emitted by the first. The reader is referred to Sparrow for a further discussion. He also describes an interchange factor for specular reflecting surfaces. The treatment can be extended without great complication to surfaces with radiance distributions which vary in angle as some power of $\cos \theta$. This approach sidesteps the angle factor catalog but is most appropriate when one surface is perpendicular to the line of contact. Several examples should illustrate this line of attack.

The relationship between the radiance of a Lambertian source and its radiant exitance is found by integrating over a hemisphere:

$$M = (dA_1)^{-1} \int_0^{\pi/2} \int_0^{2\pi} L \cos \theta_1 \cos \theta_2 \, dA_1 \, dA_2$$

In Fig. 10, dA_2 is given by $R^2 \sin \theta_1 \, d\theta_1 \, d\phi$ and $\cos \theta_2 = 1$. We assume that A_1 is a differential element that is so small that R measured from the center or the edge has the same value. Then $M = L\pi$.

This is the familiar relationship between M and L for a Lambertian body. If, however, L is not independent of the angle, we can characterize it by the sum of different powers of cosines:

$$L = \sum_m \alpha_m \cos^m \theta$$

This is a complete, orthogonal set that is equivalent to a Fourier expansion that can be used as a complete description of the angular distribution of radiance. Then, for the flux density contained in a solid angle of cone with half-angle θ one has

$$\int_0^\theta \int_0^{2\pi} \sum \alpha_m \cos^{m+1}\theta \sin \theta \, d\theta \, d\phi = 2\pi \sum (m + 2)^{-1} \alpha_m (1 - \cos^{m+2}\theta)$$

The maximum was chosen to be θ for generality. For the exitance calculation it is $\pi/2$ and

$$E = 2\pi \sum (m + 2)^{-1} \alpha_m$$

C. SIMPLIFICATIONS

It often occurs in radiometric problems that solid angles can be adequately approximated by areas divided by the square of an appropriate distance—when the area is small and the distance large. Then, if A_s is the source area, p is the distance to the source from the entrance pupil, A_0 is the area of the entrance pupil, A_d is the area of the detector or image, and

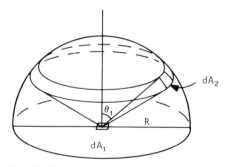

FIG. 10. Radiative transfer with a hemisphere.

q is the image distance (see Fig. 11), for a lossless, aberrationless system one has

$$\Phi = LA_sA_0/p^2 = LA_0A_d/q^2$$

It is assumed that the transmission of the optical system is 100%, that all the radiation falls on the detector, and that the medium is air so that n is 1. For a system which focuses the radiation in a medium of refractive index n, the expression is

$$\Phi = \tau LA_sA_0/p^2 = \tau LA_0A_d/q^2n^2$$

The solid angles can be identified, and the equations become

$$\Phi = LA_s\,\Omega_{0s} = LA_0\,\Omega_{s0} = \tau LA_0\,\Omega_{d0}/n^2 = LA_d\Omega_{0d}/n^2$$

Each of the solid angles can be identified by comparison with the previous equation. The first expression is used if the target area and solid angle of the optics (as seen at the target) are known; if only the properties of the optical system are known, the last expression is used. Note that for an object at infinity the last expression can be written

$$\Phi = A_d\tau L\pi/4F^2n^2$$

where F is the focal ratio. The penultimate expression utilizes the solid-angle instantaneous field of view and the optics area.

The treatment thus far is for extended sources and optical systems which put all the transmitted flux on the detector. It is worth emphasizing that the flux on the detector (for a given source radiance) is a function of the square of the focal ratio and not the optics collecting area.

If the target does not subtend the entire field of view or more—if it is not an extended target—then the situation is a little different. The flux is still given by

$$\Phi = LA_sA_0/p^2 = LA_s\Omega_{0s} = LA_0\Omega_{s0}$$

Now, however, the subtense of the detector is not the same as that of the target, so one cannot write it in terms of the characteristics of the optical system in the same way. However, if all the flux on the aperture is fo-

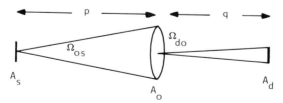

FIG. 11. Radiative transfer from source to detector.

cused onto the detector except for transmission losses, the flux on the detector is given by

$$\Phi = \tau L A_0 \Omega_{s0} = I\Omega_{0s}$$

In this case, the "point" or subresolution source case, the flux on the detector is increased by increasing the size of the collecting aperture.

In some infrared radiometer applications one needs to consider the self-radiation of the optical system. The methods by which one can calculate the detector irradiation resulting from the emission by mirrors and lenses in the system is based on the basic ideas of radiative transfer already discussed.

A single mirror with radius of curvature R and focal length $f\,(=R/2)$ has an emitted radiance given by

$$L_m = \epsilon L^{BB}(\lambda, T) = (1 - \rho)L^{BB}(\lambda, T)$$

The flux on the detector resulting from this is given approximately by

$$\Phi = L_m A_d A_0 / f^2$$

The irradiance is given by

$$E = \Phi/A_d = (1 - \rho)L^{BB}(\lambda, T)\pi D^2/4f^2$$
$$= (1 - \rho)M^{BB}(\lambda, T)/4F^2 = M_m/4F^2$$

If the mirror is at 300 K, is $F/1$, and has a reflectivity of 95%, the detector irradiance is 24×10^{-4} W cm^{-2}. If the irradiance is on a detector area of a about 100 μm on a side, the flux is 24×10^{-8} W, and this may be a detectable amount of flux.

The calculation which gives the irradiance term $M/4F^2$ assumes that the detector is at the center of curvature of the mirror. This is of course not quite right. A second approximation is to assume that the mirror is a flat disk. In that case

$$E = M/(4F^2 + 1)$$

The correct value is given by

$$E = M/(4F^2 + \tfrac{1}{2})$$

It is interesting (but probably only coincidental) that the two approximations add 0 and 1 and that the correct value is halfway between.

D. INVARIANCE THEOREMS

Invariance theorems involve brightness or radiance, throughput, and power, as well as the Lagrange and Helmholtz optical invariants.

One can show for paraxial optics that

$$n\bar{y}u - ny\bar{u} = K$$

where n is the index of refraction, y the ray height of a marginal ray, $\bar{y}$ the ray height of a chief ray, u the slope angle of a marginal ray, $\bar{u}$ the slope angle of a chief ray, and K the Lagrange invariant. The proof is shown simply on the basis of paraxial imaging equations (see for example Smith, 1966).

$$n'u' = nu + y(n' - n)/R, \qquad n'\bar{u}' = n\bar{u} + \bar{y}(n' - n)/R$$

By setting $(n' - n)/R$ equal to the rest of the expression in both equations, these are seen to be equal, and rearrangement then yields

$$n\bar{y}u - ny\bar{u} = n'y\bar{u}' - n'\bar{y}u'$$

The Lagrangian invariant is a constant throughout the optical system. One can also show that the constant propagates through the optical system.

The Lagrange invariant can be applied to the object plane, entrance pupil, and image plane. In the object and image planes y is zero. In the pupil planes, $\bar{y}$ is zero. Therefore

$$n_o y_o \bar{u}_o = n_p \bar{y}_p u_p = n_i y_i \bar{u}_i$$

where o indicates object, p indicates pupil, and i indicates image. If each of these terms is squared, then

$$n_i^2 y_i^2 u_i^2 = n_i^2 A_i \Omega_i$$

where n is the refractive index in the ith region, y_i the optical element height at the ith element, and Ω_i the solid angle subtended at the $(i - 1)$th or $(i + 1)$th element. The quantity $A\Omega$ is called by many the throughput or étendue of the optical system, and it is often designated T. In fact, it should be $n^2 A\Omega$ which is the throughput, and I will use that meaning here.

The flux through the system is given by

$$\Phi = LT = Ln^2 A\Omega$$

Further, by virtue of the results of Section II,F, the radiance is inversely proportional to the square of the refractive index, so that

$$\Phi = L_o A\Omega$$

The flux is not changed by a refractive index change, but the flux distribution is.

In summary, for an optical system with no transmission losses, for which the power or flux is conserved, in the paraxial approximation, the radiance divided by the square of the refractive index and the $A\Omega$ product times the square of the refractive index are conserved.

E. THE COSINE-FOURTH LAW

This radiometric relationship shows how the irradiance in an image varies as the field angle of the object or image increases. The falloff is proportional to the fourth power of the cosine of the angle between the line of centers and the optical axis.

The system shown in Fig. 12 indicates on-axis and off-axis elements of the source and the receiver. The irradiance is given by

$$dE = LdA_s \cos \theta_s \cos \theta_r /r^2$$

The two cosine factors are due to the inclinations of the surfaces to the line of centers. The distance r is given by $D/\cos \theta$, where $\cos \theta = \cos \theta_r = \cos \theta_s$. The result is

$$dE = LdA_s \cos^4 \theta /D^2$$

This simple result says that the irradiance between two plane surfaces falls off as the fourth power of the cosine. This result can be extended easily to a source, a lens (or mirror), and an image plane.

F. THE INVERSE SQUARE LAW

This "law" of radiometry is often discussed in terms of the flux density of a spherical wave as a function of distance or of the radiation from a

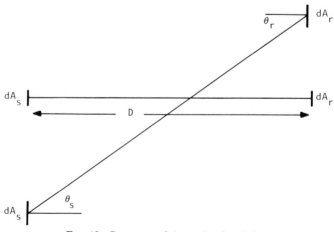

FIG. 12. Geometry of the cosine-fourth law.

point source. Each is an appropriate development for radiation emanating from a source which is small compared to its distance from the viewer, so the flux density is inversely proportional to the square of the distance. As the source becomes large, it becomes more extended and must be represented by the superposition of spherical waves. This section, by considering a special case, provides some quantification for "large" and "small."

Consider a Lambertian circular disk of radius r with an optical axis passing through its center and perpendicular to it. The flux at any point along this optical axis can be calculated from the basic radiance theorem:

$$d\Phi = L \, dA_1 \cos\theta_1 \, dA_2 \cos\theta_2 / R^2$$

$$\Phi = \int_0^R \frac{2\pi L \cos^2\theta r \, dr}{r^2 + d^2} = \pi L d^2 \int_0^R \frac{2r \, dr}{(r^2 + d^2)^2}$$

$$= \frac{\pi R^2 L}{d^2} \frac{1}{1 + (R/d)^2}$$

The evaluation is based on the fact that the two angles θ_1 and θ_2 are equal. The result shows that an inverse square law applies as long as R/d is small enough. The expression can also be written in an infinite geometric series to evaluate the contribution of $(R/d)^2$ term by term. (For an R/d of 10% the error in the inverse square assumption is only 1%.)

V. NORMALIZATION

A. INTRODUCTION

Most of the radiometric quantities one wishes to measure are distributions of one sort or another. The number of photons per unit wavelength, the power per unit area, and the power per unit solid angle are examples. These quantities are defined only for infinitesimal intervals of the independent variable. However, a finite spectral band is needed, for instance, to measure the spectral distribution of radiance—because there is no flux in an infinitesimally small band. Thus every measurement is a weighted average over a finite interval of the distribution quantity to be measured. This quantity is weighted by the variation in responsivity of the measuring instrument. The quantity reported is then usually given the epithet "normalized." The most common normalization process is spectral normalization, and the most common normalization technique is normalization to the peak. This section includes fairly general equations for normalization to the peak and a less detailed discussion of the other techniques. The independent variable is chosen to be λ, but others representing area, frequency, angle, or polarization could be used equally well.

B. BASIC ANALYSIS

Assume that a radiometer has a spectral responsivity $R(\lambda)$ and that it views a calibration source with an absolute spectral distribution of radiation $S_c(\lambda)$. The output V of the radiometer (usually a voltage) is given by

$$V_c = \int_{\Delta\lambda} S_c(\lambda)R(\lambda)\, d\lambda = S_c\, R \int_{\Delta\lambda} s_s(\lambda)r(\lambda)\, d\lambda$$

The subscript c denotes calibration. The capital letters are the peak values of spectrally varying quantities, and the lowercase letters represent the normalized distributions. For photometry, which of course deals with the response of the human eye, the luminous response of a given system is important. This can be obtained by calculating the luminous content of the source, which we can write

$$S_{eff} = \int_{\Delta\lambda} K(\lambda)S_c(\lambda)\, d\lambda = KS_c \int_{\Delta\lambda} k(\lambda)s_c(\lambda)\, d\lambda$$

Then the luminous response of the system is

$$R_{eff} = S_c R \int s_c r\, d\lambda \Big/ S_c K \int s_c k\, d\lambda$$

The quantity $K(\lambda)$ could also be the response of a lead sulfide detector, an indium antimonide detector, or the weighted output of any given instrument. Often if S_c is in terms of power, the quantity S_{eff} is called "effective watts." Of course it is necessary to specify "effective with respect to what." The quantity R_{eff} is the normalized, effective, or weighted responsivity.

Now we can assume that a measurement has been made of some unknown radiant source with an absolute spectral distribution $S(\lambda)$. The output of the measuring system is given by

$$V_m = \int_{\Delta\lambda} R(\lambda)S_m(\lambda)\, d\lambda = RS_m \int r(\lambda)s_m(\lambda)\, d\lambda$$

By substitution for R_{eff}, V_m becomes

$$V_m = R_{eff} KS_m \int s_c k\, d\lambda \int rs_m\, d\lambda \Big/ \int rs_c\, d\lambda$$

If, in addition, the effective number of watts at the aperture is known,

$$S_{eff} = \int S_m(\lambda)K(\lambda)\, d = S_m K \int s_m k\, d\lambda$$

Therefore

$$V_m = R_{eff} S_{eff} \left(\int s_c k \, d\lambda \int s_m r \, d\lambda \Big/ \int s_c r \, d\lambda \int s_m k \, d\lambda \right)$$

In order to see the significance of this expression, consider the following erroneous statement which sounds so plausible: In order to obtain the output of a given system, multiply its luminous responsivity by the luminous content of the radiation. For "luminous" one can also substitute any of the other effective quantities.

The mathematical equivalent of this statement is to assume that the ratio of the four integrals in parentheses is 1. This is true only under the following two conditions:

(a) The calibration source and unknown have the same (normalized) spectral distribution; $s_c = s_m$.

(b) The system (normalized) spectral distribution is the same as the assumed one; $r = k$.

The factor in parentheses can be used as a correction term so that it and $R_{eff} S_{eff}$ can be multiplied together to obtain the proper output. Then it is called a spectral matching factor and must be specified for both sources and both spectral responses. The preceding discussion centered around the concept of normalization to the peak. The values of S_{eff} and R_{eff} can be viewed as single peak values which when multiplied together give the right output. The entire distributed quantity has been gathered together into a single quantity and multiplied by another to give an output—as if it all occurred at the peak of the distribution. In fact, the process of writing each quantity as its peak value times a normalized distribution does just this; it normalizes to the peak.

C. BANDWIDTH NORMALIZATION

One can also normalize to the bandwidth. This is most commonly done in the calculation of effective noise bandwidth in electronic systems. Most noises which afflict radiometric instruments are not flat. They usually have some high-frequency rolloff and $1/f$ noise at low frequencies. Thus the total mean square noise is given by

$$\sigma^2 = \int_{\Delta f} |N(f)|^2 \, df$$

The radiometric system usually has low- and high-frequency cutoffs and is never quite flat in midband. Its transfer function can be written as $H(f)$. The output noise (assuming it was generated in front of the electronic filter) then is

$$\sigma_0^2 = \int_{\Delta f} |N(f)|^2 |H(f)|^2 \, df$$

It is convenient, however, to think only of white noise and of a square bandwidth, so that

$$\sigma_0^2 = |N_W|^2 \, \Delta f_{\text{eff}}$$

where N_W is the value of the white noise assumed for the system and Δf_{eff} is the effective bandwidth. Therefore

$$\Delta f_{\text{eff}} = |N_W|^2 \int |N(f)|^2 |H(f)|^2 \, df$$

If the dominant noise is the generation-recombination (gr) noise of a detector and $H(f)$ is a single time constant system with gain G, then

$$\sigma_0^2 = N_0^2 G^2 / [1 + (\omega \tau_1)^2][1 + (\omega \tau_2)^2]$$

where N_0^2 is the low-frequency value of the gr noise and τ_1 and τ_2 are the time constants of the noise rolloff and filter rolloff, respectively. If, for simplicity, τ_1 and τ_2 are made equal, then the noise output rolls off at 24 dB/octave and the difference between the square-band filter which cuts off at $\tau \, (= \tau_1 = \tau_2)$ is very small.

The situation at the other end of the frequency spectrum is quite different, however, for most detectors and transistors have an important excess noise contribution which is $1/f$ in character. The noise variance (or power) varies as $1/f^2$, and the effective noise bandwidth depends importantly on the low-frequency cutoff of the filter. It depends upon where the $1/f$ spectrum crosses the white noise spectrum and the location of the low-frequency cutoff with respect to that point. In normalization to the bandwidth one adjusts the value of the bandwidth while keeping a fixed amplitude or power level so that the integral is correct. For electronic noise it is convenient to choose the level as that for the white noise (e.g., Johnson) characteristics of the resistance, temperature, or other real parameters.

D. NORMALIZATION TO THE AVERAGE

The final normalization encountered with any regularity is normalization to the mean value. (Although it could be the rms or any other arbitrary level.) The reason this is used generally is because no peak value is unambiguously distinguishable. The development for peak normalization applies directly to normalization to the average, except that S, S_c, and R are replaced by their mean values and the normalized distributions indicated by lowercase letters are normalized to the average and may have values greater than 1.

E. EFFECTIVE VALUES

One interesting use of these effective quantities can be illustrated by considering the signal of a radiometer which uses a lead sulfide detector. The signal voltage is proportional to the responsivity of the detector and the flux on it. The latter quantity depends upon the blackbody radiation, the emissivity of the source, and the transmission of the atmosphere and optics. (Geometric factors are not germane to this discussion.) Then,

$$V = \int \epsilon(\lambda)\tau(\lambda)L^{B}(\lambda)R(\lambda) \, d\lambda$$

This can be written in terms of effective values for emissivity transmission and even radiance:

$$\epsilon_{eff} = \int \epsilon(\lambda)\tau(\lambda)L^{B}(\lambda)R(\lambda) \, d\lambda \Big/ \int \tau(\lambda)L^{B}(\lambda)R(\lambda) \, d\lambda$$

$$\tau_{eff} = \int \tau(\lambda)L^{B}(\lambda)R(\lambda) \, d\lambda \Big/ \int L^{B}(\lambda)R(\lambda) \, d\lambda$$

$$L^{B}_{eff} = \int L^{B}(\lambda)R(\lambda) \, d\lambda$$

$$V = \epsilon_{eff} \, \tau_{eff} \, L^{B}_{eff}$$

The output voltage is now found by a simple multiplication of three terms. (The widely held law of the conservation of difficulties ensures that there is about as much work in the determination of each of the effective quantities as there is in the evaluation of the integral of the product of four spectral variables.) Some system trade-off analyses are simplified by the use of these effective quantities. Notice, however, that this is a chainlike effect: Each of the effective quantities is dependent upon the others used in the calculation. One could have calculated an effective emissivity based on the responsivity or the effective radiance—and the results would be different. It is *not* true that

$$V = \epsilon_{eff} \, \tau_{eff} \, L^{B}_{eff}$$

where

$$\epsilon_{eff} = \int \epsilon(\lambda)L^{B}(\lambda)R(\lambda) \, d\lambda \Big/ \int L^{B}(\lambda)R(\lambda) \, d\lambda$$

$$\tau_{eff} = \int \tau(\lambda)L^{B}(\lambda)R(\lambda) \, d\lambda \Big/ \int L^{B}(\lambda)R(\lambda) \, d\lambda$$

$$L_{\text{eff}}^{B} = \int L^{B}(\lambda)R(\lambda)\,d\lambda$$

This is wrong by the ratio

$$\frac{\int \epsilon(\lambda)\tau(\lambda)L^{B}(\lambda)R(\lambda)\,d\lambda/\int \tau(\lambda)L^{B}(\lambda)R(\lambda)\,d\lambda}{\int \epsilon(\lambda)L^{B}(\lambda)R(\lambda)\,d\lambda/\int L^{B}(\lambda)R(\lambda)\,d\lambda}$$

F. Photodetector Effective Watts

This background provides a good opportunity for showing the relationship for the effective watts seen by a photodetector with constant quantum efficiency. The effective flux is given by

$$\Phi_{\text{eff}} = \int \Phi(\lambda)R(\lambda)\,d\lambda$$

The responsivity for a photodetector with constant quantum efficiency increases linearly with wavelength until it reaches its maximum value at the spectral cutoff:

$$R(\lambda) = (\lambda/\lambda_m)R_m$$

Therefore

$$\Phi_{\text{eff}} = R_m \lambda_m^{-1}\Phi_m \int \phi_q(\lambda)\lambda\,d\lambda$$

The flux distribution $\phi(\lambda)$ is given by

$$\phi(\lambda) = c_1\lambda^{-5}[\exp(c_2/\lambda T)-1]^{-1}$$

The equivalent photon distribution is

$$\phi_q(\lambda) = c_1\lambda^{-5}[\exp(c_2/\lambda T)-1]^{-1}/(hc/\lambda)$$

Thus

$$\phi_{\text{eff}} = \frac{hc}{\lambda_m}\int_{\Delta\lambda}\phi_q(\lambda)\,d\lambda$$

So we see that the effective energy flux is the integral of the photon rate times the energy of the photon at the maximum wavelength.

VI. RADIOMETRIC IDEAS

A. INTRODUCTION

It is probably fair to say that most radiometric problems are simple in principle but very difficult in practice. At least they are difficult to accomplish with high accuracy. One of the main problems, if not the only one of real importance, is the inclusion of every important factor. I have found that a systematic way of doing this is to write the radiometric equation including every possible variable. Then each variable can be considered in turn. Some can be eliminated, and attention can then be centered on those that are important to the specific application.

B. RADIOMETRIC EQUATION

The radiometric equation is usually written as the responsivity of the instrument, the quotient of the output signal to an input radiometric term. The output can be a voltage or a current; the input can be a radiance, power, irradiance, or intensity, and it can be in terms of either energy or photon flux rates. The differences in the radiometric input terms are considered in Section VI,C. Here I consider the dependence of the radiometric responsivity on all the parameters.

The responsivity R can be written

$$R(\lambda, f, t, \theta, \phi, x, y, z, P_B, T, s, p)$$

The responsivity is certainly a function of the wavelength of the radiation to be measured, whether it be for a quasi-monochromatic band or over a reasonable range. It can also be a function of the modulation frequency of the radiation, because the detector and electronics have specific frequency responses. The responsivity is a function of time; one aspect of this is frequency response, but another has to do with longer time periods. One example of this is a radiometer meant for a planetary mission or an orbiting spacecraft. The noise level may increase with electronic leakages in the circuit. The responsivity can vary according to the direction (θ, ϕ) from which the light comes or even from the different directions in a single F cone. Since detectors have spatial sensitivity contours, the response of the radiometer can vary with the position (x,y) of the image on the detector. The detectivity and the responsivity of some detectors, especially photodetectors which are limited by background noise, can be a function of the background flux P_B. The temperature of the radiometer can be a very influential factor. In chopper radiometers the temperature can be a second-order effect. In unchopped radiometers the case temperature and that of the surround become the reference temperature and are therefore first-order effects. All radiation is polarized, and all optical systems have a polarization response. This response can be small in well-designed, all-reflective systems with blackened detectors, but there is no reason to ignore the influence a priori. Systems with antireflection coatings, dielectric beam splitters, and interference filters must be evaluated for their polarization properties (s,p). In modern measurements, especially with laser radiation, or in astronomical measurements with distant point sources, the degree of coherence γ can be a factor which influences the accuracy of the measurement.

C. RELATIVE AND ABSOLUTE MEASUREMENTS

It has often been said that radiometric measurements are much more easily made on a relative basis than absolutely. I do not argue with the statement and believe it needs considerable interpretation. Far more measurements are relative than absolute, and probably more measurements are relative than is generally believed.

The absolute measurements of which I am aware include the lumen and spectral irradiance from about 0.1 to 10 μm made by the National Bureau of Standards (NBS), Washington, D.C. The lumen is defined as the radiation from a $\frac{1}{60}$ cm^2 area of a blackbody at the temperature of melting platinum. This measurement has an accuracy which depends on the measurement of the area, the temperature, and the quality of the blackbody. It is not related at all to the measurement of some other flux level.

Most of the standards in the spectral region 0.1–0.2 μm now depend upon the blackbody nature of synchrotron radiation. These measurements also do not depend upon another flux level. The electrically calibrated radiometer makes a comparison between electrical heating (electrical watts) and an input flux. The measurement depends upon temperature, thermal conductivity, absorptivity, etc., but no comparison is made with another flux level. I believe it can be argued logically and successfully that only the measurements made by NBS personnel are absolute. All other measurements are relative. The remaining questions involve only the accuracy required and the closeness of the relationship.

Standards organizations usually maintain secondary standards of spectral radiance and irradiance, and field measurements are generally made with tertiary standards calibrated against one of these. If a radiometer is placed aboard a satellite or probe, it will be calibrated against one of these standards before the flight and maybe against an onboard calibration source during the flight. Two accuracies can then be quoted, one with respect to the standard secondary source (often called absolute) and one with respect to the onboard calibration source. These are *both* relative calibrations; they are just relative to different quantities. The largest difference seems to be the drift in the instrument and the two sources.

There is an important distinction between measuring a radiation or temperature and the difference between two radiation or temperature levels. The distinction is minimized if the difference is taken between an unknown and one of the sources.

D. Rules of Radiometric Measurement

The two rules of radiometric measurement can be stated so simply that they seem almost trivial. In fact most radiometrists do not appreciate them until they have violated them. They are:

(1) Take into account every factor that might influence the measurement.

(2) Calibrate the radiometer in exactly the same way it will be used.

One immediate example of rule (2) includes putting a calibration source next to an unknown at a distant point along the same transmission path. A second fairly obvious example is using a calibration source that puts about the same amount of flux on the detector as the unknown. Similarly one should maintain similar polarizations, spectral distributions, angular distributions, image sizes, etc. The more the calibration source is like the unknown, the better. The more the measurement arrangement is like the calibration arrangement, the better.

E. Apparent Values

A simple but sometimes useful concept in radiometry is that of an apparent value. Often radiometric measurements are made at a distance through a partially absorbing atmosphere that has a complicated spectral transmission $\tau(\lambda)$. If the source has a spectral radiance $L_{s\lambda}$, then the irradiance at the aperture is given by

$$E_\lambda = \tau(\lambda)L_{s\lambda}\Omega$$

where Ω is the solid-angle field of view of the radiometer. In the measurement of course one determines the value of irradiance and calculates the source radiance from it:

$$L_{s\lambda} = E_\lambda / \tau(\lambda)\Omega$$

The fraction $E_\lambda / \tau(\lambda)$ is sometimes quite difficult to evaluate because $\tau(\lambda)$ is not known accurately or because both the source and atmosphere are virtually "combs" with complicated interactions. Some workers therefore report their evaluation of both $L_{s\lambda}$ and $\tau(\lambda)L_{s\lambda}$ or just the latter quantity which is the *apparent radiance*. The apparent radiance can usually be measured quite accurately, but it is directly useful only in the situation in which it was measured.

VII. RADIOMETER DESIGNS

A. INTRODUCTION

There are basically two types of radiometers: those which use a light chopper (episcotister) and those which do not. Each type has its characteristic advantages, disadvantages, and applications; and there are probably about an equal number of both in use. Most systems of either kind consist of an optical system for collecting enough flux to focus onto the transducer. Some place in the optical system is an element or a combination of elements that determine(s) the spectral responsivity. Some combination determines the angular responsivity, and similarly the temporal frequency response, etc. There are so many different ways to arrange the important optical elements that it is hard to be general about designs.

B. RADIOMETERS WITH CHOPPERS

Figure 13 shows the basic elements of a radiometer with a chopper. The chopper causes the detector to be irradiated alternately from the field of view which has in it the subject to be measured and from an internal reference which is a secondary calibration source. It provides the reference radiation level of the instrument. In theory the internal source provides a known amount and kind of radiation on the detector. The input signal to the electronics is essentially a square wave in which one level is that of the internal source and the other is unknown. Most instruments of this type then use a synchronous rectifier amplifier (SRA) to obtain a dc voltage proportional to the difference in levels, and an appropriate reference level is inserted so that the output represents an absolute value of radiation. Potential sources of error in such a system are: The detector usually views one field of view via an extra folding flat mirror; the radiation from the internal source is of a spatial distribution different than that from the scene; the internal reference may not be sufficiently accurate or uniform, the SRA may have jitter, wander, or other imperfections; and the waveform generated by the chopper may not be satisfactory. We discuss each of these in turn.

The mirror which folds the radiation onto the detector reduces the radiation from the internal source by its reflectivity $\rho(\theta_i,\phi_i,\theta_r,\phi_r,\lambda)$, emits additional radiation by virtue of its emissivity $\epsilon(\theta,\phi,\lambda)$ and its temperature, and reflects stray radiation to the detector. Because it is usually used at about a 45° angle it can also introduce polarization (which is usually small because the mirrors are metallic reflectors with a thin protective overcoat). Figure 13 shows one geometric configuration for the placement of the detector, primary, calibration source, and chopper mirror. The detector is flooded with flux from the source, but it also receives inputs from other portions of the instrument. The system is ac-coupled and only sees changes in radiation level. If the instrument stays at a constant temperature, this *may* be all right. The chopper moves, so there are also requirements for the uniformity of the radiation levels in the instrument. For measurements of high accuracy and low flux levels this arrangement is not recommended—but so much of what is satisfactory depends upon the intended use that instruments of this configuration should not be immediately condemned.

One possibility of course is to make the source fill the entire F cone of the detector, just as the primary optical element should, and a properly designed field stop should define the field of view of the detector.

Let us now consider the effect of the mirror on a cone that varies from $F = 1$ to $F = 3$, typical values for most such radiometers. The chief ray reflects at 45°, and marginal rays reflect at angles $45 \pm \alpha$ where α is the half-angle of the cone. For an $F/1$ cone the angles vary from 18° to 72°. The degree of polarization introduced varies. If the detector is

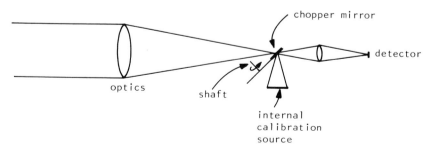

FIG. 13. A generalized chopper radiometer.

polarization-sensitive, this introduction of polarization into the internal calibration signal can be important.

C. CHOPPER CONSIDERATIONS

Two different chopper implementations are in common use: the butterfly rotating chopper and the canted mirror on the end of a tuning fork. Neither extinguishes the beam instantaneously; each extinguishes the beam gradually, and the waveform can be calculated by a convolution of the source image and the chopper blade. This process combined with the transfer function of a detector provides the signal waveform. The geometry is about as shown in Fig. 14. The shape of the chopper can be designed to taste, but the simplest is a radial sector. The cross section of the optical beam is certainly not uniform. It can be modeled, however, as a uniform, Gaussian, or bessinc $J_1(x)/x$—or as a detailed measurement may provide.

In each there is a leading edge, the duration of which is a function of the geometry and the speed of rotation. The essence of the accuracy of the SRA technique is to time an electronic gate to open at t_1, close at t_2, and measure the level V during the period t_1 to t_2. In this way the shape of the waveform can be rendered unimportant. There is a trade-off. The two

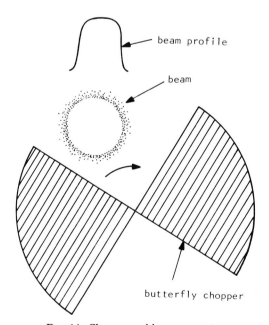

beam profile

beam

butterfly chopper

FIG. 14. Chopper and beam geometry.

gates should be opened as far from the leading and trailing edges as possible to avoid errors that accrue from including the lower signal levels, but the sampling time of V should be as long as possible to smooth the noise. It is tempting to arrange the chopper so that most of the time is spent in viewing the scene. The calibration level can be set higher so the signal-to-noise ratio is greater and less time is needed for calibration. This contradicts rule (2): The calibration should be as much like the measurement as possible. One *can* set the flux level of the calibration signal higher, but it must stay in the range of linearity and the spectral distribution must not change significantly.

Since most radiometers use thermal detectors which have time constants of about 10^{-3} s, the waveform has to rise for about 10^{-2} s and decay for the same amount of time. Therefore the calibration signal should be flat for 5×10^{-2} s or longer. Generally calibration takes about one-fourth the measurement time, so a full chopper rotation (which usually includes two calibrations and two measurements) takes about $\frac{1}{2}$ s.

D. Radiometers without Choppers

Since the modulation introduced by an optical chopper does not exist in this type of radiometer, the reference temperature or reference radiation level is determined by the detector itself. Systems of this type enjoy the advantages of lower power consumption, because they do not require a heated cavity or a chopper motor. However, they are subject to drift and to the difficulties of direct-coupled electronics. The calibration techniques and determination of the reference level can often be difficult. In some cases these determinations are not critical: for example, the measurement of solar flux levels. The radiation is of such a level and character that small drifts in the instrument and small ambient flux changes cause negligible error. This is the basis of most pyrometer and pyrheliometer designs. In cases for which these effects cannot be ignored a straightforward approach is to measure the temperature of the detector and use it as the reference level.

We had the opportunity to design a radiometer which would measure continuously the temperature of a nearby 1-m patch of ground at a remote location. The design of this radiometer serves as an illustration of the general type (Palmer, 1973). Some of its characteristics of course were determined by the specific application.

Figure 15 is a line drawing of the radiometer. The optical collector is a fairly massive polished cone of 20° half-angle. This gives good radiative coupling between the detector and the patch of ground and poor coupling to the instrument. Entering radiation next passes through a wide-band

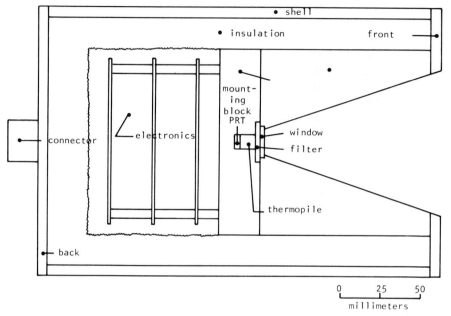

FIG. 15. Radiometer cross section.

protective window and filter (to define the 10.7–12.9 μm band) to the thermopile detector assembly. The spectral band is chosen to reject sunlight and the silica reststrahlen band; a platinum resistance thermometer (PRT) is placed as close as possible to the thermopile to measure its temperature. This is a unique and important feature. The electronics follow. The entire instrument is encased in a protective aluminum shell and painted white for thermal reasons. Foam insulation is used appropriately. A PRT* is, with aluminum-filled epoxy, affixed to the outside of the TO-5 can which houses the thermopile detectors. The cold junctions of the elements of the thermopile† are heat-sunk to the housing. In this way the reference temperature of the radiometer is measured by a PRT.

The radiometer was arranged electronically to measure temperature. In principle, if the radiometer responds equally to power at all wavelengths, the output voltage is proportional to the fourth power of the temperature and a logarithmic fourth-rooting circuit will give a direct temperature readout. This radiometer was responsive over only a small spectral interval not far from the maximum of the blackbody curve. Therefore a

---

* Model 118L from Rosemont Engineering Company.
† Model C-1 from Sensors, Inc.

calculation based on the integral over the band was made, and experimental curve fitting by trimming resistors was accomplished. The procedure for this calibration was simply to fill the field of the radiometer with a blackbody of known temperature and plot the calibration-output voltage as a function of source temperature. Resistors in the antilog circuit were trimmed until a linear responsivity was obtained.

The temperature characteristic of the radiometer was then checked. Since the detector views a portion of the radiometer, there is an output signal due to the emission from the radiometer itself. This responsivity-versus-temperature curve is shown in Fig. 16. The nature of the curve is interesting. At temperatures above 40°C the temperature rate of change in

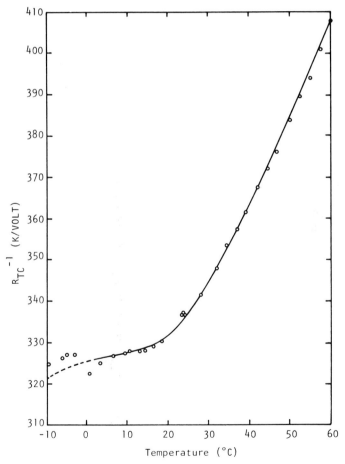

FIG. 16. (Radiometer responsivity)$^{-1}$ versus radiometer temperature.

the reciprocal responsivity is about 20 V^{-1}. This is clearly due to the input of appreciable radiation from the radiometer linearized by the antilog circuit. Below 20 K a new slope that is about zero shows there is virtually no change with radiometer temperature. The conclusion to be drawn is that the contribution of radiometer self-radiation became negligible. This curve is a function of the temperature of the calibration source.

The theoretical curve takes into account the processing circuitry. If indeed all the radiation is sensed, then one can use the Stefan–Boltzmann law for input with constants A and B to account for the different geometric effects. Since there is a fourth-rooting circuit, the output voltage is

$$V_{\text{out}} = K(AT_s^4 + BT_R^4)^{1/4}$$

The responsivity is based on the source temperature T_s and the output voltage

$$\mathscr{R} = \frac{K(AT_s^4 + BT_R^4)^{1/4}}{T_s} = KA^{1/4}\left[1 + \frac{B}{A}\left(\frac{T_R}{T_s}\right)^4\right]^{1/4}$$

From this it is easy to see that, when there is no radiometer contribution, $B = 0$ and the responsivity is given by $KA^{1/4}$, where A is a geometric constant. As T_R/T_s increases the responsivity increases. In fact, when the second term in the root is large with respect to 1, the responsivity is just proportional to T_r/T_s; the higher the radiometer temperature, the greater the (static) responsivity. Conversely, when T_r/T_s is small enough, either because T_r is small or T_s is large, the responsivity is just $KA^{1/2}$ and independent of radiometer or source temperature. More careful analysis would include integration of the Planck curve over the spectral band of the radiometer and the proper circuitry for temperature linearization.

VIII. RADIATION STANDARDS

A. Introduction

In discussing radiation standards, I discuss what I consider the only absolute radiometric measurements one ever makes. The major laboratories of the world making these measurements are the NBS in the United States, the National Physical Laboratory (NPL) in Great Britain, the Technische Physikalische Reichanstalt in Germany, the Bureau des Poids et Mesures in France, and the National Research Council in Canada. Their techniques are similar, so it should suffice to discuss the accuracies and procedures at the NBS and compare the others to it.

There are in concept two different ways to establish the fundamental

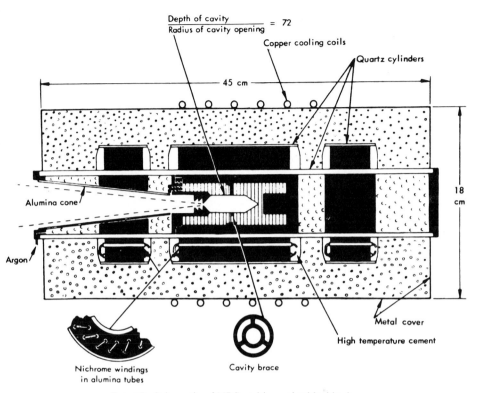

FIG. 17. Schematic of NBS melting-point blackbody.

standard of radiant power. One is to use a source which in theory provides an exact amount of flux in some spectral region that can be calculated from constructional details and other measurements. The other is to construct a receiver that can be similarly calibrated. The standard source has received much more attention over the years, but today it seems that the standard receiver is beginning to provide superior results.

The main idea of the standard source is the use of a cavity radiator which must be as close a realization of a blackbody as possible. Blackbody cavity theory predicts the shapes and sizes of such cavities with reasonable accuracy, and practicality enters the design choices as well. The cavity, in short, ought to be large and deep compared to the dimensions of the aperture. It should be isothermal, and it should be at exactly the temperature that its monitoring "thermometer" records. Although much good research has been done by the NBS on such bodies (Richmond *et al.*, 1962), the only official standard which utilizes a cavity radiator is that of luminance. The best of these designs seem to be for the melting-point standards, as shown in Fig. 17. The gold, platinum, or other metal is heated at a constant rate, and a PRT is used to measure the temperature. As the melting temperature is reached, the time–temperature curve becomes flat as the input energy is absorbed during the change of state of the metal. This provides a relatively stable and well-known temperature. The curve is level for about 30 min, thereby allowing time for thorough equilibration. The temperature is known to the extent that the pressure above and purity of the metal are known. The cavities that have been used so far have been long cylinders with conical ends or double-ended cones. Unfortunately they tend to develop temperature gradients because they have varying view factors within the cavities. The effect is hard to analyze. The emissivity is usually calculated on the basis of the cavity geometry and the assumption of isothermality.

It is noteworthy that the only standard specifically required of the NBS by charter is for the lumen. It is based on the flux from a blackbody at the temperature of freezing platinum. The cavity used is similar to that described above in the shape of a cylinder and a cone. The last time this "standard candle" was used to calibrate the secondary standards was 1931.

B. NBS STANDARDS

The primary standards of emission in the 0.3–0.8 μm range are a blackbody maintained at a known temperature of the International Practical Temperature Scale (1968). Deuterium lamps used as secondary sources are calibrated as to spectral shape with synchrotron radiation and with respect to magnitude with tungsten lamps. Tungsten strip and ribbon lamps are used in the region 0.3–2.8 μm. The relative calibrations of secondary-detector standards are made with special-window thermopiles in blackened cavities, and the absolute values are found calorimetrically. The currently available standards are listed in Table VII.

The NBS has described its technique for converting the basic standard of radiance to a standard of irradiance. The latter is often more convenient to use (Saunders, 1977). The irradiance standard is also a quartz, halogen, tungsten lamp, but the filament is a coiled coil rather than a strip. The technique for carrying out this transfer illustrates several useful techniques of radiometry, including the use of substitution, an integrating sphere, detector linearity considerations, and great care.

A monochromator and telescope are arranged as shown in Fig. 18; then the sources of radiance and irradiance are placed in the "source position" alternately and signals compared.

TABLE VII

SOURCE AND DETECTOR STANDARDS CURRENTLY AVAILABLE FROM THE NBS

Source	Wavelength region (nm)	Uncertainty (%)
A. Spectral radiance transfer source standards		
Argon mini-arc	114–140	10
Argon mini-arc	140–350	5
Deuterium lamp	165–350	10
Tungsten strip lamp	225–2400	
Tungsten strip lamp	At 225	2
Tungsten strip lamp	At 650	0.7
Tungsten strip lamp	At 2400	0.6
B. Spectral irradiance transfer source standards		
Deuterium lamp	200–350	6
Mercury lamp	253.7	5
Tungsten halogen	250–1600	
Tungsten halogen	At 250	3
Tungsten halogen	At 555	1.2
Tungsten halogen	At 1600	1.2
C. Spectral irradiance nonportable transfer source standard		
SURF storage ring	5 plus infrared	~5
D. Spectral irradiance transfer detector standards		
Vacuum ultraviolet windowless diode	5–125	5–10
Vacuum ultraviolet window diode	115–253.7	5–10
Ultraviolet-B diode	200–320	10
Silicon photovoltaic	257	5
Silicon photovoltaic	364	5
Silicon photovoltaic	420–700	2
Silicon photovoltaic	700–1150	5
E. Spectral irradiance high-level primary detector standard		
Electrically calibrated pyroelectric	257–2000	1
	2000–14000	2
NET (1 sec) = 1 W (in air), 100 nW (in *vacuo*)		

The use of an integrating chamber in the "source position" with the sources alternately illuminating it is preferred to make all the geometric aspects of the two measurements identical. When this is done, the output for the radiance source is

$$V = \int_{\Delta\lambda} \int_A \int_\Omega L_\lambda R(\lambda) \cos \theta \, d\Omega \, dA \, d\lambda$$

For the irradiance source it is

$$V = \int_{\Delta\lambda} \int_A E_\lambda R(\lambda) \, dA \, d\lambda$$

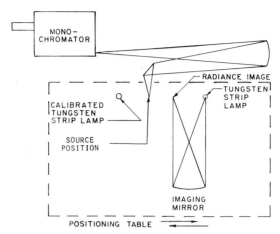

FIG. 18. Measurement setup for calibrating the radiance of the image.

The null or equal-signal condition provides that these are equal. If the responsivity is independent of angle of incidence and the projected solid angle is the same over the receiver area A, then L, E, and R can be "pulled through" the $d\Omega$ integral and

$$\int_{\Delta\lambda} \int_A E_\lambda R(\lambda)\, dA\, d\lambda = \int_{\Delta\lambda} \int_A L_\lambda R(\lambda)\, dA\, d\lambda \int_\Omega \cos\theta\, d\Omega$$

Then

$$\overline{E}_{\Delta\lambda} = \overline{L}_{\Delta\lambda}\Omega_p$$

where $\overline{E}_\lambda$ and $\overline{L}_\lambda$ are effective radiometric quantities and Ω_p is the projected solid angle. For the experimental arrangement shown here and used by the NBS the solid angle did not vary by more than 0.01%. This still leaves one with the situation of verifying that

$$E_{\Delta\lambda A} = L_{\Delta\lambda A}\Omega_p$$

where the weighted averages are over space and spectrum. They must be evaluated separately. In theory of course the effective value can be a function of both, even over a very small wave band. A mapping of the aperture to establish uniformity is necessary, and this was done. The problem then is to calculate Ω_p and to make sure $\overline{E}_{\Delta\lambda\,A}$ and $\overline{L}_{\Delta\lambda\,A}$ are separable so that the weighting is only over λ. Then, using the spectrometer, a spectral bandpass as narrow as possible is used. The uniformity with angle must be established experimentally, and it was. Therefore

$$\overline{E}_{\Delta\lambda} = \overline{L}_{\Delta\lambda}\Omega_p$$

The integrating sphere is used first to project the radiance source flux onto the entrance pupil of the monochromator via the mirror telescope. Then the sphere is rotated so that the entrance port of the sphere is irradiated by the source of irradiance. The flux which enters the sphere is the product of the port area and the flux density at the aperture. By translating the sphere (or the irradiance source) one can also determine the area distribution of the source. One way to view the operation of the integrating sphere is as a blackbody-type cavity radiator with an input. The radiation undergoes very many reflections off the walls. The wall material is made to be as diffuse and as reflective as possible. Then very little radiation is lost, and the radiation is made as uniform in angle, area, polarization, and coherence as possible.

The NBS workers report their uncertainties in the realization that "each of the following error sources contributed 0.01% or less: Nonlinearity of the detector-amplifier system, wavelength error, stray light, lamp polarization, lamp current measurements, and the integrating sphere response to the different solid angles. . . . Each of the following contributed 0.05% or less: The projected solid angle determination, uniformity of mirror M1, and spectral light scattering due to the slit function wings. Finally the non-uniformity of the radiance sources over the viewing error contributes an uncertainty varying from 0.12% in the uv to 0.05% in the ir."

The rms total error resulting from these sources is 0.17–0.20%. This is considerably smaller than most measurements in any part of the spectrum, but it is just the transfer from a radiance source to a source of irradiance—and it does not include a major error when more than one source is used. This drift error of the lamps is estimated at 0.8%, which dominates the 0.2% error evaluated above. The authors also list the total errors with respect to S1 units and in terms of reproducibility (see Table VIII). They represent the best that one can do today in this spectral region.

C. NPL STANDARDS

According to Gillam (1977) the NPL undertakes the following main types of calibration:
(1) Spectral radiance and irradiance 0.2–0.8 μm.
(2) Luminous intensity and luminous flux.

TABLE VIII

Standard Uncertainties

Quantity	Wavelength (μm)							
	0.25	0.35	0.45	0.555	0.6546	0.800	1.300	1.600
Radiance								
Absolute	1.66	1.20	0.93	0.65	0.65	0.46	0.28	0.27
Reproducibility	1.06	0.76	0.61	0.30	0.26	0.22	0.12	0.18
Irradiance								
Absolute	1.8	1.3	1.1	0.86	0.84	0.69	0.85	0.85
Reproducibility	1.26	0.95	0.81	0.64	0.60	0.56	0.81	0.82

(3) Color temperature.
(4) Luminance temperature.
(5) Total radiant emission.
(6) Detector spectral sensitivity 0.240–2.8 μm.
(7) Detector sensitivity to broad-band radiation (with nonselective detectors).

IX. RADIATION PYROMETRY (RADIOMETRIC MEASUREMENTS OF TEMPERATURE)

A. Introduction

The word "pyrometry" originates from the Greek word for fire (*pyr*) and the subject usually deals with the measurement of high temperatures. This section deals with such radiometric measurements of temperature whether or not they are high. There are three ways to make such measurements: measuring the total amount of radiation, measuring the spectral distribution of the radiation, and measuring the radiation at a single wavelength. Each of these techniques and the names generally ascribed to them are discussed below.

B. Radiation Temperature

The name universally applied to the temperature determined from the total radiation emitted from an object is "radiation temperature." One formal definition is:

The radiation temperature of an object is the temperature of a black-body that gives the same quantity of total integrated radiation.

The total radiation (radiant exitance) from a blackbody is given by the Stefan–Boltzmann law:

$$M = \sigma T^4$$

The total radiant exitance from an arbitrary body is given by

$$M = \int_0^\infty \epsilon(\lambda) c_1 \lambda^{-5}(e^x - 1)^{-1} \, d\lambda = \epsilon_{\text{eff}} \sigma T^4$$

where

$$\epsilon_{\text{eff}} = \int \epsilon(\lambda) c_1 \lambda^{-5}(e^x - 1)^{-1} \, d\lambda / \sigma T^4$$

Therefore the radiation temperature T_r can be found by equating σT_r^4 and $\epsilon_{\text{eff}} \sigma T^4$ so that

$$T_r = \epsilon_{\text{eff}}^{1/4} \, T$$

where T is the thermodynamic or "true" temperature of the body. It is easy to show that the relative difference between the radiation temperature and the true temperature is $1 - \epsilon_{\text{eff}}^{1/4}$.

Few, if any, instruments really measure the irradiance over an infinite spectral band, and the definition is only true for such a measurement. The instrument parameter can be included in the integral as a responsivity:

$$T_r = \int_0^\infty \epsilon(\lambda) R(\lambda) M(\lambda) \, d\lambda = (\epsilon R)_{\text{eff}} \, \sigma T^4$$

$$(\epsilon R) = \int_0^\infty \epsilon(\lambda) R(\lambda) M(\lambda) \, d\lambda / \sigma T^4$$

$$T_r = [(\epsilon R)_{\text{eff}}]^{1/4} \, T$$

It is impossible in general to proceed any further in the analysis of the error in this measurement of temperature unless further information is provided. This can be in terms of the relative spectral distributions or absolute levels, but note that $\epsilon(\lambda)$ and $R(\lambda)$ cannot be separated.

C. BRIGHTNESS TEMPERATURE

This measure of temperature, also called radiance temperature, is based on the amount of radiation in a narrow spectral band. The radiance temperature of a body is the temperature of a blackbody that gives the same radiance in a narrow spectral band as the body.

This can be written in mathematical terms:

$$L_\lambda^{BB}(T_B) = \epsilon\, L^{BB}(T)$$

where T_B is the brightness temperature, T the true temperature, and ϵ the emissivity in the narrow spectral band. The detailed description in terms of the Planck equation is

$$\exp(c_2/\lambda T_B) - 1 = [\exp(c_2/\lambda T) - 1]\epsilon^{-1}$$

The usual approximation, one which permits analytic, closed-form results, is that Wien's law applies, so that

$$\frac{c_1}{\lambda T_B} = \frac{c_2}{\lambda T} - \ln \epsilon, \qquad T = T_B\left(\frac{\lambda T_B}{c_2}\ln \epsilon + 1\right)^{-1}$$

Under this assumption the relative difference between T and T_B is

$$-(\lambda T_B/c_2)\ln \epsilon = -x_B \ln \epsilon$$

It is interesting to note that the relative error is a function not only of the emissivity but also of the wavelength and brightness temperature itself and is in fact a function of the normalized radiation variable x_B.

If Planck's law is retained, then the true temperature can be written

$$T = (\lambda/c_2)\ln\{\epsilon[\exp(c_2/\lambda T_B) - 1] + 1\}$$

The relative error is given by

$$1 - (T_B/c_2)\ln\{\epsilon[\exp(c_2/\lambda T_B) - 1] + 1\}$$

D. Distribution Temperature

This measure of temperature has several aliases and several variants. It is sometimes called distribution temperature, ratio temperature, and color temperature. I prefer retention of "color temperature" for the quantity that involves the chromaticity diagram and the visual sensation of color. The distinction between "ratio temperature" and "distribution temperature" is better made after a short mathematical treatment of the subject. The ratio temperature can be defined as follows:

The ratio temperature is the temperature of a blackbody that gives the same ratio of radiances at two distinct wavelengths as the body in question.

In mathematical terms this can be written

$$\frac{L^B(\lambda_1, T_d)}{L^B(\lambda_2, T_d)} = \frac{\epsilon(\lambda_1)L^B(\lambda_1, T)}{\epsilon(\lambda_2)L^B(\lambda_2, T)}$$

A meaningful expression can be obtained if the Wien approximation can be used:

$$T = T_d \left[1 - \frac{T_d \ln \epsilon(\lambda_1)/\epsilon(\lambda_2)}{c_2 \left[(1/\lambda_1) - (1/\lambda_2)\right]} \right]$$

If the Wien approximation does not hold, and it is necessary to use the Planck expression, only an implicit relationship connects T_d with T:

$$\frac{\exp (c_2/\lambda_2 T) - 1}{\exp (c_2/\lambda_1 T) - 1} = \frac{\epsilon(\lambda_2)[\exp (c_2/\lambda_2 T_d) - 1]}{\epsilon(\lambda_1)[\exp (c_2/\lambda_1 T_d) - 1]}$$

When two narrow bands are used, I prefer the term "ratio temperature." It is also possible to use three or more bands and make a best fit to the distribution.

It is possible to extend this analysis to more complicated situations. For instance, if the object is seen through an atmosphere with spectral transmittance $\tau(\lambda)$, the ratio of irradiances in the two narrow spectral bands is found to be

$$\frac{E(\lambda_1, T)}{E(\lambda_2, T)} = \frac{\tau(\lambda_1)\epsilon(\lambda_1)\lambda_2^5(e^{x_2} - 1)}{\tau(\lambda_2)\epsilon(\lambda_2)\lambda_1^5(e^{x_1} - 1)}$$

The voltage ratio is the ratio of each irradiance times the spectral responsivity at that wavelength, and one finds

$$\frac{e^{x_2} - 1}{e^{x_1} - 1} = \frac{\tau(\lambda_2)\epsilon(\lambda_2)R(\lambda_1)E(\lambda_1)}{\tau(\lambda_1)\epsilon(\lambda_1)R(\lambda_2)E(\lambda_2)} = \frac{\tau_2\epsilon_2 V_1}{\tau_1\epsilon_1 V_2}$$

One can solve this implicitly or use the Wien approximation. A second extension is to assume that only part of the field of view of the radiometer is filled with the temperature T_m to be measured and the rest with some other background temperature T_b. Then the voltage ratio is given by

$$\frac{V_1}{V_2} = \frac{R_1 E_1}{R_2 E_2}$$

$$= \frac{R_1\tau_1\epsilon_1\lambda_2^5\{[\exp (c_2/\lambda_1 T_m) - 1]^{-1}F + [\exp (c_2/\lambda_1 T_b) - 1]^{-1}(1 - F)\}}{R_2\tau_2\epsilon_2\lambda_1^5\{[\exp (c_2/\lambda_2 T_m) - 1]^{-1}F + [\exp (c_2/\lambda_2 T_b) - 1]^{-1}(1 - F)\}}$$

where F is the fraction of the field of view subtended by the object to be measured and $1 - F$ is the rest.

The accuracy this measure can attain depends upon the values of the different parameters. It is not unreasonable to expect 10% error for realistic values.

The use of three colors and best-fit techniques seems like a straightforward extension of what has been discussed here. That extension is left for the reader to apply to his own special problems.

APPENDIX:
DERIVATION OF EXPANSIONS FOR INTEGRALS

The radiance expression in terms of wavelength is

$$L_\lambda = c_1\pi^{-1}\lambda^{-5}[\exp(x) - 1]^{-1} = 2c^2h\lambda^{-5}[\exp(x) - 1]^{-1}$$

But

$$L_\lambda \, d\lambda = L_x \, dx$$

$$L_x = L_\lambda \frac{d\lambda}{dx} = - L_\lambda \frac{c_2}{T}\frac{1}{x^2}$$

Therefore

$$\int_0^\lambda L_\lambda \, d\lambda = \int_\infty^x L_x \, dx = 2c^2h \left(\frac{c_2}{T}\right)^4 \int_\infty^x x^3 \, [\exp(x) - 1]^{-1} \, dx$$

The photon radiance expression is

$$L_{q\lambda} = 2c\lambda^{-4}[\exp(x) - 1]^{-1}$$

By similar substitution it is easy to show that

$$\int_0^\lambda L_{q\lambda} \, d\lambda = 2c \left(\frac{c_2}{T}\right)^3 \int_\infty^x x^2 \, [\exp(x) - 1]^{-1} \, dx$$

These two expressions can be handled as a pair by noting their great similarity. Now the denominator exponential term can be recognized as the sum of a geometric series so that

$$\int_0^\lambda L_\lambda \, d\lambda = \int_\infty^0 L_x \, dx = 2c \left(\frac{c_2}{T}\right)^3 \int_\infty^x \sum_{m=1}^\infty x^2 e^{-mx} \, dx$$

The summation and integration operations can be interchanged and the integral integrated by parts:

$$\int_0^\lambda L_\lambda \, d\lambda = 2c^2h \left(\frac{T}{c_2}\right)^4 \sum_{m=1}^\infty m^{-4}e^{-mx}[(mx)^3 + 3(mx)^2 + 6mx + 6]$$

$$\int_0^\lambda L_{q\lambda} \, d\lambda = 2c \left(\frac{T}{c_2}\right)^3 \sum_{m=1}^\infty m^{-3}e^{-mx}[(mx)^2 + 2mx + 2]$$

The temperature derivatives are only a little more difficult:

$$\frac{\partial L_\lambda}{\partial T} = \frac{xe^x L_\lambda}{T(e^x - 1)} = \frac{2c^2h}{T} \left(\frac{T}{c_2}\right)^4 \int_\infty^x \frac{x^4 \, e^x}{(e^x - 1)^2} \, dx$$

$$\frac{L_{q\lambda}}{T} = \frac{xe^x L_{q\lambda}}{T(e^x - 1)} = \frac{2c}{T} \left(\frac{T}{c_2}\right)^3 \int_\infty^x \frac{x^3 \, e^x}{(e^x - 1)^2} \, dx$$

The integrals have terms $e^x(e^x - 1)^{-2}$ rather than $(e^x - 1)^{-1}$, but this is no problem. It can be written

$$\frac{e^x}{(e^x - 1)^2} = \frac{e^x}{e^{2x} - 2e^x + 1} = \frac{1}{e^x - 2 + e^{-x}}$$

Then long division shows

$$
e^x - 2 + e^{-x} \overline{\big)\,1} \quad \begin{array}{l} e^{-x} + 2e^{-2x} + 3e^{-3x} + \cdots \\[4pt] \hline \end{array}
$$

$$
\begin{array}{r}
1 - 2e^{-x} + e^{-2x} \\ \hline
2e^{-x} - e^{-2x} \\
2e^{-2x} - 4e^{-2x} + 2e^{-3x} \\ \hline
3e^{-2x} - 2e^{-3x}
\end{array}
$$

It is just the sum $\sum_{m=1}^{\infty} m\, e^{-mx}$. Therefore

$$\int \frac{\partial L_\lambda}{\partial T}\, d\lambda = \frac{2c^2 h}{T}\left(\frac{T}{c_2}\right)^4 \sum m^4 e^{-mx}[(mx)^4 + 4(mx)^3$$
$$+ 12(mx)^2 + 24mx + 24]$$

$$\int \frac{\partial L_{q\lambda}}{\partial T}\, d\lambda = \frac{2c}{T}\left(\frac{T}{c_2}\right)^3 \sum m^{-3} e^{-mx}[(mx)^3 + 3(mx)^2 + 6mx + 6]$$

REFERENCES

Geist, J. (1976). Trends in the development of radiometry, *Opt. Eng.* **15**, 537–540.

Gillam, E. J. (1977). *Appl. Opt.* **16**, 300–301.

Harvey, J. E. (1976). Light scattering characteristics of optical surfaces, Ph.D. Dissertation, University of Arizona Tuscon.

Jones, R. (1963). Terminology in photometry and radiometry, *J. Opt. Soc. Am.* **53**, 1314–1315.

Judd, D. B. (1967). *J. Opt. Soc. Am.* **57**, 445.

Kirchhoff, G. (1860). On the relation between the radiating and absorbing powers of different bodies for light and heat, *Phil. Mag. J. Sci.* **20**.

Nicodemus, F. E. (1969). Optical resource letter on radiometry, *J. Opt. Soc. Am.* **59**, 243.

Nicodemus, F. E. (1970). Reflectance nomenclature and directional reflectance and emissivity, *Appl. Opt.* **9**, 1474–1475.

Nicodemus, F. E. Radiometry, Selected Reprints, American Institute of Physics, New York.

Palmer, J. M. (1973). A radiometer for remote measurements of earth surface temperatures, M.S. Thesis, University of Arizona, Tuscon.

Richmond, J. C., Harrison, W. N., and Shorten, F. J. (1962). An approach to thermal emittance standards, Symposium on Measurement of Thermal Radiation Properties, Dayton, Ohio, Sept. 5–7, pp. 403–423 (NASA SP-31, Washington, D. C.), reprinted in NBS Special Publication 300, Vol. 7, Precision Measurement and Calibration: Radiometry and Photometry.

Saunders, R. D., and Shumaker, J. B. (1977). Optical radiation measurements: The 1973 NBS scale of spectral irradiance, U. S. Government Printing Office, Washington, D.C.

Smith, W. J. (1966). "Modern Optical Engineering," McGraw-Hill, New York.

Sparrow, E. M., and Cess, R. D. (1970). "Radiation Heat Transfer," Brooks/Cole, Belmont, California.

CHAPTER 6

The Calculation of Image Quality

WILLIAM B. WETHERELL

Optical Systems Division
Itek Corporation
Lexington, Massachusetts

I. INTRODUCTION

Most optical systems have as their principal function the formation of aerial images, either for detection and recording by an image sensor such as photographic film, or for direct visual examination. The prospective user of such an image-forming optical system requires that its image quality be specified in terms related to its application. The engineers designing the optical system must be able to relate its image quality to design parameters they can control. The field of image quality analysis is concerned with the development of quantitative merit functions which can be used to calculate image quality in user terms from lens design parameters defined in engineering terms.

It is the purpose of this chapter to review the foundations of modern image quality analysis, to show how image quality may be calculated from basic optical design parameters, and to discuss those aspects of optical systems which most affect image quality. The point of view throughout is that of an optical engineer concerned with the design of optical systems whose image quality approaches the limits imposed by diffraction. Most of the chapter is devoted to the optical portions of the imaging process; only limited attention is given to image sensors, although they are a vital factor in determining the performance limits of the complete image-forming system.

A. HISTORICAL BACKGROUND

Modern concepts of image quality have very diverse roots. Traditionally, image quality merit functions have been defined by the user communities. Since optical instruments have many different scientific, industrial, medical, and military applications, the result has been a chaotic array of special merit functions and nomenclature that has tended to disguise similar measures of image quality behind radically different terminologies. Developments during the middle half of the current century based on two-dimensional Fourier analysis have unified many of these root sources into a common structure, showing how both new and traditional merit functions can be related quantitatively to optical system parameters. Use of the newer image quality merit functions which have arisen from these developments is not universal, being confined primarily to applications requiring the highest image quality. There also remains some diversity in nomenclature, as reflected in the text.

The basis of modern image quality analysis is the perception of the image-forming process as a two-stage Fourier transformation. Both object and image are perceived as two-dimensional distributions of light of varying brightness. When a lens forms an image of an object, the object distribution of light is in effect transformed from the linear domain into an object spectrum in the spatial frequency domain, passed through a spatial frequency bandpass filter, and retransformed into the linear domain to become the image distribution of light. Viewed in this framework, the lens system is a low-pass spatial frequency filter. Its effect on image quality can therefore be specified in terms of a spatial frequency bandpass curve. The latter is termed the *optical transfer function* (OTF) in current nomenclature.

The OTF is the core of modern image quality theory. Later in the chapter, more precise definitions will be given, and we will show how the OTF is computed from basic lens design parameters and how it relates to other image quality merit functions. For the moment, we wish to show how the concept developed historically.

The OTF was not invented in one step. It is the result of a series of independent developments whose true significance was not fully recognized until the second quarter of the present century. The earliest step in the process was the invention of the diffraction grating by Rittenhouse, in 1785, and by Fraunhofer, independently, several decades later (Hecht and Zajac, 1974). The diffraction grating shows that a periodic structure diffracts collimated light, transforming it into an "angular spectrum" of plane waves. The angular divergence of each diffracted plane wave, or *sideband,* from its original direction of travel is a function of the spatial frequency of the grating and the wavelength of the light. If an object can be perceived to be composed of a spectrum of amplitude gratings of different spatial frequency, the object spectrum mentioned earlier can be seen as an analogy for the angular spectrum of plane waves arising from diffraction.

The next step was made by Abbe in 1873 (Abbe, 1873; Martin, 1966).* Abbe was studying the theory of microscopes, with the hope of designing instruments capable of seeing finer details. In examining diffraction gratings with a microscope, using collimated light for illumination, he discovered that no image of the grating was formed unless diffracted light from at least the first-order sidebands passed through the microscope. Since each diffracted sideband formed a point source of light at the rear focal point of the lens, Abbe viewed image formation as a double-diffraction process: first, diffraction by the grating, and second, interference between wavefronts originating at the point sources formed by the different sidebands. Thus the smallest grating period, or highest spatial frequency, which can be seen through a microscope is a function of the sine of the cone angle subtended by the aperture of the microscope objective at the object. The numerical aperture (NA) is now used to specify this cone angle.

Abbe's analysis was thought to apply only to illuminated targets, or imagery by coherent light, in present nomenclature. In 1896, Lord Rayleigh (Martin, 1966; Rayleigh, 1896, 1964a) showed that the same forms of analysis could be applied to self-luminous objects (incoherent imagery). He examined image formation for both illuminated and self-luminous objects in the form of arrays of equally spaced point and line sources, using Fourier techniques to determine the resolution limit (he coined the term *"resolving power"*) in each case. He found a factor of 2 difference in the resolution limit between coherent and incoherent imagery. More importantly, in historical perspective, he demonstrated the applicability of Fourier techniques to image quality analysis.

---

* Chapter VI of Martin (1966) reviews the history of early work on the microscope.

Both Abbe and Rayleigh were concerned primarily with limiting resolution. The modern concept of the OTF is concerned with the quality of reproduction at each spatial frequency below the limiting resolution. The transition from limiting resolution to the OTF is hard to date specifically. It seems to have occurred in a number of steps during a period 30–50 years after Lord Rayleigh's paper was published, without any overt connection with that paper. In most of these cases, an extension of analysis techniques applied to audio-frequency electric circuits was involved.

Frequency analysis is a natural development in audio systems, because of the nature of human hearing. Its extension to electric circuit analysis is also natural, since electric circuits were used often to reproduce sound and since they involved time sequences of events. The extension of frequency analysis to optical imagery began in the late 1920s and early 1930s with the development of two technologies: (1) the recording of sound on film for talking pictures and (2) the sequential scanning of images for transmission over wire (phototelegraphy) and over radio waves (television). What may be the earliest published OTF appeared without fanfare in a paper by MacKenzie (1928). It is actually an audio-frequency response curve for a complete variable-density sound on film recording system, showing the effects of the recording slit width on high-frequency response and how to compensate for the resultant performance losses.

Other early examples of more direct interest appear in papers by Mertz and Gray (1934) and Frieser (1935). The former is concerned with phototelegraphy and shows the effects of different forms of scanning apertures on frequency response, expressed in terms of an "equivalent transfer admittance" function. This is a landmark paper which can still be read with value today. Frieser's paper includes the earliest published "sine wave response" curves for photographic film in essentially the same form in which they are now plotted.

The full application of Fourier techniques to optical imagery did not occur until after World War II. The most historically significant postwar publications were a book by Duffieux (1946, 1965) and a series of papers by Schade (1951, 1952, 1953, 1955). The former strongly influenced the development of image quality theory in Europe, and the latter strongly influenced American developments, particularly in the television industry.

B. FORMAT

The material in this chapter is presented in an order which builds from geometrical concepts of image formation and image quality and shows how these tie in with concepts based on two-dimensional Fourier analysis. Section II formally defines the terms "object," "image," and "image quality" as used in image quality analysis. Section III introduces the concept of the geometrically perfect lens and discusses its properties. Section IV discusses aberrations in geometrical ray tracing and shows their relationship to wavefront error. Section V generalizes the concept of wavefront error in the form of the complex pupil function, which is the most complete description of lens performance in terms directly connected to lens design parameters.

Section VI introduces the point spread function, one of the two most complete measures of image quality in user terms. Several corollary merit functions are also defined, along with techniques for calculating the point spread function (PSF) by Fourier transformation of the complex pupil function. Section VII formally defines the OTF, the second complete measure of image quality in user terms. Corollary merit functions are also discussed, along with techniques for calculating the OTF by Fourier transformation of the PSF and by autocorrelation of the complex pupil function. Section VIII discusses image sensors in terms of their relationship to optical image quality and describes measures of image quality which integrate the image sensor and the optics.

Sections IX, X, and XI return to the optical system and discuss how image quality is degraded by wavefront error, amplitude transmittance variations across the pupil, and image motion. The results of this discussion are used in Section XII to close the chapter with a brief discussion of several models of merit functions which may be of value in preliminary design analyses.

Because of the diffuse origins of modern image quality concepts, there is considerable diversity in nomenclature and symbology. Where standards exist (Inglestam, 1961) or have been suggested, as in symbols for radiometric quantities (Muray *et al.*, 1971), they are used. Otherwise, the author's idiosyncrasies are adhered to, as is customary. Where first defined, each optical term is printed in italics. In many cases, common variants found in the literature are given.

Considerable emphasis is given to one-number merit functions, discussing their advantages and shortcomings. The field of image quality analysis, in common with other fields of intellectual endeavor, is beset by an affliction the present author likes to call *"unimania"* —the belief that highly complicated processes can be compared fully and accurately using a single one-real-number merit function. In optics, this takes the form of overreliance on merit functions such as Strehl definition, rms wavefront error, and limiting resolution.

Unimania arises from an overzealous application of Ockham's razor in the search for a simple, unambiguous, universal merit function for comparing competing entities. The problem arises not in the one-number merit functions themselves but in their extension to areas of application where they lose relevance. One-number merit functions are in fact indispensable if used correctly. Their use should be confined to two circumstances: (1) cases where the merit function is an exact measure of performance for the particular application and (2) cases where the two optical systems being compared are identical in all respects except that measured by the merit function. In other cases, the use of a single one-number merit function may be misleading, to varying degrees. Using two different one-number merit functions simultaneously, such as low-contrast and high-contrast limiting resolution, can extend their range of application.

Most one-number merit functions for optical systems were developed for image quality degradation resulting from wavefront error. Other factors such as image motion and aperture obstructions also degrade image quality, but in subtly different fashions. We discuss some examples where using a merit function developed for wavefront error when other forms of image degradation are present can give misleading results. The application most often used for this purpose is the detection of a point source against a uniform background.

II. OBJECT, IMAGE, AND IMAGE QUALITY

The words "object" and "image" in optical terminology have formalized meanings which differ slightly from common usage. Both refer to distributions of light at surfaces, rather than real objects. This formalism is a mathematical convenience, making it possible to specify image quality in quantitative terms by comparing the two distributions of light. The use of surfaces rather than volumes arises from the nature of image sensors.

A. IMAGE SURFACE AND OBJECT SURFACE

An *image sensor* is a device for detecting and recording patterns of light. The sensing element which converts light energy into another, recordable form of energy may be termed the *image transducer*. The image transducer is a surface which may be flat or curved, but which has insufficient thickness at each point to affect image quality. (Some image sensors may have several image transducer surfaces, as in color film. Each surface is treated as a separate image sensor in these cases.) The *image surface* is the surface in lens system coordinates which coincides with the image transducer. The pattern of light projected onto this surface by the lens system is termed the *aerial image*.

Figure 1 shows the coordinate system for a rotationally symmetric *objective* lens system. It is used as the basis of all discussions in this chapter. Mathematically, an objective is a device for transforming the coordinates (x,y,z) in *object space* into the coordinates (x',y',z') in *image space*. Object space coincides with the real scene coordinates, extends in all directions to infinity, and is characterized by an index of refraction n. Image space is a mathematical construct, the transformation of object space. It also extends to infinity in all directions and is characterized by an index of refraction n'. The common denominator between the two coordinate systems is the *optical axis,* which is the axis of rotational symmetry of the objective and which coincides with the z and z' axes. The origin for (x,y,z) is at the front focal point F, and the origin for (x',y',z') is at the rear focal point F'. The objective is characterized by its front and rear focal lengths f and f', which are equal when $n = n'$. Primed quantities always refer to image space.

The position of the image surface is defined by the image transducer. For each image surface, there is one associated surface in object space, termed the *object surface,* whose location is defined by a point-for-point transformation of the image surface coordinates. These two surfaces are referred to as *conjugate image surfaces.* Either or both may be flat or curved, depending on the transformation characteristics of the objective. It is a common practice to define one surface as flat, the other being said

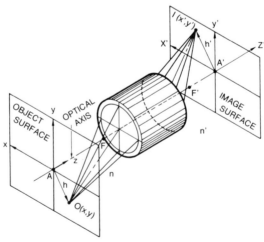

FIG. 1. Object space and image space coordinate systems for a rotationally symmetric lens system.

to show *field curvature* if it is the object surface, or *image curvature* if it is the image surface. The position of the axial intercepts A and A' are related by the Newtonian imaging equation

$$zz' = -ff' \tag{1}$$

where z is negative and z' is positive in the example in Fig. 1. Both coordinate systems are right-handed, with z and z' positive to the right.

In general, the discussion is confined to points within the conjugate image surfaces, so that the z and z' coordinates can be ignored. The exception is a brief discussion of depth of focus in Section IV,D. Most attention is paid to the conjugate pair of *object point* $O(x,y)$ and *image point* $I(x',y')$. Their positions are denoted by the two vectors, the *object height* h and the *image height* h'. The *lateral magnification* $m = h'/h$ defines the relative scale of the image and object surfaces. The optical axis, h, and h' all lie in a single plane termed the *meridional plane*. It is used to define the y and y' axes in the following discussions.

The object surface is assumed to be self-luminous and incoherent, having a radiance distribution $L_0(x,y)$. It is also assumed to be monochromatic, although polychromatic cases are commented on where appropriate. Lateral variations in $L_0(x,y)$ and how they are reproduced in the aerial image form the subject of image quality. The portion of the object surface of most interest to the observer is termed the *target*, and the rest is referred to as the *background*. In theoretical analyses, the background radiance is usually assumed to be uniform. Variation in the background radiance is called *clutter*.

Variation in $L_0(x,y)$ across the target is characterized by the *target contrast* C_t:

$$C_t = \text{maximum target } L_0/\text{minimum target } L_0, \tag{2}$$

which is a quantity of particular significance in image quality analysis. Where the range of radiance in background clutter far exceeds that in the target, the term *dynamic range* is applied. The image quality requirements of optical systems used for long-dynamic-range imagery, such as solar coronagraphy, differ substantially from those used for short-dynamic-range imagery. It is in this area that one-number image quality merit functions can be most abused, as in the use of high-contrast limiting resolution to compare systems which will be used for low-contrast imagery.

B. LINFOOT'S IMAGE QUALITY FACTORS

If the wavelength of light were infinitely short and the lens system perfect, the object radiance distribution $L_0(x,y)$ would be transformed into an ideal image irradiance distribution $E_0(x',y')$, whose only significant change would be that of lateral magnification. In practice, the real image irradiance distribution $E_i(x',y')$ differs from $E_0(x',y')$ as a result of the lateral spreading of light by diffraction and by imperfections in the design and fabrication of the lens system. Image quality measures the degree to which $E_i(x',y')$ differs from $E_0(x',y')$.

Linfoot (1956) has defined three quantities for comparing the image distribution to the object distribution. These are the *relative structural content* T_L,

$$T_L = \int E_i^2(x',y')\, dx'\, dy' \Big/ \int\int E_0^2(x',y')\, dx'\, dy' \tag{3}$$

the *image fidelity* Φ_L,

$$\Phi_L = 1.0 - \frac{\int\int [E_0(x',y') - E_i(x',y')]^2\, dx'\, dy'}{\int\int E_0^2(x',y')\, dx'\, dy'} \tag{4}$$

and the *correlation quality* Q_L,

$$Q_L = \int\int E_0(x',y')E_i(x',y')\, dx'\, dy' \Big/ \int\int E_0^2(x',y')\, dx'\, dy' \tag{5}$$

The integration limits are nominally $-\infty$ to $+\infty$, but integration over the image format is usually sufficient. Note that

$$Q_L = (T_L + \Phi_L)/2 \tag{6}$$

The relative structural content corresponds most closely to what is usually meant by image quality, measuring, in effect, the change in the variance of $L_0(x,y)$ as it is recorded. Relative structural content has limitations in cases where the optical and image-sensing systems can introduce spurious detail, as will be made evident later in the chapter. It is also insensitive to coordinate mapping errors arising from distortion. Image fidelity is extremely sensitive to mapping errors and is useful in applications where accurate mapping is critical. Correlation quality indicates the balance where both structural content and mapping fidelity are important.

Linfoot's quality factors are convenient merit functions where the object surface characteristics are well known in advance, in that they are one-number merit functions which are conveniently compared and directly relevant. For more general usage, some modification is required. In most real situations, $E_0(x',y')$ contains detail finer than can be reproduced by any real optical system. A more useful merit function would compare the actual optical system performance to that of the theoretically best possible lens system. A modification of Linfoot's relative structural content is developed for this purpose in Section VII,D. Before doing so, the standard of comparison must be defined and tools for calculating image irradiance developed.

III. PROPERTIES OF THE PERFECT LENS

The ideal image irradiance distribution $E_0(x',y')$ was defined in terms of a geometrically perfect lens and light of infinitely short wavelength. If a geometrically perfect lens is constructed, its performance will still be limited by diffraction and the wavelength of light. The longer the wavelength, the poorer the image quality. Thus a performance standard based on diffraction in a geometrically perfect lens is perforce a monochromatic standard. Also, the longer the wavelength, the greater the departure from geometrical perfection which can be tolerated in the lens design.

Analysis of the departure from geometrical perfection is the province of geometrical optics and ray tracing and is the basis of lens design concepts of image quality. The foundations of lens design theory lie in the properties of the geometrically perfect lens.

A. DEFINITION OF THE GEOMETRICALLY PERFECT LENS

If the lens in Fig. 1 were geometrically perfect, it would meet three conditions formulated by Maxwell in 1858 (Boutry, 1962):

(1) All rays from object point $O(x,y)$ which traverse the lens must pass through the image point $I(x',y')$.

(2) Every element of the plane normal to the optical axis which contains $O(x,y)$ must be imaged as an element of the plane normal to the optical axis which contains $I(x',y')$.

(3) The image height h' must be a constant multiple of the object height h, no matter where $O(x,y)$ is located in the object plane.

Violations of the first condition are known as *aberrations*, or in more general terms, *image degradation*. Violations of the second condition were introduced earlier as field curvature and image curvature. Violations of the third condition are termed *distortion*.

Violations of the second and third conditions are of secondary importance in discussing image quality. There are imaging systems in which they are critical, as in mapping cameras. There are also some imaging systems in which condition (1) is badly violated unless specific amounts of image curvature or distortion are present. A failure to come within tolerance limits of meeting Maxwell's first condition is critical in almost all cases.

The first condition may be restated in terms of physical optics: In a geometrically perfect lens, light takes identically the same time to travel from $O(x,y)$ to $I(x',y')$ along all ray paths traversing the lens. Lens designers use the *optical path length* (OPL) instead of the transit time. The OPL is defined by the line integral along any ray path s:

$$\text{OPL} = \int_{O(x,y)}^{I(x',y')} n(s)\ ds, \tag{7}$$

where $n(s)$ is the index of refraction at each point along the ray path s, and $ds^2 = dx^2 + dy^2 + dz^2$. The OPL is measured in units of length and equals the path length in a vacuum over which light would travel in the same time as along the actual path length in the lens.

If Maxwell's first condition is met, the OPL will be identical along all ray paths. Departures from the first condition are then measured in units of *optical path difference* (OPD), where the OPD measures the difference in the OPL between the given ray and a standard ray. If the OPD is a significant fraction of one wavelength, image quality is reduced below the limit set by diffraction. The OPD, which can be calculated directly from lens design parameters by ray tracing, is the foundation of lens design image quality analysis.

Lenses which meet Maxwell's first condition exactly are termed *stigmatic*. Perfectly stigmatic designs exist, but they are generally stigmatic for only one pair of on-axis conjugate image points. The province of lens design is to find optical configurations that greatly extend the region over which conjugate image points will be within a specified tolerance of being stigmatic. The conditions required for extending the region of stigmatic performance can be defined analytically (Boutry, 1962, Chap. IV).

Assume that the lens shown in Fig. 2 is exactly stigmatic at the conjugate image points A and A'. If it is to be stigmatic at the axial conjugates B and B' as well, it must satisfy the *Herschel condition:*

$$n\ dz\ \sin^2(u/2) = n'\ dz'\ \sin^2(u'/2) \tag{8}$$

If the lens is to be stigmatic at the off-axis conjugates C and C', it must satisfy the *Abbe sine condition:*

$$n\ dy\ \sin u = n'\ dy'\ \sin u' \tag{9}$$

In general, these two conditions cannot be met exactly and simultaneously except where $|u'| = |u|$, e.g., in a unit magnification relay. Since image transducers are two-dimensional and most lenses are designed for a specific pair of conjugate image surfaces, the Abbe sine condition is

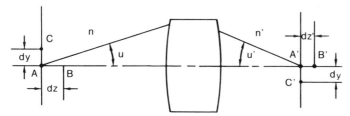

FIG. 2. Object and image parameters used to define the conditions for extended stigmatic performance.

usually given priority. Historically, the sine condition has played a large part in the development of lens design techniques, with the quantity *offense against the sine condition* (OSC) being used as an image quality criterion. OSC data on lenses are still published in some texts (Cox, 1964).* A lens which meets the sine condition and is stigmatic on axis is termed *aplanatic* and has acceptable image quality in the image surface region near the axis.

B. PARAXIAL PROPERTIES OF THE PERFECT LENS

In the *paraxial* region, where $\sin u \simeq u$ and $\sin u' \simeq u'$, both the Herschel condition and the Abbe sine condition are satisfied simultaneously. Thus a paraxial model of a lens system behaves like a perfect stigmatic lens. The ray aberrations and OPD of a real lens can be approximated using a power series of aberration coefficients whose terms are computed by tracing two paraxial rays through the lens. Thus the paraxial lens model is a fundamental part of lens design image quality analysis.

Figure 3 illustrates the more important paraxial optical parameters with a section in the meridional plane of the objective lens in Fig. 1. The meridional plane has been set coincident with the y and y' axes, and $O(y)$ and $I(y)$ now represent the object and image points. The *stop* is the physical aperture within the lens system which most restricts passage of light from object point to image point. The two paraxial rays used for image quality analysis are the *principal ray* (PR in Fig. 3), which passes through the intersection of the optical axis AA' and the stop surface at S, and the *marginal ray* (MR in Fig. 3), which extends from A to A' in the meridional plane and just grazes the edge of the stop. Other important parameters include the marginal ray angles u and u' and the principal ray angles u_p and u'_p.

---

* The extensive and useful list of lens designs in Cox (1964, pp. 557–661) uses OSC as one measure of image quality.

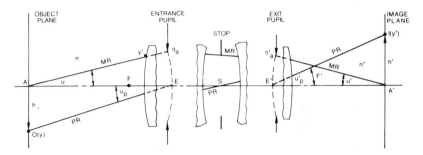

FIG. 3. Basic lens parameters for image quality analysis.

When discussing image quality in general terms, it is desirable to eliminate any need to refer to real lens surfaces. This can be done by replacing the lens with its *entrance pupil* and *exit pupil*. The entrance pupil is the image of the stop formed by all lens elements between the object plane and the stop and is found by extending PR in Fig. 3 forward from $O(y)$ to the point E where it intersects the optical axis. The exit pupil is the image of the stop formed by all lens elements between the stop and the image plane and is found by extending PR in Fig. 3 backward from $I(y')$ to the point E' where it intersects the optical axis. Object space now extends from the object surface to the entrance pupil, and image space extends from the exit pupil to the image plane. The exit and image pupils are conjugate image surfaces, so that ray intercept points and ray directions at the entrance pupil can be transferred to the exit pupil with suitable magnification. Coordinates in the entrance pupil are given as (ξ, η), and coordinates in the exit pupil as (ξ', η'), with the η and η' axes lying in the meridional plane.

In the paraxial region, the Abbe sine condition reduces to

$$hnu = h'n'u' \tag{10}$$

Equation (10) is referred to as the *Lagrange equation,* the *Smith–Helmholtz equation,* and several other variants. The quantity hnu is known as the *Lagrange invariant,* or simply the *optical invariant.*

Equation (10) may be generalized to apply to any surface in the optical system by writing it in a form involving both the marginal ray and the principal ray. If, in Fig. 3, MR is specified at each surface by an angle u and a height y measured relative to the optical axis, and if PR is specified by an angle u_p and a height y_p,

$$\text{optical invariant} = y_p nu - ynu_p \tag{11}$$

Equation (10) follows directly from Eq. (11), if it is noted that $y = 0$ at the object and image planes.

Since $y_p = 0$ at the entrance and exit pupils, Eq. (10) might also be written

$$hnu = -\eta_a n u_p = -\eta'_a n' u'_p = h'n'u' \tag{12}$$

where η_a and η'_a are the heights of the marginal ray marking the pupil rims. Very few real lenses fall completely into the paraxial domain, and so a quasiparaxial form of Eq. (12) is usually more convenient. If $\sin u'$ is substituted for u' and $\tan u_p$ for u_p, Eq. (12) leads to

$$h'n'\sin u' \simeq -\eta_a n \tan u_p \tag{13}$$

The sine implies that the exit pupil is a sphere centered at A' and the entrance pupil is a sphere centered at A. The tangent implies that both object and image surfaces are planes. In more complete analyses, the exit pupil sphere is assumed to be centered at $I(y)$, and its position varies with image height.

If the object point is on the optical axis at infinity and the lens obeys the sine condition, $y_1 = \eta_a$, $u' = u'(\infty)$, and the rear focal length

$$f' = \eta_a / \sin u'(\infty) = D_p / 2 \sin u'(\infty) \tag{14}$$

where D_p is the entrance pupil diameter. It follows from Eq. (13) that

$$h' = -(n/n')f' \tan u_p \tag{15}$$

Equation (15) is the standard definition for a *distortion-free lens*.

The parameter most closely related to image degradation due to diffraction is the cone angle u'. The cone angle is more commonly specified by either the NA or the *focal ratio* F (also called the *F number* and the *relative aperture*). The NA is given by

$$\text{NA} = n' \sin u' \tag{16}$$

If the lens satisfies the Abbe sine condition, the classical definition of the focal ratio is

$$F = f'/D_p = 1/[2 \sin u'(\infty)] = n'/2\text{NA} \; (\infty) \tag{17}$$

The classical definition of focal ratio may then be extended to finite conjugates by defining an *effective focal ratio* in terms of $\sin u'$.

The reader interested in a more extensive discussion of paraxial optics is directed to Kingslake (1965a) or a standard text such as Smith (1966) or Hecht and Zajac (1974).

IV. ABERRATIONS AND WAVEFRONT ERROR

In lens design terminology, departures from Maxwell's first condition are specified by *ray aberrations* and *wave aberrations*. Ray aberrations denote the linear displacement from the ideal of the points at which rays intersect the image surface. Wave aberrations measure the OPD of each ray compared to that of the principal ray. Both types of aberrations are computed by ray tracing the lens system. For detailed performance analysis of finished lenses, exact ray tracing is used. During the earlier stages of lens design, paraxial ray tracing is frequently preferred, with aberrations being represented by polynomial functions whose coefficients may be computed by summing contributions from each lens surface.

Lens design procedures have been reviewed by Kingslake in an earlier volume (Kingslake, 1965b). Aberration theory is discussed here in terms of the aberration polynomials referenced to the exit pupil, based in large part on material published by Rimmer (1962). Detailed ray tracing procedures are not discussed. The interested reader is directed to MIL-HDBK-141 (1962).

A. RAY ABERRATIONS

Figure 4 shows the image space portion in Fig. 3 in perspective. The nominal location of $I(y')$ is $(0,h')$, as defined by the principal ray. If aberrations are present, any other ray leaving the exit pupil from an off-axis

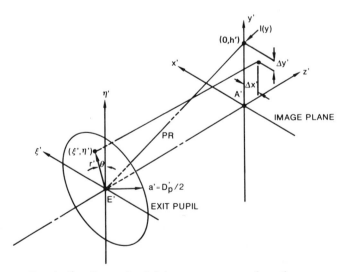

FIG. 4. Coordinates for defining transverse ray aberrations.

position, such as (ξ',η'), is likely to miss $(0,h')$, striking the image surface at, for example, $(\Delta x', h' + \Delta y')$. The quantities $\Delta x'$ and $\Delta y'$ are termed *transverse ray aberrations*. These transverse ray aberrations are defined by *ray aberration polynomials* which consist of the sums of two series of terms containing fixed constants and powers of ξ', η', and h'.

1. *The Ray Aberration Polynomials*

The ray aberration polynomials are most conveniently written in terms of normalized coordinates. The paraxial image height h' is usually normalized:

$$\bar{H} = h'/h'_{max} = \tan u_p / \tan u_{p\,max} \tag{18}$$

where $u_{p\,max}$ is the field of view half-angle. Polar coordinates are used for the exit pupil in writing the polynomials. If r' and θ are as defined in Fig. 4, with D'_p the exit pupil diameter and

$$\xi' = r' \sin \theta, \qquad \eta' = r' \cos \theta \tag{19}$$

the normalized polar pupil coordinates are (ρ, θ), where

$$\rho = 2r'/D'_p \tag{20}$$

Note that the normalized coordinates $\bar{H}$, ρ, and θ apply equally well to object space and image space.

The normalized Cartesian pupil coordinates (X, Y) are also used, to present ray aberration data graphically:

$$X = 2\xi'/D'_p = \rho \sin \theta, \qquad Y = 2\eta'/D'_p = \rho \cos \theta \tag{21}$$

The ray aberration polynomials can now be defined by two series of terms of the form

$$_aC_{bc}\, H^a \rho^b \cos^c \theta \tag{22}$$

where $_aC_{bc}$ is called the *aberration coefficient*, and the sum $a + b$, which is always odd, is the *order number*. In their most general form, the equations for $\Delta x'$ and $\Delta y'$ are written

$$\Delta x' = \mu_0 \sum\, _aC_{bc}\bar{H}^a\rho^b \cos^c \theta \sin \theta$$
$$\Delta y' = \mu_0 \sum\, _aC_{bc}\bar{H}^a\rho^b \cos^c \theta \cos \theta \tag{23}$$

where $\mu_0 = 1/(2 n' \sin u')$. The sign convention is that of Rimmer (1962) where u' as shown in Fig. 3 is positive. Specific forms of notation used by various workers differ substantially (Buchdahl, 1968; Cruickshank and Hill, 1960; Rimmer, 1962). The major differences are a result of combining terms and trigonometric manipulation.

2. Ray Aberration Nomenclature

Ray aberrations are divided into classes according to their order number. In the most common American nomenclature, these classes are *first-order aberrations* $(a + b = 1)$, *third-order aberrations* $(a + b = 3)$, *fifth-order aberrations* $(a + b = 5)$, and so forth. The third-order aberrations are also termed *Seidel aberrations,* for histroical reasons. Typical modern computer programs for lens design use first-, third-, and fifth-order aberrations, plus seventh-order spherical aberration. The ray aberrations of lenses containing substantial amounts of seventh- and higher-order aberrations are usually calculated by exact ray tracing.

The classification of aberrations by type is more than just a mathematical convenience. Different types of aberrations affect image quality differently. In terms of Linfoot's image quality factors, aberrations having rotational or quadrilateral symmetry affect relative structural content but not image fidelity. Aberrations exhibiting asymmetry affect image fidelity strongly, even when tolerable in terms of relative structural content. Since it is frequently possible to balance one aberration against another in designing a lens, understanding which affects image quality in the manner most significant to the final application can be important. The differences in effect can be illustrated by discussing first- and third-order aberrations alone.

In first-order ray aberrations, either a or b is 1, and the other is 0. The term $_0c_{10}\rho$ is called *defocus* and is rotationally symmetric. The term $_1c_{00}\bar{H}$ represents a variation in magnification or focal length. Monochromatically, this can occur only as a time-varying effect due to environmental changes or variation in focal length in a zoom lens. The only other possible first-order term is $_0c_{11}\rho \cos \theta$ which represents a displacement of the entire image surface. It cannot occur if rotational symmetry is maintained in the lens system but can occur if a component is tilted or a refracting element has a wedge angle; that is, in the case of fabrication errors. Thus if attention is confined to monochromatic, time-invariant systems, the first-order ray aberration polynomials are

$$\Delta x'(1) = \mu_0 \,_0c_{10}\rho \sin \theta, \quad \Delta y'(1) = \mu_0 \,_0c_{10}\rho \cos \theta \quad (24)$$

In polychromatic systems, all three first-order aberration coefficients can vary with wavelength, but variations in $_0c_{11}$ still require departures from rotational symmetry. Variations in defocus with wavelength are termed *axial color*. Variations in magnification are termed *lateral color*.

Third-order ray aberration nomenclature is well standardized, although notation is not. The most common American form of the third-order ray aberration polynomials is (Rimmer, 1962):

$$\Delta x'(3) = \mu_0[B'\rho^3 \sin \theta + F'\rho^2\bar{H} \sin 2\theta + (C' + \Pi')\rho\bar{H}^2 \sin \theta]$$
$$\Delta y'(3) = \mu_0[B'\rho^3 \cos \theta + F'\rho^2\bar{H}(2 + \cos 2\theta) \qquad (25)$$
$$+ (3C' + \Pi')\rho\bar{H}^2 \cos \theta + E'\bar{H}^3]$$

The third-order aberrations and their coefficients are *spherical aberration B'*, *coma F'*, *astigmatism C'*, *field curvature* Π', and *distortion E'*. Spherical aberration is circularly symmetric and affects relative structural content principally. Spherical aberrations of all orders are the only monochromatic aberrations except defocus which affect on-axis image quality in rotationally symmetric lenses. Coma is symmetric only in $\Delta x'$ and affects both relative structural content and image fidelity. Astigmatism is quadrilaterally symmetric at best focus. When coupled with appropriate image curvature and defocus, astigmatism produces the classical line segment images of a point source, with the lines parallel or perpendicular to the meridional plane. Astigmatism is most annoying in document reproduction but does not greatly affect image fidelity. Third-order distortion, one form of violation of Maxwell's third condition, affects only image fidelity and not relative structural content.

All third- and higher-order aberrations can vary with wavelength. In most lens design programs, only the chromatic variations of first-order aberrations are called out specifically. *Spherochromatism,* variation in spherical aberration with wavelength, is enough of a problem with special lens designs such as the Schmidt telescope to have received a name.

Designs corrected only for spherical aberration are stigmatic only on-axis. Designs corrected for coma and spherical aberration obey the sine condition and can therefore be called aplanatic. OSC is thus a measure of coma for a lens corrected for spherical aberration. If a lens is stigmatic for an infinitely distant axial object point, its paraxial focal length f'_p is given by

$$f' \to f'_p = y_1/u' \qquad \text{as} \quad y_1 \to 0 \qquad (26)$$

where y_1 is the height of the meridional ray on the first optical surface. The OSC may then be defined (Cox, 1964) numerically as

$$\text{OSC}_\infty(y_1) = (y_1/\sin u') - f'_p \qquad (27)$$

In a lens which satisfies the sine condition, $f' = f'_p$ for all zones of radius $r = y_1$ in the entrance pupil. If corrected for spherical aberration as well, it is called *aplanatic*. If in addition it is corrected for astigmatism, it is termed *anastigmatic*.

In most of this chapter, it is assumed that aberrations are both monochromatic and time-invariant. It is also assumed that the net ray aberrations $\Delta x'$ and $\Delta y'$ vary slowly enough with $\bar{H}$ so that they may be assumed constant within tolerances over a significant region of the image surface. This condition of shift invariance is termed *isoplanatism,* and the region of the image over which shift invariance holds is called the *isoplanatic patch.*

B. WAVE ABERRATIONS

Wave aberrations are measured in units of OPD, as defined earlier. To restate the definition formally, for the ray $(\rho,\theta;\bar{H})$ and the principal ray $(0,0;\bar{H})$,

$$\text{OPD}(\rho,\theta;\bar{H}) = \text{OPL}(\rho,\theta;\bar{H}) - \text{OPL}(0,0;\bar{H}) \tag{28}$$

The OPD expressed in units of length varies with wavelength only if the lens system contains refracting elements which have power or are located in noncollimated light. The effects of OPD on image quality vary inversely with wavelength, when measured against the diffraction-limited performance of a perfect lens. We therefore define the *wavefront aberration function*

$$W_a(\rho,\theta;\bar{H}) = \text{OPD}(\rho,\theta;\bar{H})/\lambda_v = \text{OPD}(\rho,\theta;\bar{H})/n'\lambda \tag{29}$$

where λ_v is the wavelength in vacuum and λ is the *wavelength in the local medium* [λ is so used throughout this chapter].

1. Computing the OPD from Ray Aberrations

Hopkins (1950) shows that the OPD can be calculated directly from the transverse ray aberrations $\Delta x'$ and $\Delta y'$. Figure 5 shows the geometry for the calculation. For simplicity, the on-axis image point at A' is used. The same calculation can be done for any image point if E'A' is matched to the principal ray, the exit pupil surface is set up as a sphere centered at the paraxial image point, and the coordinates are properly normalized (Hopkins, 1974).*

The curve E'B represents a section of the exit pupil surface lying in a plane tilted at an angle θ to the η' axis. The point B has the coordinates (r',θ), where

$$r' = R \sin \gamma = \rho R \sin u' = \rho R/2F \tag{30}$$

and $R \sin u'$ marks the boundary of the exit pupil. E'P is a section of the aberrated wavefront, and BPM represents the ray path normal to the aberrated wavefront at P. (M is not necessarily in the same plane as E',

---

* Hopkins (1974, p. 2) gives an extensive discussion of the normalization requirements.

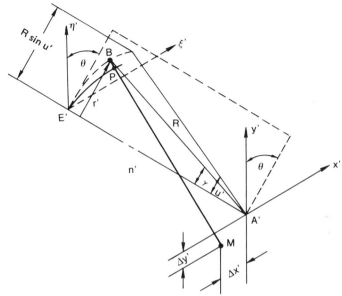

FIG. 5. Coordinates for computing the OPD from the transverse ray aberrations $\Delta x'$ and $\Delta y'$.

A', B, and P, but so long as BP $\lll$ PM, the assumption that BPM is a straight line produces negligible error.) Hopkins (1950) shows that

$$\text{OPD} = -(n'/R) \int_0^{r'} (\Delta x' \sin \theta + \Delta y' \cos \theta) \, dr' \tag{31}$$

In the normalized pupil coordinates (ρ, θ),

$$\text{OPD} = -n' \sin u' \int_0^{\rho} (\Delta x' \sin \theta + \Delta y' \cos \theta) \, d\rho \tag{32}$$

2. Wave Aberration Coefficients

Since the OPD is computed directly from $\Delta x'$ and $\Delta y'$, it follows that the equations for OPD or the wavefront aberration function can be expanded into the same form of aberration polynomial. The resulting *wave aberration coefficients* have the same general form as the ray aberration coefficients, differing primarily in being one unit higher power in ρ. Terms in the polynomial take the form

$$_dC_{ef} \, \bar{H}^d \, \rho^e \cos^f \theta \tag{33}$$

The term $_dC_{ef}$ is the wave aberration coefficient. Its relationship to the corresponding ray aberration coefficient may be noted by the subscripts: $d = a$, $e = b + 1$, and $f = c$. The order number $d + e$ is one unit larger than for the matching ray aberration. Despite this, the common American practice is to refer to the aberrations by their ray aberration order numbers. Table I lists the aberration designation, name, coefficient notation, and OPD contribution for the first-, third-, and fifth-order aberrations, plus seventh-order spherical aberration. The names and symbols used derive primarily from Sayanagi (1966). Alternative order designations found in keystone references (Buchdahl, 1968; Hopkins, 1950) are also shown.

3. Wave Aberration Polynomial

In expanding Eq. (29) into the *wave aberration polynomial*, it is convenient to reduce the number of terms by setting $\bar{H}$ at a fixed value. The new wave aberration coefficients have the general form

$$W_{ef} = {}_dC_{ef}\bar{H}^d/\lambda \tag{34}$$

where W_{ef} sums all the terms having the subscripts e and f. In the process,

TABLE I

OPD Contributions and Nomenclature for the Principal
Wave Aberration Terms Used in Current Design Practice

Order designation			Coefficient[a,b]	Name[a]	OPD contribution
1st[a]	—	—	${}_0C_{20}$	Defocus	${}_0C_{20}\rho^2$
3rd[a]	1st[b]	Primary[c]	${}_0C_{40}$	Spherical aberration	${}_0C_{40}\rho^4$
			${}_1C_{31}$	Coma	${}_1C_{31}\bar{H}\rho^3 \cos\theta$
			${}_2C_{22}$	Astigmatism	${}_2C_{22}\bar{H}^2\rho^2 \cos^2\theta$
			${}_2C_{20}$	Field curvature	${}_2C_{20}\bar{H}^2\rho^2$
			${}_3C_{11}$	Distortion	${}_3C_{11}\bar{H}^3\rho \cos\theta$
5th[a]	2nd[b]	Secondary[c]	${}_0C_{60}$	Zonal spherical aberration	${}_0C_{60}\rho^6$
			${}_1C_{51}$	Zonal coma	${}_1C_{51}\bar{H}\rho^5 \cos\theta$
			${}_3C_{33}$	Arrows	${}_3C_{33}\bar{H}^3\rho^3 \cos^3\theta$
			${}_2C_{42}$	Wings	${}_2C_{42}\bar{H}^2\rho^4 \cos^2\theta$
			${}_2C_{40}$	Lateral spherical aberration	${}_2C_{40}\bar{H}^2\rho^4$
			${}_3C_{31}$	Lateral coma	${}_3C_{31}\bar{H}^3\rho^3 \cos\theta$
			${}_4C_{22}$	Lateral astigmatism	${}_4C_{22}\bar{H}^4\rho^2 \cos^2\theta$
			${}_4C_{20}$	Lateral field curvature	${}_4C_{20}\bar{H}^4\rho^2$
			${}_5C_{11}$	Lateral distortion	${}_5C_{11}\bar{H}^5\rho \cos\theta$
7th[a]	3rd[b]	Tertiary[c]	${}_0C_{80}$	Spherical aberration	${}_0C_{80}\rho^8$

[a] K. Sayanagi (1966).
[b] H. H. Hopkins (1950).
[c] H. A. Buchdahl (1968).

all the distortion terms are merged into a wavefront tilt term, and image curvature terms are merged into one defocus term. Table II lists the new wavefront error coefficients, the terms which combine to make each, and a name for each. The altered nomenclature reflects the fact that terms are summed from several different order numbers. Given the new coefficients, Eq. (29) expands to become

$$W_a(\rho,\theta) = W_{20}\rho^2 + W_{40}\rho^4 + W_{60}\rho^6 + W_{80}\rho^8 + W_{22}\rho^2 \cos^2 \theta$$
$$+ W_{42}\rho^4 \cos^2 \theta + W_{31}\rho^3 \cos \theta + W_{51}\rho^5 \cos \theta + W_{33}\rho^3 \cos^3 \theta$$
$$+ W_{11}\rho \cos \theta \tag{35}$$

An important property of $W_a(\rho,\theta)$ is whether it is even or not even. If it is even, only relative structural content will be degraded. If it is not even, image fidelity will also be degraded. Any function can be written as a sum of an even function $[f(x,y) = f(-x,-y)]$ and an odd function $[f(x,y) = -f(-x,-y)]$. In the case of Eq. (35), the individual terms are either even or odd. If the subscript f in Eq. (34) is zero or even, the term is even. If f is odd, the term is odd. When $f = 0$, as for defocus and spherical aberration, the term is rotationally symmetric. When f is even, as with astigmatism, the term has quadrilateral symmetry. All terms are bilaterally symmetric with respect to the meridional plane.

C. Graphical Presentation of Aberration Data

Transverse ray aberration data is commonly presented by plotting *ray fans* or *spot diagrams* (Herzberger, 1947). Each takes a uniformly spaced set of rays striking the entrance pupil of the lens and plots their interception points on the image surface. Ray fans lie along a single line in the

TABLE II
WAVE ABERRATION COEFFICIENTS IN REDUCED FORMAT

Coefficient	Name	Terms
W_{20}	Defocus	$({}_0C_{20} + {}_2C_{20}\bar{H}^2 + {}_4C_{20}\bar{H}^4)/\lambda$
W_{11}	Tilt	$({}_3C_{11}\bar{H}^3 + {}_5C_{11}\bar{H}^5)/\lambda$
W_{40}	Primary spherical	$({}_0C_{40} + {}_2C_{40}\bar{H}^2)/\lambda$
W_{60}	Secondary spherical	${}_0C_{60}/\lambda$
W_{80}	Tertiary spherical	${}_0C_{80}/\lambda$
W_{22}	Primary astigmatism	$({}_2C_{22}\bar{H}^2 + {}_4C_{22}\bar{H}^4)/\lambda$
W_{42}	Secondary astigmatism	${}_2C_{42}\bar{H}^2/\lambda$
W_{31}	Primary coma	$({}_1C_{31}\bar{H} + {}_3C_{31}\bar{H}^3)/\lambda$
W_{51}	Secondary coma	${}_1C_{51}\bar{H}/\lambda$
W_{33}	Elliptical coma (arrows)	${}_3C_{33}\bar{H}^3/\lambda$

pupil and are presented by plotting $\Delta x'$ or $\Delta y'$ versus distance along the line in the pupil. OPD data are also presented as ray fan plots. For spot diagrams, a two-dimensional grid of rays is traced through the lens system. The spot diagram plots $\Delta x'$ versus $\Delta y'$ for each ray, using a spot to indicate where the ray strikes the image surface.

Figure 6 shows transverse aberration ray fan plots for third-order aberrations. The *meridional* or *tangential ray fans* and the *sagittal ray fans* shown are the fans most commonly used. Tangential ray fans are spaced along the η' axis of the exit pupil. They lie entirely in the meridional plane and are fully represented by plotting $\Delta y'$ versus Y. The sagittal ray fans are spaced along the ξ' axis of the exit pupil and are customarily represented by plotting $\Delta x'$ versus X. This is not totally accurate since, in the case of coma, sagittal rays are displaced in $\Delta y'$.

$\Delta x'$ versus X is always an odd function if the lens is rotationally symmetric. The tangential ray fan components may be either odd or even, except on-axis, where the tangential and sagittal fans are identical. The tangential ray fan must always be plotted over the full pupil from $Y = -1$ to $Y = +1$. Customarily, only half of the sagittal fan is plotted, from $X = 0$ to $X = +1$. The axial image is usually represented by a sagittal half-fan.

Distortion displaces all rays the same amount. It is represented by the separation between the paraxial principal ray and the real principal ray. Ray fans are plotted with their origin at the real principal ray image surface intersection. Distortion is indicated by marking the location of the paraxial principal ray on the $\Delta y'$ axis of the tangential ray fans (PPR in Fig. 6).

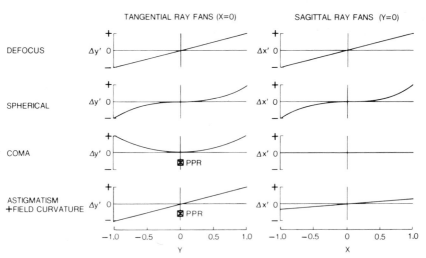

FIG. 6. Transverse ray aberration plots for sagittal and tangential (meridional) ray fans for third-order aberrations.

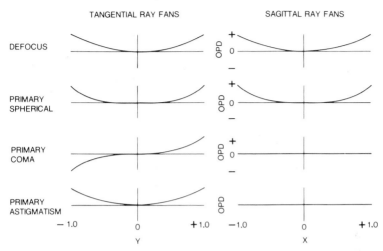

FIG. 7. OPD plots for sagittal and tangential (meridional) ray fans for third-order aberrations.

Figure 7 shows OPD ray fans for the same third-order aberrations as in Fig. 6. OPD ray fans differ from transverse aberration ray fans only in the substitution of OPD for $\Delta x'$ and $\Delta y'$ and in the absence of a mark for the paraxial principal ray. There is also a complete reversal in which terms are even and which are odd.

Ray fans represent only a fraction of the pupil area and may be very misleading as indicators of image quality. *Skew rays,* which intersect the exit pupil at points not on either axis, must be accounted for in any complete measure of image quality. Graphically, skew rays are represented by a spot diagram. Figure 8 shows a typical example, drawn in conjunction with the corresponding tangential and sagittal ray fans to show their relationship. The same scale is used for both ray fans and the spot diagram. It should be noted that some of the fine-structural detail appearing in the spot diagram is an artifact of the grid used to space rays in the exit pupil. It should also be noted that, in this instance, the ray fans are reasonably good indicators of performance, the spot diagram showing a strong central core and relatively weak flare "wings" beyond what is shown by the ray fans.

D. WAVEFRONT ERROR AND PERFORMANCE CRITERIA
BASED ON IT

The OPD and other measures of wavefront error are widely used as measures of image quality by lens designers. Several one-number merit functions and image quality criteria are based on wavefront error.

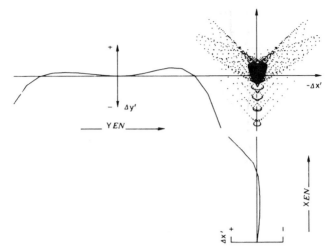

FIG. 8. A spot diagram compared to the sagittal and tangential ray fans for the same image point.

1. *Wavefront Error Specification*

There are three different forms wavefront error specifications can take: (1) If only one type of aberration is present, the wavefront error coefficient W_{ef} can be used; (2) the *peak-to-peak wavefront error* can be used in more general cases; (3) the *rms wavefront error* or *variance* can be used for all cases.

The wavefront error is by definition zero at E′ in the exit pupil. The wavefront error coefficient W_{ef} measures the value of the wavefront error contribution at $\rho = 0$, $\theta = 0$ ($X = 0$, $Y = 1.0$.)

Peak-to-peak wavefront error (also termed *peak-to-valley wavefront error*) is determined by fitting two concentric spheres to the wavefront, as shown in Fig. 9. The two spheres may be centered on the paraxial image point A′ or at the point of best focus, but the wavefront must lie entirely between them and touch each at at least one point. The peak-to-peak wavefront error is then defined by the radii of the two spheres:

$$W_{pp} = (R_1 - R_2)/\lambda \tag{36}$$

In recent years, the rms wavefront error

$$\omega = \left[\frac{1}{\pi} \int_0^{2\pi} \int_0^1 W^2(\rho,\theta)\rho \, d\rho \, d\theta \right]^{1/2} \tag{37}$$

has come into widespread use among those involved with high-precision optical systems. The definition of Eq. (37) assumes that $W(\rho,\theta)$ is mea-

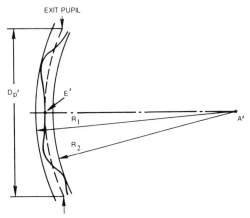

FIG. 9. Reference spheres for computing peak-to-peak wavefront error.

sured against the best-fit reference sphere. Another definition is based on the variance V, where

$$\omega^2 = V = \frac{1}{\pi} \int_0^{2\pi} \int_0^1 W^2(\rho,\theta)\rho \, d\rho \, d\theta - \frac{1}{\pi^2} \left[\int_0^{2\pi} \int_0^1 W(\rho,\theta)\rho \, d\rho \, d\theta \right]^2 \quad (38)$$

Equation (38) is used as the definition for rms wavefront error elsewhere in this chapter.

Lerman (1969) has derived the variance for the wavefront aberration polynomial of Eq. (35) for the case of a lens with a circular central obstruction of diameter ratio ϵ. The results can be found in Appendix A.

2. Image Quality Criteria Based on Wavefront Error

The classical criterion for image quality made in terms of wavefront error is the *Rayleigh quarter-wave criterion*. As originally phrased by Rayleigh (1964b, p. 434), "aberration begins to be decidedly prejudicial when the wave-surface deviates from its proper place by about a quarter of a wave-length."

Lord Rayleigh based his criterion on an analysis of the effects of third-order spherical aberration, but it is frequently assumed to apply for a quarter-wavelength of any aberration. As stated, the criterion is specified in terms of peak-to-peak wavefront error. The actual image degradation for a given peak-to-peak wavefront error varies slightly from aberration type to aberration type. One-quarter wavelength peak-to-peak of third-order spherical aberration equals about 0.075 wavelength rms wavefront error. In this form, the Rayleigh criterion is more generally applicable.

The term "*diffraction-limited*" is generally associated with the Rayleigh quarter-wave criterion. More generally, any lens system for which $\omega \leqq 0.075$ wavelength rms may be considered to have its performance limited primarily by diffraction at its pupil boundary. The term "*near diffraction-limited*" is sometimes used rather loosely to describe lenses whose performance is excellent but does not quite live up to the above standard. Near diffraction-limited may be taken to imply an rms wavefront error $0.075 < \omega < 0.150$, but it should be emphasized that there is no formally accepted definition of the term. It should also be emphasized that wavefront error is only one factor degrading image quality and that other factors such as aperture obscuration and image motion must be accounted for in any general image quality performance criterion.

3. Depth of Focus and Depth of Field

The discussion of aberrations and wavefront error given above has been confined to the nominal image surface. *Depth of focus* and *depth of field* measure the tolerance on axial displacement of the image sensor and the real object within which image quality will remain acceptable. If the image sensor is moved a distance $\Delta z'$ away from the best focus image surface, a wavefront error is added to the wave aberration polynomial:

$$W(\Delta z') = (1 - \cos u')\Delta z'/\lambda$$
$$= \frac{\Delta z'}{\lambda}\left(\frac{1}{8F^2} + \frac{1}{128F^4} + \frac{1}{1024F^6} + \frac{5}{32,768F^8} + \cdots\right) \quad (39)$$

The terms of the expansion can be identified with defocus and the various orders of spherical aberration. It can also be seen that the magnitude of each term is strongly dependent on the focal ratio F. In general, only the first term is used:

$$W_{20}(\Delta z') = \Delta z'/8\lambda F^2 \quad (40)$$

If Eq. (40) is used in place of Eq. (39), the residual error will be 7.2% for $F = 1.0$, 1.6% for $F = 2.0$, etc. More terms of Eq. (39) should be used for focal ratios substantially below $F = 1.0$.

The most commonly used criteria for setting limits on the *depth of focus* are $W_{20}(\Delta z') = 0.25$ and, for more precise optical systems, $W_{20}(\Delta z') = 0.125$. The corresponding values for depth of focus are

$$\Delta z' = \pm 2\lambda F^2 \quad (quarter\text{-}wave\ criterion)$$
$$\Delta z' = \pm \lambda F^2 \quad (eighth\text{-}wave\ criterion) \quad (41)$$

If the tolerance for defocus is given in terms of an rms wavefront error ω and the lens system has a central obstruction of diameter ratio ϵ,

$$\Delta z' = \pm 16(3)^{1/2}\lambda\omega F^2/(1 - \epsilon^2) \tag{42}$$

For the tolerance $\omega = 0.075$ wavelength rms, the numerical coefficient becomes ± 2.08. The difference between this and the quarter-wave criterion is insignificant unless the central obstruction is large.

The *depth of field* is the region in object space fore and aft of the nominal object surface corresponding to the depth of focus. It can be computed by substituting $z' + \Delta z'$ and $z' - \Delta z'$ into Eq. (1). When the lens is focused at infinity, the near boundary of the depth of field is given the special name *hyperfocal distance*, z_h:

$$\begin{aligned} z_h &= D_p^2/2\lambda & \textit{(quarter-wave criterion)} \\ z_h &= D_p^2/\lambda & \textit{(eighth-wave criterion)} \end{aligned} \tag{43}$$

V. THE COMPLEX PUPIL FUNCTION

Up to this point, image quality has been discussed solely in terms directly related to lens design. The only form of image degradation considered has been wavefront error caused by lens design aberrations. Other factors within the optical system can degrade image quality, including wavefront error from other causes, aperture obstructions, and variation in transmittance from ray path to ray path. To incorporate all these factors into a single equation and to provide the transition step from lens design theory to two-dimensional Fourier analysis theory in image quality analysis, we introduce the *complex pupil function* $\mathscr{P}(\rho,\theta)$:

$$\mathscr{P}(\rho,\theta) = P(\rho,\theta)e^{i2\pi W(\rho,\theta)} \tag{44}$$

The two components of the complex pupil function are the *wavefront error pupil function* $W(\rho,\theta)$ and the *amplitude transmittance pupil function* $P(\rho,\theta)$. Each of these components is a function of $\bar{H}$ and λ, as well as ρ and θ. Since $\bar{H}$ and λ are usually constants, they have been omitted from the notation for simplicity.

A. THE PUPIL FUNCTION COMPONENTS

1. *The Wavefront Error Pupil Function* $W(\rho,\theta)$

For each ray (ρ,θ), $W(\rho,\theta)$ specifies OPD/λ, measuring the phase retardation or advancement of the wavefront relative to a reference wavefront in fractions of a wavelength. The phase retardation is $2\pi W(\rho,\theta)$, in radians. The term $W(\rho,\theta)$ is a generalization of the wave aberration function $W_a(\rho,\theta)$ and incorporates the latter. It also incorporates aberrations due to component misalignments and wavefront error due to component fabrication tolerance limits.

The manner in which wavefront error from different sources is treated depends upon how it varies with time and the characteristic lateral dimensions of its spatial structure. Wavefront error components which vary rapidly with time or which have very fine lateral structure are usually treated with statistical models and are considered separate from the pupil function. The manner in which different forms of wavefront error arise and are treated is discussed in Section IX. For present purposes, $W(\rho,\theta)$ is assumed to be time-stationary, isoplanatic, and monochromatic.

2. The Amplitude Transmittance Pupil Function $P(\rho,\theta)$

The shape of the pupil boundary, the shape and locations of any opaque obstructions within the pupil boundary, and any variation in attenuation of light from ray path to ray path in the nominally clear areas of the pupil are defined by $P(\rho,\theta)$. The term "amplitude transmittance" is used to indicate that $P(\rho,\theta)$ measures attenuation in electric field strength, rather than energy flow. "Transmittance," when used without the qualifier "amplitude," always refers to attenuation of energy flow, as is traditional in applied optics. Transmittance is proportional to $P^2(\rho,\theta)$.

Variation in transmittance is more important than peak transmittance in image quality theory. The maximum value of $P^2(\rho,\theta)$ is therefore defined to be 1.0, while $P(\rho,\theta)$ can vary from -1 to $+1$, a negative value being interpreted as equivalent to the matching positive value coupled with a phase shift $2\pi W(\rho,\theta) = \pi$.

The two forms of $P(\rho,\theta)$ most commonly encountered in image quality analysis are the unobstructed circular pupil and the circular pupil with a centered circular obstruction of diameter D_0. For the former,

$$
\begin{aligned}
P(\rho,\theta) &= 1.0, &\quad \rho \leq 1.0 \\
&= 0.0, &\quad \rho > 1.0
\end{aligned}
\tag{45}
$$

For the latter, if $\epsilon = D_0/D_p$,

$$
\begin{aligned}
P(\rho,\theta) &= 0.0, &\quad 0 \leq \rho \leq \epsilon \\
&= 1.0, &\quad \epsilon < \rho \leq 1.0 \\
&= 0.0, &\quad \rho > 1.0
\end{aligned}
\tag{46}
$$

The term "perfect lens" implies that $P(\rho,\theta) = 1.0$ and $W(\rho,\theta) = 0.0$ everywhere within the pupil boundary. When unqualified, perfect lens also implies a circular pupil boundary.

There are important classes of pupil functions where $P(\rho,\theta)$ has values other than 0.0 and 1.0 at some points within the pupil boundary. In laser beam projector optics, the laser beam itself has a Gaussian intensity profile which imposes an effective variation in amplitude transmittance on the pupil. A variation in amplitude transmittance may also be imposed deliberately to improve image quality in a particular manner. This latter technique is called *apodization*. The sources and effects of amplitude transmittance variations are discussed in Section X.

B. PUPIL TRANSMITTANCE

When computing the image quality of an imaging system containing both an optical system and an image sensor, it is necessary to know the effective transmittance of the optics in addition to $P(\rho,\theta)$ and $W(\rho,\theta)$. The effective transmittance

$$\tau = \tau_m \times \tau_p \tag{47}$$

where τ_m is the *maximum transmittance*, measured at the point where $P(\rho,\theta) = 1.0$ and τ_p is the *pupil transmittance* as defined by the equation

$$\tau_p = \frac{1}{\pi} \int_0^{2\pi} \int_0^1 [P(\rho,\theta)]^2 \rho \, d\rho \, d\theta \tag{48}$$

In most of the following sections, it is assumed that $\tau_m = 1.0$ and $\tau = \tau_p$. The distinction between τ_p and τ_m is important. τ_m affects only the average image irradiance level. When $\tau_p < 1.0$, both image irradiance level and image irradiance distribution are affected.

For the unobstructed pupil of Eq. (45), $\tau_p = 1.0$. For the obstructed pupil of Eq. (46),

$$\tau_p = 1 - \epsilon^2 \tag{49}$$

C. MEASURING THE PUPIL FUNCTION

One important property of the pupil function is that it can be measured by interferometry. In interferometry, a test wavefront is compared to a standard wavefront by adding them coherently and observing the interference fringes. Consider, for example, the wavefront U_1 represented by the pupil function, and a wavefront U_2 corresponding to the reference wavefront centered at point A' in Fig. 5:

$$U_1 = P(\rho,\theta) \cos 2\pi W(\rho,\theta) + iP(\rho,\theta) \sin 2\pi W(\rho,\theta)$$
$$U_2 = P(\rho,\theta) \tag{50}$$

If these are added to form $U = U_1 + U_2$, and U is imaged onto photographic film, the recorded image irradiance $E = UU^*$, where the asterisk indicates the complex conjugate, is

$$E(\rho,\theta) = 2P^2(\rho,\theta)[1 + \cos 2\pi W(\rho,\theta)] \tag{51}$$

Interferometric testing of lens components has been done routinely for years. Interferometric testing of complete lens systems is used for final alignment and performance evaluation in most if not all high-precision optical systems. The results are reliable, and the pupil function so measured can be used to compute the lens system point spread function or OTF as shown in Sections VI and VII. Malacara (1978) and Houston (1974) contain papers on many aspects of interferometric testing as it is performed today.

VI. THE POINT SPREAD FUNCTION AND ITS COROLLARIES

The image of a point source object formed by a lens system is known as the *point spread function* (PSF) of the lens. Other names for the PSF include *impulse response, Green's function,* and the *Fraunhofer diffraction pattern.* It is one of the two most complete functions for describing the performance of an optical system and can be extended to include the effects of image motion, atmospheric turbulence, and other factors external to the optical system. The image irradiance distribution $E_i(x',y')$ can be calculated for any object distribution, in principle, by convolving the geometrically perfect image distribution $E_o(x',y')$ with the PSF. In practice, this is a tedious process for any real object more complex than separated point sources. Single and multiple discrete point sources are encountered often enough in astronomy and in optics for laser beam manipulation for the PSF to have substantially more than pedagogic value.

A. Computing the PSF from the Complex Pupil Function

The PSF can be computed from the complex pupil function by squaring its Fourier transform. The amplitude PSF $\mathscr{A}(\Delta x', \Delta y')$ is given directly by the Fourier integral

$$\mathscr{A}(\Delta x', \Delta y') = C \int_{-\infty}^{+\infty} \int_{-\infty}^{+\infty} \mathscr{P}(\xi',\eta')$$
$$\times \exp\left[-\frac{2\pi i}{\lambda R}(\Delta x'\xi' + \Delta y'\xi')\right] d\xi'\, d\eta' \tag{52}$$

where C is a normalization constant to be discussed below and the coordinates are those of Fig. 5. The energy distribution $E_i(\Delta x', \Delta y')$ is then given by

$$E_i(\Delta x', \Delta y') = \mathscr{A}(\Delta x', \Delta y') \, \mathscr{A}^*(\Delta x', \Delta y') \qquad (53)$$

where the asterisk indicates the complex conjugate. Readers interested in the derivation of Eq. (52) from the scalar diffraction theory are directed to Walther (1965) or other standard texts (Hecht and Zajac; Born and Wolf; Goodman, 1968).

Equation (52) can be written in the normalized polar coordinates used with the pupil function. For an exit pupil radius of a' and a focal ratio F

$$\xi'/R = (a'\rho \sin\theta)/R = \rho \sin\theta \sin u' = (\rho \sin\theta)/2F$$
$$\eta'/R = (a'\rho \cos\theta)/R = \rho \cos\theta \sin u' = (\rho \cos\theta)/2F \qquad (54)$$

The polar image coordinates for the spread function are the radius q' and the azimuth angle ψ, measured from the y' axis. Thus

$$\Delta x' = q' \sin\psi, \qquad \Delta y' = q' \cos\psi \qquad (55)$$

Equation (52) then becomes

$$\mathscr{A}(q',\psi) = (a')^2 C \int_0^{2\pi} \int_0^1 \mathscr{P}(\rho,\theta)$$
$$\times \exp\left[\frac{-\pi i p q'}{\lambda F} \cos(\theta - \psi)\right] \rho \, d\rho \, d\theta \qquad (56)$$

When the complex pupil function is rotationally symmetric, Eq. (56) reduces to the Hankel transform

$$\mathscr{A}(q') = 2\pi(a')^2 C \int_0^1 \mathscr{P}(\rho) J_0\left(\frac{\pi \rho q'}{\lambda F}\right) \rho \, d\rho \qquad (57)$$

where J_0 is a Bessel function of the first kind of zero order. Equation (57) is convenient for numerical analysis, where it can be applied, because it requires a fraction of the computation time of Eq. (52) or (56). The present author has found it particularly useful in studying the effects of fine structure in the pupil function, using rotationally symmetric models for $W(\rho)$ and $P(\rho)$. Some results of this work are discussed in Sections IX and X.

In the case of a perfect lens with a circular pupil, Eqs. (57) and (53) lead to

$$E_{ip}(q') = E_{ip}(0) \left[\frac{2J_1(\pi q'/\lambda F)}{(\pi q'/\lambda F)}\right]^2 \qquad (58)$$

where $E_{ip}(0)$ is the image irradiance at $q' = 0$ and J_1 is a Bessel function of the first kind of order one. Equation (58) is used as the basis for normalizing the PSF.

B. The Normalized PSF and Strehl Definition

Both the lateral dimensions and the peak image irradiance of the PSF are strongly dependent on wavelength λ. To compare the PSF of a real optical system to that of a perfect lens of matching focal length and focal ratio, it is convenient to normalize both radius and image irradiance units. The image irradiance at the center of the normalized PSF is an image quality merit function in its own right.

1. Radius Normalization

There are three forms of radius normalization found in the literature. One uses the numerical value of the argument of the Bessel function in Eq. (58) directly. A second uses λF as the radius unit. The third uses the radius of the first minimum of Eq. (58) as the radius unit. In Fraunhofer diffraction theory, Eq. (58) is known as the *Airy diffraction pattern,* and the central maximum is called the *Airy disk.* The third form of normalization is the form used in this chapter, and the units are called *Airy radius units* δ'. In the first form of normalization, radius units can be called *Bessel radius units B'*. Here λF equals one period at the incoherent cutoff defined by Rayleigh (1896, 1964a) and is called *cutoff* or *Rayleigh units c'*. (None of these terms has official standing.)

The first minimum of Eq. (58) occurs when $\pi q'/\lambda F = 3.8317$. The radius of the minimum in linear units is

$$q' = 1.2197\lambda F \tag{59}$$

In angular units referenced to object space, $\alpha_0 = q'/f'$:

$$\alpha_0 = 1.2197\lambda/D_p \tag{60}$$

In the three forms of radius normalization, the first minimum occurs when $B' = 3.8317$, $c' = 1.2197$, and $\delta' = 1.0$. To convert to linear units,

$$q' = \lambda F B'/\pi = \lambda F c' = 1.2197\lambda F \delta' \tag{61}$$

2. Relative Intensity Normalization

The form of relative intensity normalization depends upon the use to be made of the PSF data. In this chapter, the performance of lenses whose image quality has been degraded by wavefront error, aperture obscurations, apodization, and image motion is being compared to that of a geometrically perfect lens. In most of this chapter, the standard for normalization is therefore the image irradiance at $q' = 0$ for the perfect lens, $E_{ip}(0)$.

For other purposes, other normalization criteria may be more applicable. Normalizing to 1.0 at $q' = 0$ is a common practice. Where the interaction between different factors degrading image quality is being studied, e.g., image motion and aperture obstructions, it may be desirable to use $E_i(0)$ for the PSF degraded by one of the two factors acting alone as the normalization standard. Both of these practices are legitimate if their consequences are understood. If, for example, changes in image irradiance at large q' are being studied, normalization to 1.0 at $q' = 0$ can be misleading, since $E_i(0)$ is changed far more rapidly by most factors degrading image quality than the image irradiance at large q'.

The normalization constant C of Eqs. (52), (56), and (57) is

$$C = (1/\lambda R)\sqrt{\Phi/A'} \tag{62}$$

where A' is the pupil area and Φ is the radiant flux entering the pupil. For a perfect lens with a circular aperture and unit transmittance, Eq. (62) leads to a peak image irradiance:

$$E_{ip}(0) = \pi^2 D_p^2 E_a / 16\lambda^2 F^2 \tag{63}$$

where E_a is the entrance pupil irradiance from the on-axis point source object.

Equation (63) may be used directly to normalize the PSF when $\tau_p = 1.0$. When $\tau_p < 1.0$, there are two possible forms of relative intensity normalization, both of which are used in the literature. The first assumes that Φ is the total flux entering the pupil from the point source, and the second assumes that Φ is the total flux reaching the image surface. In the first case, normalization is accomplish by dividing $E_i(q',\psi)$ by $E_{ip}(0)$. In the second case, $E_i(q',\psi)$ is divided by $\tau_p E_{ip}(0)$. The second form of normalization is used here, and the quantity *relative intensity* $I(q',\psi)$ is defined as

$$I(q',\psi) = E_i(q',\psi)/\tau_p E_{ip}(0) \tag{64}$$

The rationale for preferring the second form of normalization is that it corresponds exactly to the standard normalization of the OTF (see Section VII,B,3) and that it distinguishes between the effects of amplitude transmittance variations on image structure and on average image irradiance. Thus, for computation of the average image irradiance, τ_p can be treated exactly as τ_m, which is universally ignored in normalizing the PSF.

Star field photography serves as an illustrative example. In stellar photography, the point source is seen against a continuous background, as shown in Fig. 10. Typically, the exposure time is extended until the background reaches a set exposure level. In a lens where $\tau_p < 1.0$, both the peak irradiance and the background irradiance are reduced below the corresponding values for a perfect lens. The most legitimate form of comparison between them, for star field photography, is thus

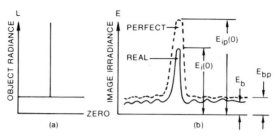

FIG. 10. Images of a point source object seen against a uniform background, formed by a perfect lens and by a real lens for which $\tau_p < 1.0$.

$$\frac{E_i(q',\psi)}{E_{ip}(0)}\frac{E_{bp}}{E_b} = \frac{E_i(q',\psi)}{\tau_p E_{ip}(0)} = I(q',\psi) \tag{65}$$

The general form of the relative intensity equation is thus

$$I(q',\psi)$$
$$= \frac{1}{\pi^2\tau_p}\left|\int_0^{2\pi}\int_0^1 \mathcal{P}(\rho,\theta)\exp\left[\frac{-\pi i\rho q'\cos(\theta - \psi)}{\lambda F}\right]\rho\,d\rho\,d\theta\right|^2 \tag{66}$$

When the pupil is rotationally symmetric,

$$I(q') = (4/\tau_p)\left|\int_0^1 \mathcal{P}(\rho)J_0(\pi\rho q'/\lambda F)\rho\,d\rho\right|^2 \tag{67}$$

For a circular pupil, the actual image irradiance is computed from the relative intensity by

$$E_i(q',\psi) = \pi^2 D_p^2 E_a\tau_m\tau_p I(q',\psi)/16\lambda^2 F^2 \tag{68}$$

3. Strehl Definition

The relative intensity at $q' = 0.0$ is commonly used as a measure of image quality and is known as *Strehl definition* $\mathcal{D}$, *Strehl ratio*, or *Strehl intensity*. ["Definition" is used here in the sense "distinctness of outline or detail." Strehl's original term was *Definitionshelligkeit* (Strehl, 1902).] For Strehl definition calculations, Eq. (66) reduces to

$$\mathcal{D} = \frac{1}{\pi^2\tau_p}\left|\int_0^{2\pi}\int_0^1 \mathcal{P}(\rho,\theta)\rho\,d\rho\,d\theta\right|^2 \tag{69}$$

In using Strehl definition for lenses with central obstructions or other forms of amplitude transmittance variations, it should always be recalled that the actual peak irradiance is proportional to $\tau_p\mathcal{D}$ and the background irradiance is proportional to τ_p.

4. *The Polychromatic PSF*

The PSF given above is monochromatic. Where the wavelength range for the point source object is more than a few percent of its mean wavelength, a polychromatic PSF can be computed by summing individual PSFs for incremental regions. Such a summation must be done in the $E_i(q',\psi)$ domain, with appropriate spectral weighting, since the peak irradiance varies as the inverse square of wavelength, in accordance with Eq. (68). Strehl definition loses its relevance for the polychromatic PSF, without extensive redefinition. In practice, simplified PSF models scaled for the mean wavelength are usually used. Precision is seldom required in a polychromatic PSF. It in any event varies with small changes in the spectral distribution.

C. Encircled Energy and Other Corollaries of the PSF

1. *Encircled Energy*

The principal corollary of the PSF is the *encircled energy* [EE(q')]. The encircled energy function measures the fraction of the total energy in the PSF which lies within the radius q'. For rotationally symmetric PSFs, the encircled energy function is centered on the real PR. For nonrotationally symmetric PSFs, the encircled energy function may be shifted laterally to coincide with the irradiance centroid.

The total energy reaching the image surface is equal to $\tau_m\tau_p E_a$ times the area of the entrance pupil. To make the total energy in the PSF unity for a circular pupil.

$$E_a = 4/\tau_m\tau_p\pi D_p^2 \tag{70}$$

From Eqs. (68) and (70),

$$EE(q') = \frac{\pi}{4\lambda^2 F^2} \int_0^{2\pi} \int_0^{q'} I(q',\psi)q' \, dq' \, d\psi \tag{71}$$

If the radius is in Airy radius units,

$$EE(\delta') = 0.3719\pi \int_0^{2\pi} \int_0^{\delta'} I(\delta',\psi)\delta' \, d\delta' \, d\psi \tag{72}$$

2. *Excluded Energy*

If the energy in the outer ring structure is to be examined in detail, it is sometimes convenient to replace the encircled energy function with *excluded energy* [XE(q')], where

$$XE(q') = 1 - EE(q') \qquad (73)$$

Excluded energy is used in evaluating apodization techniques for suppressing the ring structure in Section X,C.

3. Displaced Energy

Encircled energy and excluded energy describe the particular PSF. If we wish to compare the energy distribution in a real optical system to its perfect counterpart, one method would be to subtract the encircled energy function of one from the other. This is called *displaced energy* [DE(q')]:

$$DE(q') = EE_p(q') - EE(q') \qquad (74)$$

where $EE_p(q')$ is the encircled energy function for the perfect lens, to be defined in Section VI,D. Positive values of $DE(q')$ represent an outward displacement of energy and indicate image quality degradation. Negative values, indicating inward displacement, are found with some forms of apodization discussed in Section X.

The maximum value of $DE(q')$ usually occurs at a radius just under $\delta' = 1.0$ and measures the total energy diffracted out of the central maximum. It can be used as a one-number merit function similar to Strehl definition. It is a more sensitive measure of image degradation than Strehl definition for factors such as aperture obstructions and is less sensitive for factors such as image motion.

4. Zonal Energy Increment

The present author has found the concept of the *zonal energy increment* (ZEI) useful in examining details of the redistribution of energy in a degraded PSF. The ZEI measures the amount of energy added to or subtracted from the perfect lens PSF in the zone between radii q_1' and q_2':

$$\begin{aligned} ZEI(\Delta q') &= [EE(q_2') - EE(q_1')] - [EE_p(q_2') - EE_p(q_1')] \\ &= DE(q_1') - DE(q_2') \end{aligned} \qquad (75)$$

where $q_2' > q_1'$ and $\Delta q' = q_2' - q_1'$. Negative values of ZEI indicate energy scattered out of the zone, and positive values indicate energy scattered into the zone. Note that

$$\sum_{q'=0}^{\infty} ZEI(\Delta q') \equiv 0.0 \qquad (76)$$

ZEI is in effect a scatter function which measures both where the energy being scattered comes from and where it goes. It is used in Sections

IX, X, and XI to compare the manner in which different factors such as wavefront error, aperture obstructions, and image motion affect the PSF.

5. Half-Power Diameter

A quantity often used as a measure of image quality is the diameter of the PSF at 50% of its peak relative intensity. For a perfect lens, this *half-power diameter* d_h occurs at $\delta' \simeq 0.424$ or $d_h \simeq 1.04\lambda F$.

The half-power diameter is not useful in comparing diffraction-limited and near-diffraction-limited optical systems. The shape of the central maximum varies little with increasing wavefront error at these levels of performance. A central obstruction in fact reduces the half-power diameter.

The half-power diameter is most useful when the image quality is limited by external factors such as image motion or atmospheric turbulence, where it correlates directly with the degrading factors. It is also useful where large amounts of aberration are present and $\omega > 0.2$ wavelength rms.

D. THE PERFECT LENS PSF AND ENCIRCLED ENERGY

For the perfect lens, $W(\rho,\theta) = 0.0$ and $P(\rho,\theta) = 1.0$ at all points within the pupil. The forms of the PSF and its corollaries then depend upon the shape of the pupil boundaries.

For a perfect lens with a circular pupil, the relative intensity

$$I_p(q') = [2J_1(\pi q'/\lambda F)/(\pi q'/\lambda F)]^2 \qquad (77)$$

There is also an analytical solution to the encircled energy function:

$$EE_p(q') = 1 - J_0^2(\pi q'/\lambda F) - J_1^2(\pi q'/\lambda F) \qquad (78)$$

For a perfect lens with a rectangular pupil,

$$I(\Delta x', \Delta y') = \left[\frac{\sin(\pi\Delta x'/\lambda F_x)}{(\pi\Delta x'/\lambda F_x)}\right]^2 \left[\frac{\sin(\pi\Delta y'/\lambda F_y)}{(\pi\Delta y'/\lambda F_y)}\right]^2 \qquad (79)$$

where the exit pupil boundaries lie at $\xi' = \pm a'$ and $\eta' = \pm b'$, and

$$F_x = R/2a' = 1/(2\sin u'_x), \qquad F_y = R/2b' = 1/(2\sin u'_y) \qquad (80)$$

No analytic solutions for encircled energy exist in this case.

The Strehl definition is unity for perfect lenses without obstructions. If there is an obstruction, $\mathscr{D}$ can be calculated using Eq. (69). When $W(\rho,\theta) = 0.0$, $\mathscr{P}(\rho,\theta) = P(\rho,\theta)$ and it is informative to rewrite Eq. (69), expanding τ_p using Eq. (48):

$$\mathcal{D} = \left| \int_0^{2\pi} \int_0^1 P(\rho,\theta)\rho \, d\rho \, d\theta \right|^2 \bigg/ \int_0^{2\pi} \int_0^1 P^2(\rho,\theta)\rho \, d\rho \, d\theta \qquad (81)$$

In the case of an obstructed aperture, where $P(\rho,\theta)$ is either zero or unity, $P^2(\rho,\theta) = P(\rho,\theta)$ and it follows from Eq. (81) that $\mathcal{D} = \tau_p$. If the amplitude transmittance is neither unity nor zero at some points within the pupil, it follows from Eq. (81) that $\mathcal{D} > \tau_p$.

For the circular pupil with a centered circular obstruction,

$$\mathcal{D} = \tau_p = 1 - \epsilon^2 \qquad (82)$$

In general, if A_0 is the total area of the pupil obstructions and A_p is the area within the outer boundary of the pupil, then

$$\mathcal{D} = \tau_p = (A_p - A_0)/A_p \qquad (83)$$

regardless of the shapes of the pupil boundary and the obstructions.

E. Graphical Presentation of PSF Data

The rotationally symmetric PSF and its corollaries are most conveniently presented by plotting radial sections from the center to a chosen limiting radius. Figure 11 is a computer-generated profile of $I_p(\delta')$ and

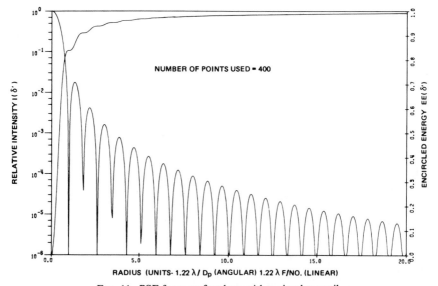

Fig. 11. PSF for a perfect lens with a circular pupil.

$EE_p(\delta')$ within the first 20 Airy radius units of the perfect lens PSF. Relative intensity is plotted on a logarithmic scale to emphasize details of the PSF structure outside the Airy disk. The secondary maxima are known as *sidebands, bright rings,* or *diffraction fringes* in optics, and the minima are known as *dark rings*. The maxima are known as *side lobes* in antenna theory. The minima all go to zero, and the nonzero minima in Fig. 11 are computer artifacts due to the finite radius sampling increment.

It may be noted in passing that the zero minima are a result of the scalar theory used in deriving Eq. (52). At very fast focal ratios, $F < 1.0$, a vector theory should be used for accuracy. The differences produced by the more complete vector analysis are not dealt with here and are seldom of practical significance. The interested reader is directed to the papers by Richards and Wolf (1959) and Boivin *et al.* (1967).

Figure 12 plots $I(\delta')$ and $EE(\delta')$ against δ' for four different values of defocus W_{20}. The Strehl definition drops rapidly with increasing defocus, reaching zero for $W_{20} = 1.0$ wavelength OPD. The minima in the ring structure fill in, but note that there is relatively little net outward transfer of energy past 6–7 Airy radius units. Figure 12 also indicates the advantages of normalizing relative intensity to the Strehl definition, rather than setting $I(0) \equiv 1.0$. Aside from the problem when $I(0) = 0.0$, the normalization in Fig. 12 more accurately reflects changes in the relative intensity in the outer rings.

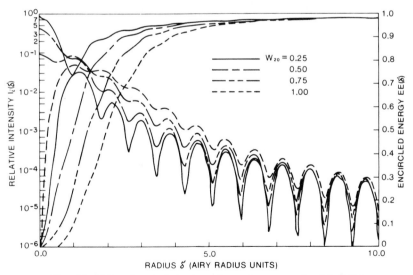

FIG. 12. PSF and encircled energy for various amounts of defocus.

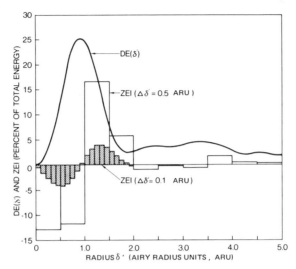

FIG. 13. Displaced energy and ZEIs for a perfect lens with a central obstruction of diameter ratio $\epsilon = 0.45$.

Figure 13 plots both displaced energy and ZEIs for a perfect lens with a centered circular obstruction of diameter ratio $\epsilon = 0.45$. The diameter ratio was selected to give a Strehl definition of 80%. The maximum value of displaced energy indicates that 25% of the total energy in the PSF has been transferred from the Airy disk to the ring structure, appearing mainly in the first bright ring. ZEIs show in more detail where the energy comes from (negative values) and goes (positive values.) Note that positive slopes of displaced energy correspond to a loss of energy from the zones and that negative slopes correspond to a gain of energy.

When the PSF is not rotationally symmetric, other forms of graphical presentation must be used. For quantitative analysis, contour plots of constant relative intensity are probably the most useful. For more qualitative pedagogic needs, perspective plots of relative intensity versus $\Delta x'$ and $\Delta y'$ are convenient. Figure 14 is an example of the latter, showing the PSF for 2.4 wavelengths OPD of third-order coma.

F. PERFORMANCE CRITERIA BASED ON THE PSF

1. Rayleigh Two-Point Resolution

The *Rayleigh two-point resolution criterion* states that two PSFs may be considered just resolved when the central maximum of one falls on the first minimum of the other. This criterion was originally formulated for

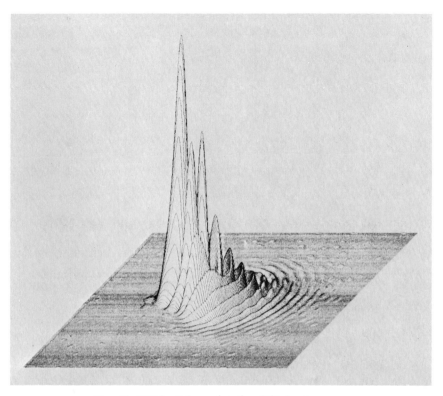

FIG. 14. PSF for 2.4 wavelengths of third-order coma.

line objects, for use in spectroscopy, but it is generally applied to point source objects as well. For point sources, the Rayleigh resolution limit is reached when the separation between the peaks of the PSFs is $s = 1.22\lambda F$.

The arbitrariness of this criterion was recognized by Rayleigh (1964b, p. 420), who states: "This rule is convenient on account of its simplicity; and it is sufficiently accurate in view of the necessary uncertainty as to what is meant by resolution." In fact, Rayleigh two-point resolution is little affected by small amounts of wavefront error and is improved by a central obstruction. There is no direct connection between the Rayleigh two-point resolution criterion and the Rayleigh quarter-wave criterion.

2. *Sparrow Two-Point Resolution*

Two incoherent PSFs separated by the Rayleigh criterion add to give the combined intensity profile shown in Fig. 15a. The irradiance midway

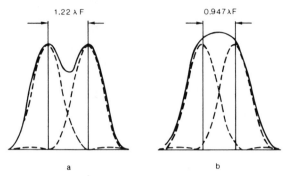

FIG. 15. Classical criteria for two-point resolution. (a) Rayleigh; (b) Sparrow.

between the two peaks dips below the two peaks. If the separation is reduced to $s = 0.947\lambda F$, the dip will just disappear. This separation is known as the *Sparrow two-point resolution criterion*. To state the criterion formally, if the combined irradiance distribution is $E(s,t)$ and the individual PSFs are centered at $(s/2,0)$ and $(-s/2,0)$, then s meets the Sparrow criterion if

$$\frac{\partial^2}{\partial s^2} I(s,t) = 0.0 \text{ at } (0,0) \qquad (84)$$

The Sparrow criterion is just as arbitrary as the Rayleigh criterion and is no more affected by small amounts of wavefront error. If applied to a series of equally spaced line spread functions of equal brightness, it will indicate the point at which they just merge into a continuum.

3. *Maréchal's Criterion for Strehl Definition*

One common criterion for diffraction-limited performance is that the Strehl definition equal or exceed 80%. This is known as *Maréchal's criterion*. For third-order spherical aberration, it coincides with Rayleigh's quarter-wave criterion.

Maréchal considered the effects of wavefront error only ($\tau_p = 0.0$). He shows that, for small amounts of wavefront error (Maréchal, 1947):

$$\mathscr{D} \geq (1 - 2\pi^2\omega^2)^2 \qquad (85)$$

Other common variants of Eq. (85) include

$$\mathscr{D} \geq 1 - 4\pi^2\omega^2 \qquad (86)$$

and

$$\mathscr{D} \simeq \exp[-(2\pi\omega)^2] \qquad (87)$$

A derivation of Eq. (87) is given in Section IX,C. Depending on the choice of Eq. (85), (86) or (87), Maréchal's criterion is just met when $\omega = 0.071$, $\omega = 0.073$, or $\omega = 0.075$ wavelength rms, respectively.

Maréchal's criterion can be extended to cover aperture obstructions and image motion by using the form of the Strehl definition given earlier. We discuss the implications of this on optical design parameters in Section XII.

VII. THE OPTICAL TRANSFER FUNCTION AND ITS COROLLARIES

The PSF was introduced earlier as being one of the two most complete functions for specifying image quality and as being inconvenient to use when the object surface irradiance distribution is continuous. The second complete function for specifying image quality, the *complex optical transfer function* OTF, is specifically tailored for use with continuous-tone imagery. The PSF and OTF are mathematically interchangeable, since each can be calculated by taking the Fourier transform of the other.

Any image irradiance distribution on a surface can be represented as a spectrum of sinusoidal irradiance gratings of varying period, orientation, and brightness amplitude. The *object spectrum* S_o and the *image spectrum* S_i are computed by taking the Fourier transforms of the geometrically ideal and real-image irradiance distributions $E_o(x',y')$ and $E_i(x',y')$, and the transformations are reversible. Each component of S_i can be computed by multiplying the corresponding component of S_o by a constant factor which is characteristic of the optical system. The OTF is the spectrum of these constant factors for the optical system.

The versatility of the OTF lies in two circumstances. First, the process of taking the Fourier transform of $E_o(x',y')$, multiplying it by the OTF, and taking the reverse Fourier transform to obtain $E_i(x',y')$ is less difficult in practice than the mathematically equivalent process of convolution with the PSF. Second, the OTFs of a series of separable factors causing image degradation (e.g., optical system, image motion, and image sensor) can be combined by simple multiplication to give a complete imaging system OTF. Generating the system PSF requires convolution of the separate PSFs.

The OTF is a comparatively new concept which is not in universal use at all levels of image quality analysis. It has been adapted most readily in fields where older testing techniques involved the use of test objects having relatively simple spectra, such as the three-bar resolution charts used in aerial reconnaissance. Much of the early work in applying OTFs was done in the television industry, where it is a logical extension of audio system electronic circuit analysis techniques.

A. Definitions of OTF Terminology

Figure 16 shows an object in the form of a simple grating, along with cross-section plots of the object and image irradiance distributions E_o and E_i. It is useful in visualizing the definitions of OTF terminology.

1. Target Orientation Angle α

Target orientation is defined by a vector pointing in the direction of modulation of the grating, normal to the lines making up the grating. The direction of the vector is represented by the angle α, measured relative to the y' axis. Note that there is an ambiguity of 180° in the direction of the vector; the angles α and $\alpha + \pi$ refer to the same grating.

2. Modulation

The irradiance amplitude of the grating is represented by its object and image modulations M_o and M_i, defined as

$$M_o = [E_{o\ max} - E_{o\ min}]/[E_{o\ max} + E_{o\ min}]$$
$$M_i = [E_{i\ max} - E_{i\ min}]/[E_{i\ max} + E_{i\ min}] \tag{88}$$

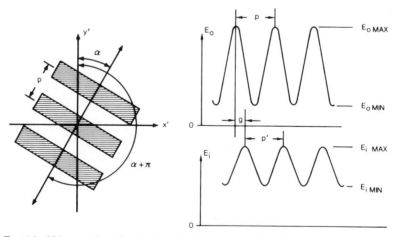

FIG. 16. Object grating of period p and orientation angle α. Here E_o and E_i are the geometrically ideal and real image irradiance distributions for a sinusoidal object grating. The real image is shifted from its ideal position a distance g and has a period p'.

Modulation is a normalized quantity and thus is unitless. The use of modulation eliminates any need for considering average irradiance level, at least in linear systems. Optical systems are linear, and the image grating is sinusoidal if the object grating is sinusoidal. Some image sensors, such as photographic film, are nonlinear. They introduce harmonic distortion, the amount being a function of the exposure level. Techniques for treating nonlinear sensors are not included in this chapter.

3. *Spatial Frequency*

The grating periods are p, referred to the object surface, and p', referred to the image surface. The corresponding spatial frequencies are

$$\nu = 1/p, \qquad \nu' = 1/p' \tag{89}$$

If the lens system is distortion-free and has a magnification m, $\nu' = \nu/m$. If the lens has distortion or is anamorphic, then $m_x \neq m_y$, and both the spatial frequency and the direction of the image grating may be altered. In this case,

$$\nu' = \nu \left(\frac{\sin^2 \alpha}{m_x^2} + \frac{\cos^2 \alpha}{m_y^2} \right)^{1/2} \tag{90}$$

and

$$\alpha' = \text{arc tan} \left(\frac{m_x \sin \alpha}{m_y \cos \alpha} \right) \tag{91}$$

Spatial frequency is expressed in units of *cycles per unit length*, cycles per millimeter being the most common form. For an infinite conjugate object or image, *cycles per unit angle* can be used, with radians, milliradians, or arc seconds the common angular units.

The terminology for spatial frequency has varied considerably, *lines per millimeter, line-pairs per millimeter,* and *television lines per millimeter* being the most common variants. The term *"cycle"* has been adopted by most modern workers to avoid the confusion between the optical practice of calling a period one line and the television practice of calling a period two lines. Note that *1 cycle = 1 optical line = 1 line pair = 2 television lines = 1 period p.* Another unit which is common in television parlance is the *line number N*, defined as the number of television lines per picture height; N_{max} corresponds to the number of raster lines used to scan the image, and N has obvious advantages in comparing different-sized camera and display formats.

The increased use of diode arrays as image sensors has brought two other terms into use, *pixel* and *resel*. A pixel is a square whose sides equal the center-to-center spacing of the diode array. The term resel refers to $\frac{1}{2}$ cycle at some spatial frequency, usually the limiting resolution spatial frequency to be defined in Section VIII,C. It derives from information theory, which states that two samples must be taken per cycle if a frequency is to be detected unambiguously. The two terms are sometimes used interchangeably.

Each grating component in a spectrum is defined by the spatial frequency coordinates (ν,α). For use in Cartesian coordinates, the Cartesian spatial frequencies ν_x and ν_y are defined as

$$\nu_x = \nu \sin \alpha; \qquad \nu_y = \nu \cos \alpha \tag{92}$$

Note that (ν_x,ν_y) and $(-\nu_x,-\nu_y)$ refer to the same grating, as do (ν,α) and $(\nu,\alpha + \pi)$.

4. Cutoff Frequency and Frequency Normalization

In electronic terminology, a lens is a low-pass filter which does not transfer information on any spatial frequency higher than the cutoff spatial frequency ν_c'. In image space coordinates,

$$\nu_c' = 1/\lambda F \quad \text{(cycles/unit length)} \tag{93}$$

In object space coordinates, for an infinitely distant object surface,

$$\nu_c = D_p/\lambda \quad \text{(cycles/radian)} \tag{94}$$

For general discussions of image quality, it is convenient to use normalized spatial frequency units. The *reduced* or *normalized spatial frequency* ν_n is defined here as

$$\nu_n = \nu'/\nu_c' = \nu/\nu_c \tag{95}$$

By this definition, $0.0 \leq \nu_n \leq 1.0$. Some texts use a normalized scale running from 0 to 2, based on the coherent cutoff frequency $1/2\lambda F$, or on the procedure for calculating the OTF from the pupil function, as discussed below.

5. Modulation Transfer Function

Each component of the object spectrum S_o is characterized by four quantities, spatial frequency, orientation, modulation, and position. We have discussed how spatial frequency and orientation can be altered when distortion is present. The *modulation transfer function* (MTF) $T(\nu',\alpha)$ measures the change in modulation:

$$T(\nu',\alpha) = M_i(\nu',\alpha)/M_o(\nu,\alpha) \tag{96}$$

Since (ν',α) and $(\nu',\alpha + \pi)$ represent the same component grating, the MTF must by definition be an even function.

6. *Phase Transfer Function**

The image grating may be shifted in position as well as reduced in modulation. This shift is measured by the *phase transfer function* (PTF) $\Phi(\nu',\alpha)$, which expresses the displacement in radians. If the linear displacement is g and the period of the image grating is p', as shown in Fig. 16,

$$\Phi(\nu',\alpha) = 2\pi g/p' \tag{97}$$

The direction of the displacement is associated with the vector defining the orientation of the target but does not have the same 180° ambiguity. Mathematically, the ambiguity is eliminated by noting that $g(\nu,\alpha) = -g(\nu,\alpha + \pi)$, making the PTF an odd function by definition.

7. *The Complex OTF*

The MTF and PTF are combined to form the *complex* OTF $\mathcal{T}(\nu',\alpha)$:

$$\mathcal{T}(\nu',\alpha) = T(\nu',\alpha)\exp[i\Phi(\nu',\alpha)] \tag{98}$$

The OTF reduces to the MTF whenever the PTF is zero. OTFs for separable image quality degradation factors such as wavefront error and image motion can be cascaded by simple multiplication: $\mathcal{T}_1, \mathcal{T}_2, \mathcal{T}_3, \ldots$ multiply to give

$$\mathcal{T}(\nu',\alpha) = [T_1 \times T_2 \times T_3 \times \cdots]\exp i[\Phi_1 + \Phi_2 + \Phi_3 + \cdots] \tag{99}$$

OTFs for lenses in a relay system are not separable. Such coherently coupled lenses must be ray-traced as a unit to determine their combined OTF. Incoherently coupled optics, such as fiber optics faceplates, are separable.

OTF nomenclature has been standardized recently (Inglestam, 1961), but many variants are found in the older literature. *Contrast transfer function* and *frequency response* can be found in some optical literature. Linfoot (1956) used the term *"transmission factor,"* which is avoided now to eliminate confusion with the other optical definition for transmission. Mertz and Gray (1934) used *"equivalent transfer admittance."* The term *"sine-wave response"* is still in use by people working with image

---

* For a more complete discussion of the significance of the phase transfer function, see Shack (1974).

sensors, when describing sensor characteristics. The term *modulation transfer factor* is used when only one spatial frequency is involved.

B. COMPUTING THE OTF

Figure 17 shows the mathematical relationships among the complex pupil function, the OTF, and the PSF. When starting from the complex pupil function, the OTF may be calculated directly by autocorrelation, or indirectly by first calculating the PSF and taking its Fourier transform. Which route is taken depends upon the needs of the user and the form of the pupil function. If the pupil function is rotationally symmetric and the Fourier transforms become Hankel transforms, calculation via the PSF offers some advantages. The PSF may also be calculated from the OTF, since Fourier transformation is a reversible process. The latter is useful when working with models for the OTF. For purposes of exposition, calculation via autocorrelation of the complex pupil function is treated first.

1. Computing the OTF by Autocorrelation of the Pupil Function

If the complex pupil function of Eq. (44) is written in normalized Cartesian coordinates, the OTF is given by the autocorrelation integral

$$\mathcal{T}(s,\alpha) = \int_{-1}^{1} \int_{-1}^{1} \mathcal{P}^* (X + k, Y + j) \, \mathcal{P} (X - k, Y - j) \, dX \, dY$$

$$\Big/ \int_{-1}^{1} \int_{-1}^{1} [P(X,Y)]^2 \, dX \, dY \quad (100)$$

Comparison to Eq. (48) shows the denominator to be equal to $\pi \tau_p$. Figure 18 defines the coordinates and shows the significance of the numerator; the centers of $\mathcal{P}$ and its complex conjugate are separated a distance s in the direction α, and then the integration is performed over their common area. Note that $k = (s/2) \sin \alpha$ and $j = (s/2) \cos \alpha$.

If the complex pupil function is expanded into its trigonometric form,

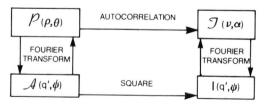

FIG. 17. Mathematical relationships among the pupil function $\mathcal{P}$, the PSF I, and the OTF $\mathcal{T}$.

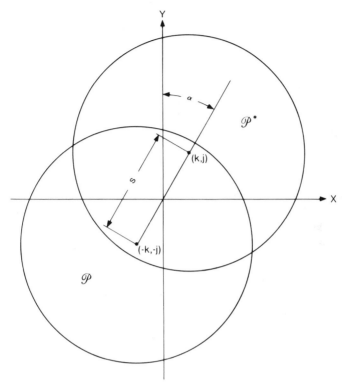

Fig. 18. Coordinates for the autocorrelation integral, showing the pupil shifts s. Here X and Y are the normalized pupil coordinates defined in Eq. (21).

$$\mathscr{P}(X,Y) = P(X,Y)\cos[2\pi W(X,Y)] + iP(X,Y)\sin[2\pi W(X,Y)] \quad (101)$$

the OTF can be written in the form

$$\mathscr{T}(s,\alpha) = A + iB \quad (102)$$

where

$$A = \frac{1}{\pi\tau_p}\int_{-1}^{1}\int_{-1}^{1} P(X+k, Y+j)P(X-k, Y-j)$$
$$\times \cos 2\pi[W(X+k, Y+j) - W(X-k, Y-j)]\, dX\, dY \quad (103)$$

and

$$B = \frac{1}{\pi\tau_p}\int_{-1}^{1}\int_{-1}^{1} P(X+k, Y+j)P(X-k, Y-j)$$
$$\times \sin 2\pi[W(X+k, Y+j) - W(X-k, Y-j)]\, dX\, dY \quad (104)$$

The MTF is then given by

$$T(s,\alpha) = (A^2 + B^2)^{1/2} \qquad (105)$$

and the PTF is given by

$$\Phi(s,\alpha) = \text{arc tan } (B/A) \qquad (106)$$

Equation (100) is normalized both in terms of modulation $[\mathcal{T}(0,0) \equiv 1.0]$ and in terms of spatial frequency. Examination of Fig. 18 indicates that $0 \leq s \leq 2$, and therefore $\nu_n = s/2$. In some references, the MTF is plotted directly as a function of s, which is the second reason for using a reduced frequency scale of 0 to 2 (see Section VII,A,4).

The PTF disappears completely when $B = 0$, and the OTF devolves into the MTF. Thus the condition for a lateral displacement of a sinusoidal image grating component is that $B \neq 0.0$. Examination of Eq. (104) shows the conditions for the existence of B to be:

(1)   If $W(\rho,\theta) = 0.0$ at all points, $B = 0.0$.
(2)   If $W(\rho,\theta)$ and $P(\rho,\theta)$ are both even functions, $B = 0.0$.
(3)   If $W(\rho,\theta) \neq 0.0$ and either $W(\rho,\theta)$ or $P(\rho,\theta)$ is not even, $B \neq 0.0$.

It is for this reason that odd components in the wavefront aberration function are a matter of concern; they cause lateral displacements of components of the image spectrum.

2. Computing the OTF from the PSF

The PSF is defined in its most general terms by Eqs. (52) and (53). The unnormalized OTF is computed by taking the Fourier transform of Eq. (53). The normalized OTF is given by

$$\mathcal{T}(\nu_x',\nu_y')$$

$$= \frac{\displaystyle\int_{-\infty}^{\infty}\int_{-\infty}^{\infty} E_i(\Delta x', \Delta y') \exp[-2\pi i(\nu_x'\Delta x' + \nu_y'\Delta y')] \, d\Delta x' \, d\Delta y'}{\displaystyle\int_{-\infty}^{\infty}\int_{-\infty}^{\infty} E_i(\Delta x', \Delta y') \, d\Delta x' \, d\Delta y'} \qquad (107)$$

The denominator in Eq. (107) is both the unnormalized OTF at $\nu_x' = 0$, $\nu_y' = 0$ and the total energy in the PSF. Thus the practice of normalizing the OTF to 1.0 at its origin is equivalent to normalizing the PSF to unit total energy at the image surface, as was done in Eq. (66).

Equation (107) can be rewritten in terms of the normalized PSF using polar coordinates:

$$\mathcal{T}(v',\alpha) = \frac{\pi}{4\lambda^2 F^2} \int_0^{2\pi} \int_0^{-\infty} I(q',\psi)$$

$$\times \exp[-2\pi i v'q' \cos(\psi - \alpha)]q' \, dq' \, d\psi \quad (108)$$

Since the PSF and the OTF are a Fourier transform pair, it follows that

$$I(q',\psi) = \frac{4\lambda^2 F^2}{\pi} \int_0^{2\pi} \int_0^{\infty} \mathcal{T}(v',\alpha) \exp[2\pi i v'q' \cos(\psi - \alpha)]v' \, dv' \, d\alpha \quad (109)$$

In the important case where the PSF and thus the MTF are rotationally symmetric, Eqs. (108) and (109) reduce to the Hankel transforms:

$$T(v') = \frac{\pi^2}{2\lambda^2 F^2} \int_0^{\infty} I(q')J_0(2\pi v'q')q' \, dq' \quad (110)$$

and

$$I(q') = \frac{2\lambda^2 F^2}{\pi^2} \int_0^{\infty} T(v')J_0(2\pi v'q')v' \, dv' \quad (111)$$

For an optical system, the OTF is a bounded function, but the PSF is not. (Cases where the opposite is true will be dealt with shortly.) In any numerical computation of the OTF from the PSF, the integration cannot be carried to $q' = \infty$. If the OTF computed from a truncated PSF is normalized to 1.0 at $v = 0$, it will be in error at all higher spatial frequencies. The preferred procedure in this case is to normalize to a value equal to the total energy in the truncated PSF. This form of normalization is accomplished automatically in Eqs. (108) and (110).

The author has used a computer program based on Eqs. (66) and (110) to examine the effects of fine structure in the pupil function on the PSF and the MTF (Wetherell, 1974a). Results from studying the properties of this program will give a feeling for the effects of truncating the PSF on the accuracy of the MTF. Figure 19 plots the error in the MTF versus spatial frequency for a truncation radius $\delta'_t = 3.0$ Airy radius units. The error function shown is for a perfect lens with a circular pupil and gives the ratio of the numerically calculated MTF T to the analytically calculated MTF T_i. Credible accuracy is achieved even with very small truncation radii if the MTF is normalized to the encircled energy at the truncation radius. In the spatial frequency range of most common interest, between $v_n = 0.1$ and $v_n = 0.8$, the error is on the order of one-fifth of the value of the excluded energy. In the example shown, the error is less than 1% over this spatial frequency range, despite the small truncation radius. For a lens with wavefront error, the objective in selecting the truncation radius

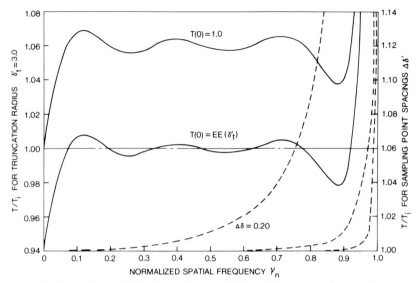

FIG. 19. Errors in the MTF of a perfect lens calculated by numerical integration of Eq. (110): Curves (——) are for a truncation radius $\delta_t' = 3.0$ Airy radius units ($\Delta\delta' = 0.05$) and show the advantage of normalizing $T(0)$ to equal $EE(\delta_t')$; Curves (---) show the effect of varying the radius sampling increment: $\Delta\delta' = 0.20, 0.10$, and 0.05 ($\delta_t' = 100.0$).

should be to reduce the excluded energy to less than, say, five times the acceptable error at middle spatial frequencies. Equation (110) automatically normalizes $T(0)$ to equal $EE(\delta_t')$.

Figure 19 also shows the error produced by varying the radius increment $\Delta\delta'$ at which the PSF is sampled for the Hankel transform. This error is confined primarily to the higher spatial frequencies. Ten to 20 samples per Airy radius unit is sufficient for most practical purposes.

Since the PSF and OTF are a Fourier transform pair, symmetry characteristics in the OTF imply corresponding properties in the PSF. The OTF is either real (PTF = 0.0) or Hermitian (MTF even and PTF odd). The PSF is therefore always real. When the OTF is Hermitian, the PSF cannot be even. Otherwise, the PSF is always even. When $W(\rho,\theta) = 0.0$ for all points, the PSF is an even function, no matter how irregular the pupil boundary or amplitude transmittance pupil function.

3. Computing Strehl Definition from the OTF

Strehl definition can be computed from the OTF by use of Eq. (109) or (111). Where only the Strehl definition is of interest, Eq. (109) reduces to

$$\mathcal{D} = \frac{4\lambda^2 F^2}{\pi} \int_0^{2\pi} \int_0^\infty T_c(\nu',\alpha)\nu' \; d\nu' \; d\alpha \tag{112}$$

4. The MTF for the Perfect Lens

Consider a pupil function in which $W(\rho,\theta) = 0.0$ at all points and $P(\rho,\theta)$ is either 1.0 or 0.0 at all points. The OTF then devolves into the MTF, and Eq. (100) is reduced to calculating the common clear area between the two displaced pupils divided by the total pupil area. This is true no matter what the pupil boundary shape, and no matter the number and shape of obstructions within the pupil boundary. Such calculations can be made numerically, graphically, or experimentally, using masks, a uniform light source, and a photometer.

For a perfect lens with a circular pupil,

$$\begin{aligned} T_i(\nu_n) &= (2/\pi)[\text{arc cos } \nu_n - \nu_n \sin (\text{arc cos } \nu_n)] \\ &= (2/\pi)[\text{arc cos } \nu_n - \nu_n \sqrt{1 - \nu_n^2}] \end{aligned} \tag{113}$$

$T_i(\nu_n)$ is used as the standard of comparison in judging the effects on image quality of wavefront error, aperture obstructions, and other factors.

For the rectangular pupil described in Section VI,D,

$$T(\nu_{nx},\nu_{ny}) = (1 - \nu_{nx})(1 - \nu_{ny}) \tag{114}$$

where ν_{nx} and ν_{ny} are normalized spatial frequencies defined in terms of cutoff spatial frequencies $\nu'_{cx} = 1/\lambda F_x$ and $\nu'_{cy} = 1/\lambda F_y$ [see Eq. (80)].

Analytic solutions for a perfect lens with a circular pupil and a centered circular obstruction are more complex. O'Neill's (1956)* solution can be found in Appendix B.

5. Polychromatic MTF

It is possible to define the MTF for a lens operating in white light. To do this, it is necessary to know the *spectral radiance distribution* $L(\lambda)$ of the object, the *spectral radiant sensitivity* $S(\lambda)$ of the image sensor, and the *spectral transmittance* $\tau(\lambda)$ of the complete optical system from object surface to image surface. All three quantities may be normalized to 1.0 at a common reference wavelength λ_0.

In computing the *polychromatic* MTF $T_p(\nu_0)$, the normalized spatial frequency ν_n is replaced by $\nu_0\lambda$, where ν_0 is the *specific spatial frequency:*

---

* The erratum on p. 1096 of O'Neill (1956) contains a vital correction for the equation for the MTF.

$$\nu_0 = \nu' F = n' \nu'/2NA \qquad (115)$$

The polychromatic MTF is the weighted average of the monochromatic MTFs at each wavelength, computed by

$$T_p(\nu_0) = \int \tau(\lambda) L(\lambda) S(\lambda) T(\nu_0 \lambda) \, d\lambda \Big/ \int \tau(\lambda) L(\lambda) S(\lambda) \, d\lambda \qquad (116)$$

The limits on integration are usually defined by the wavelength limits at which the product $\tau(\lambda)S(\lambda)$ drops low enough to be ignored. Levi (1969) has published polychromatic MTFs for the perfect lens for a series of combinations of sources and detectors.

The polychromatic MTF is a useful tool for estimating the performance potential of a lens in white light, but care must be taken in its application. It is completely applicable only where $L(\lambda)$ does not vary across the object surface.

C. GRAPHICAL PRESENTATION OF OTF DATA

Graphical presentation of OTF data presents the same problems as graphical presentation of PSF data and is solved in much the same way. For a rotationally symmetric MTF, a single section plotting MTF versus ν' from zero to ν'_c is sufficient. If the MTF is not rotationally symmetric, multiple sections can be plotted and constitute the preferred approach where the data must be used quantitatively. Contour plots and perspective drawings are also used.

Figure 20 shows sections of five MTFs, plotted on linear graphs (log–log plots are preferred by some). The uppermost curve is for a perfect lens, and the other four are for lenses with increasing amounts of defocus W_{20}. It is noted that two of the defocus curves show negative values of MTF. This phenomenon is known as *spurious resolution* and indicates a contrast reversal in the image grating. While this might be thought of as equivalent to a PTF of π radians, there is a sign ambiguity which negates such an interpretation. In a target consisting of, for example, five equally spaced bright lines on a dark background, spurious resolution reduces the apparent number of bright lines to four, spaced midway between their proper positions.

Figure 21 shows a perspective drawing of the MTF of a lens with 2.4 wavelengths of third-order coma, matching the PSF in Fig. 14. It shows the strong variation with azimuth angle α which is possible in real lens MTFs. It also indicates the possible shortcomings of the common practice of plotting sections 90° apart. While the complete MTF is 50% redundant, two orthogonal sections can be disastrously incomplete. If the MTF is ro-

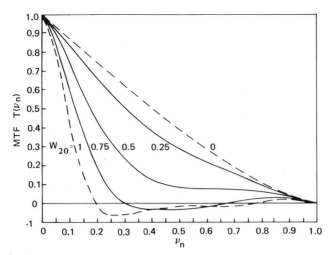

FIG. 20. MTFs for various amounts of defocus W_{20} measured in wavelengths OPD, showing spurious resolution.

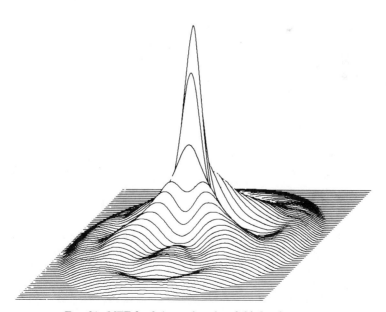

FIG. 21. MTF for 2.4 wavelengths of third-order coma.

tationally symmetric, only one section is needed. For the MTF in Fig. 21, three sections are needed, at $\alpha = 0°$, $45°$, and $90°$. If the wavefront error pupil function is asymmetric, four sections are the minimum required, at $\alpha = 0°$, $45°$, $90°$, and $135°$.

D. COROLLARIES OF THE OTF

1. *The* MTF *Degradation Function*

In most image quality analyses, the MTF is used in place of the OTF. During the early stages of designing a new optical system, when its basic parameters are being selected, various image-degradation models are used to predict an expected MTF. It is convenient in these situations to divide the system MTF into a perfect lens MTF $T_i(\nu_n)$ and an MTF *degradation function* $T_d(\nu_n, \alpha)$, such that

$$T(\nu_n, \alpha) = T_d(\nu_n, \alpha) \times T_i(\nu_n) \qquad (117)$$

where T_d summarizes the effects of all factors degrading the image quality and can itself be broken up into MTF degradation functions for each factor, e.g., T_ω for wavefront error and T_σ for image motion.

The MTF degradation function can be defined empirically in any real situation, but the value of the concept lies in the validity of the implied assumption that the different factors degrading image quality are statistically independent. There are circumstances where this assumption breaks down, and some of these are cited later in the chapter. The assumption of statistical independence is sufficiently accurate for preliminary image quality analysis if not overextended.

Even empirically produced MTF degradation functions are useful in comparing how different factors differ in the manner in which they degrade the MTF. The MTF degradation functions for different lens aberrations have been reproduced in Figs. 22 and 23 (Solomon and Wetherell, 1973). Figure 22 shows data for rotationally symmetric aberrations. The nonrotationally symmetric aberrations in Fig. 23 require more than one curve for a complete picture of the MTF degradation function. In the case of astigmatism, if the image surface is at either the sagittal or tangential focus, one of the $0°$ or $90°$ curves will be unity at all spatial frequencies. In Figs. 22 and 23, all aberrations have the same rms wavefront error, to facilitate comparison. It is noted that the degree of MTF reduction at any given spatial frequency differs considerably with type of aberration. Only the defocus and astigmatism curves have similar shapes.

Lerman (1969b) has generated a rotationally symmetric MTF degradation function model for a mix of aberrations, to be used in preliminary

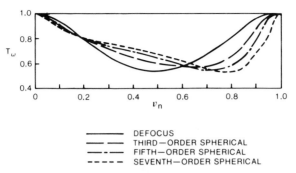

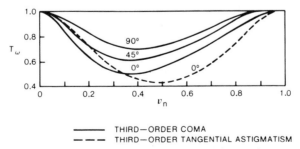

FIG. 22. The MTF degradation function for 0.1 wavelength rms of rotationally symmetric aberrations through seventh-order spherical.

analyses of optical systems. The model is represented by the five curves in Fig. 24. In Section IX,C, an MTF degradation function model for random wavefront error is discussed.

2. Noise Equivalent Relative Structural Content

Linfoot's (1956) concept of relative structural content was described in Section II,B. In the form shown in Eq. (3), it is tied too intimately to the specific object irradiance distribution to be of use in general analysis of optical system image quality. For the latter purpose, it is convenient to shift to the spatial frequency domain, replacing $E_0(x',y')$ and $E_i(x',y')$ by their Fourier transforms $S_0(\nu_x',\nu_y')$ and $S_i(\nu_x',\nu_y')$. Within the isoplanatic patch,

$$S_i(\nu_x',\nu_y') = \mathcal{T}(\nu_x',\nu_y')S_0(\nu_x',\nu_y') \tag{118}$$

The irradiance distributions in Eq. (3) are squared and are replaced by the power spectrum $|S_0(\nu_x',\nu_y')|^2$. For more general analysis, the actual object

FIG. 23. The MTF degradation function for 0.1 wavelength rms of nonrotationally symmetric aberrations.

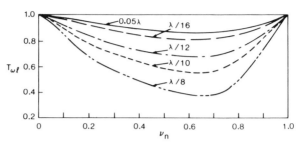

FIG. 24. Lerman model for the average MTF degradation function for a mix of Seidel aberrations plus fifth- and seventh-order spherical aberrations (S. H. Lerman, 1969b).

power spectrum is replaced by the statistical mean power spectrum $<|S_0(\nu'_x,\nu'_y)|^2>$. Equations (3) and (118) combine to give

$$T_L = \int \int T^2(\nu'_x,\nu'_y)<|S_0(\nu'_x,\nu'_y)|^2> \, d\nu'_x \, d\nu'_y \Big/ \int \int <|S_0(\nu'_x,\nu'_y)|^2> \, d\nu'_x \, d\nu'_y \tag{119}$$

Equation (119) represents the total spectral power content of the image divided by the total spectral power content of the object and is still tied to a particular spectral power distribution. More generally, the quantity of interest is the fraction of the maximum possible spectral power content of any object which can be reproduced in the image. The maximum possible spectral power content is defined by white noise, for which $<|S_0(\nu'_x,\nu'_y)|^2> = 1.0$ at all spatial frequencies. In this case, Eq. (119) reduces to

$$T_L^* = \int_0^{2\pi} \int_0^{\infty} T^2(\nu',\alpha)\nu' \, d\nu' \, d\alpha \tag{120}$$

in polar coordinates, where T_L^* is termed the *noise equivalent relative structural content.*

To compare the noise equivalent relative structural content of a real lens to that of the corresponding perfect lens, T_L^* can be normalized:

$$Q^* = \frac{T_L^* \text{ (real)}}{T_L^* \text{ (perf)}} = \frac{12\pi\lambda^2 F^2}{3\pi^2 - 16} \int_0^{2\pi} \int_0^{\infty} T^2(\nu',\alpha)\nu' \, d\nu' \, d\alpha \tag{121}$$

Q^* has a special relevance to the problem of detecting a dim point source against a continuous bright background, which is discussed in Section VIII,B.

3. Relative Edge Response

A problem encountered in aerial reconnaissance is that of determining whether or not the complete imaging system is performing to specifications by examining the output data. Since it is seldom convenient to arrange for sinusoidal bar charts to be incorporated into real scenes, techniques for estimating the MTF from the edges of real objects have evolved (Scott *et al.*, 1963).

Figure 25 illustrates a typical edge object and its image. The object is assumed to have a sharp edge. In the recorded image, the edge has been smeared out into a shoulder of varying slope. The image can be scanned by a microdensitometer with a slit aperture, and the output data differentiated to determine the slope profile. The slope profile is actually the line spread function for the imaging system and can be converted into a one-directional MTF by Fourier transformation. By using a number of edges with different orientations, the complete imaging system MTF can be estimated. Finding the optical system MTF is then a matter of backing out the MTF contributions due to the microdensitometer and image sensor, which is not a trivial job; photographic images contain grain noise. Noise is emphasized by differentiation and can be suppressed only by averaging the signal before differentiation. In edge scanning, this requires the use of long scanning slits, or averaging measurements over a number of edges.

Details of edge measurement processes are not of direct concern here; suffice it to say that edge response is considered a useful measurement of image quality. *Edge response* is specified as the maximum slope of the edge image, which equates with the peak of the line spread function. The *relative edge response* $\mathscr{E}$ is defined by analogy to the Strehl definition as the ratio of the real lens edge response to the perfect lens edge response. It is computed by integrating the area under the one-directional MTF:

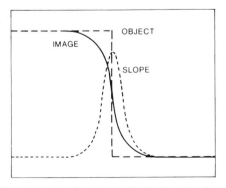

FIG. 25. Image of an edge, showing the profiles in the direction of scan of the object, image, and slope (line spread function).

$$\mathscr{E} = \int_0^1 T(\nu_n) \, d\nu_n \Big/ \int_0^1 T_i(\nu_n) \, d\nu_n$$

$$= (3\pi/4) \int_0^1 T(\nu_n) \, d\nu_n \tag{122}$$

4. Schade's Equivalent Passband

Otto Schade has defined a concept which he terms the *equivalent passband* N_e^*, which can be thought of as a one-directional analog of T_L^* in Eq. (120). The equivalent passband measures the amount of noise a scanning imaging system such as a television camera generates when viewing an object which is pure white noise. It is thus a measure of the maximum possible signal content of the image. The real imaging system produces the same noise as an idealized system whose MTF is 1.0 for $0 \leq N \leq N_e^*$ and 0.0 for $N > N_e^*$, where N and N_e^* were spatial frequencies in line numbers in Schade's original article. Schade's (1952) equation for N_e^*, translated into cycles per millimeter or the equivalent, is

$$N_e^* = \int_0^\infty T^2(\nu') \, d\nu' \tag{123}$$

Here Q^* is used in preference to the equivalent passband for reasons shown in Section VIII,B, and N_e^* is introduced partially for historic reasons, since Schade's analysis of imaging systems was a key step in the search for more relevant merit functions for image quality. It can also be used to help illustrate the subtle differences in the way different image quality degradation factors affect superficially similar merit functions.

5. Spatial Frequency Weighting in $\mathscr{D}$, $\mathscr{E}$, N_e^*, and Q^*

The strehl definition $\mathscr{D}$, relative edge response $\mathscr{E}$, equivalent passband N_e^*, and noise equivalent relative structural content Q^* are all one-number merit functions which can be computed from the MTF. Each uses the MTF in a different manner. The significance of this difference may be better understood by noting the weight each places on different regions in the spatial frequency domain.

If the lens MTF is expanded in terms of the MTF degradation function, as in Eq. (117), the relative spatial frequency weighting of the four merit functions can be seen by examining the integrands in Eqs. (112), (121), (122), and (123):

for $\mathscr{E}$, the integrand is $T_d(\nu_n) \times T_i(\nu_n)$;
for $\mathscr{D}$, the integrand is $T_d(\nu_n) \times T_i(\nu_n)\nu_n$;
for N_e^*, the integrand is $T_d^2(\nu_n) \times T_i^2(\nu_n)$;
for Q^*, the integrand is $T_d^2(\nu_n) \times T_i^2(\nu_n)\nu_n$.

Assume $T_d(\nu_n)$ is the same in all cases. Then the relative weighting of the MTF degradation function can be compared by examining the portion of the integrand to the right of the multiplication sign. The latter quantities have been normalized and plotted as a function of spatial frequency in Fig. 26. (Strictly speaking, the fact that the MTF degradation function is squared in two of the four cases should be considered for quantitative comparisons. For qualitative comparisons, Fig. 26 should be satisfactory.)

Spatial frequencies below $\nu_n = 0.1$ are heavily weighted by relative edge response and equivalent passband but are almost ignored by the other two merit functions. Q^* weights low to middle spatial frequencies most heavily, while Strehl definition weights middle to high spatial frequencies most heavily. The MTF degradation functions in Figs. 22 and 23 suggest that different forms of wavefront error affect the merit functions differently. More significant quantitative differences will be shown later for image motion, which affects Strehl definition most strongly, and central obstructions, which affect Q^* most strongly.

Spurious resolution, the presence of negative modulation at some spatial frequencies, affects the four merit functions differently. Spatial frequencies above the first zero crossing of the MTF are not accurately reproduced in the image, due to phase (positional) ambiguities. If these values of the MTF are included when calculating the four merit functions, they will not accurately reflect the image quality. Examination of the integrands shows that N_e^* and Q^* will be unduly optimistic, while edge response and Strehl definition will be unduly pessimistic. In the case of Strehl definition, inclusion of the negative modulation values has a physical significance, since the central maximum of the PSF can reduce to zero, as was shown in Fig. 12, for one wavelength of defocus. This zero value is misleading as a measure of image quality, however, since the defocussed

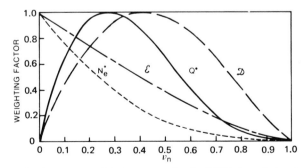

FIG. 26. Spatial frequency weighting curves for the MTF degradation function for the Strehl definition, relative edge response, equivalent passband, and noise equivalent relative structural content.

lens still accurately reproduces spatial frequencies below the first zero crossing of the MTF. If the four merit functions are to represent image quality, the integration by which each is calculated should be terminated at the first zero crossing of the MTF.

E. Aperture Processes

The term *"aperture"* has a somewhat different meaning to television engineers than it does to optical engineers. Optical engineers use "aperture" and "pupil" more or less interchangeably and apply the terms exclusively to lens pupils and diffracting apertures. When making finer distinctions, optical engineers use the term *"clear aperture"* to define the physical boundaries at the edge of a mounted lens element, and *"effective aperture"* to define the pupil. Television engineers extend the meaning of aperture to include the PSF, or more generally the *sampling aperture* which scans the image.

The reasons for this redefinition are historic, arising from early analyses of facsimile machines which used pinhole apertures to scan and reconstruct images transmitted over wires. The concept was extended to television, where the sampling aperture is generated with an electron beam, and is now being applied to diode arrays, where the individual diodes form the sampling apertures.

From the point of view of image quality analysis, lens pupils and sampling apertures perform essentially the same function; both act as low-pass spatial frequency filters. Both may be represented by either a PSF or an OTF, and both are treated in the same manner in computing image quality for the complete imaging system. Otto Schade (1951, 1952, 1953, 1955), whose research pioneered the application of Fourier techniques in image quality analysis, uses the term *"aperture process"* to include both lens and sampling apertures. Aperture process is a useful term which covers all elements in an imaging system which act to restrict the spatial frequency bandpass.

There are distinctions between lens pupils and sampling apertures. In lenses, the OTF is bounded and the PSF is unbounded. The opposite is true for sampling apertures. Lens pupil OTFs and PSFs vary as a function of wavelength, while sampling aperture OTFs and PSFs are wavelength-independent. Sampling apertures do not suffer from the effects of wavefront error, but their OTFs can be modified by varying the transmittance within the sampling aperture, which is comparable to apodizing a lens pupil.

Two common examples of sampling apertures are the square or rectangular aperture of a diode in a diode array, and the circular aperture of an idealized fiber in a fiber optics faceplate. If we define the rectangular sampling aperture by its normalized PSF,

$$
\begin{aligned}
I(x',y') &= 1.0, \quad |x'| < a \quad \text{and} \quad |y'| < b \\
&= 0.0, \quad |x'| > a \quad \text{or} \quad |y'| > b
\end{aligned}
\tag{124}
$$

then its normalized MTF is

$$
T(\nu'_x,\nu'_y) = \frac{\sin(\pi a \nu'_x)}{\pi a \nu'_x} \frac{\sin(\pi b \nu'_y)}{\pi b \nu'_y}
\tag{125}
$$

If the circular sampling aperture is defined by the normalized PSF

$$
\begin{aligned}
I(r') &= 1.0, \quad r < a \\
&= 0.0, \quad r > a
\end{aligned}
\tag{126}
$$

then the normalized MTF is

$$
T(\nu') = 2J_1(2\pi a \nu')/2\pi a \nu'
\tag{127}
$$

Note that both MTFs are unbounded, both can assume negative values, and neither is a function of wavelength. The magnitude of the negative values can be suppressed by tapering the transmission of the sampling aperture from center to edge (Mertz and Gray, 1934; Schade, 1951, 1952, 1953, 1955).

VIII. INTEGRATION WITH IMAGE SENSORS

It is possible to discuss image quality entirely in terms of the properties of the optical system and to rate the merit of the optical system according to a purely optical criterion, such as wavefront error or Maréchal's criterion for the Strehl definition. In subsequent sections, purely optical merit functions are used to compare the effects of different factors degrading image quality. But the optical system is only part of the complete imaging system, and no discussion of image quality is complete without some consideration being given to image sensors.

The topic of image sensors and image displays is very broad. For the purposes of the present chapter, only those aspects which bear on the choice of optical image quality merit functions are examined. In particular, attention is confined to the concept referred to by Overington (1976) as *"acquisition"* and its relationship to optical concepts of detection and resolution, and to several specific imaging situations which illustrate these relationships. The reader interested in more detailed information on image sensors, image displays, and observer performance is directed to

Biberman (1973); Biberman and Nudelman (1971), Dainty and Shaw (1974), Overington (1976), Rose (1973), and Shaw (1976).

A. Acquisition, Detection, and Signal-to-Noise Ratio

The most common imaging systems have an output in the form of a recorded image presented to a human observer for direct visual inspection. The displayed image may be a photographic print or a television cathode ray tube screen, or it may take one of many other forms. In all cases, the purpose is to present visual stimuli to the observer, which convey some information about the object whose image is being viewed. It is this process of acquiring information through visual stimuli which Overington refers to as *acquisition*. Image quality analysis is concerned with the fraction of the available information which can be acquired using the given imaging system.

Overington divides acquisition into three subcategories, *detection, recognition,* and *identification.* Detection involves the observer becoming aware that a local variation in image structure is a result of a real variation in object structure and not an artifact of the imaging system. Recognition involves the first level of sorting detected object variations into classes, e.g., artificial versus natural features. Identification involves recognition of sufficient detail within a feature to specify that it is a particular one of a class.

Clearly, each higher level of acquisition requires an improvement in image quality. Not as obviously, each higher level requires an increase in the amount of light detected. The reasons for this are tied to the photon statistical nature of light and the properties of image transducers which convert the light image into an energy form which can be recorded.

Light energy comes in discrete increments, and it interacts with image transducers in discrete steps. In effect, an image transducer defines the time of arrival and position on the image surface of each detected photon. More accurately, a real image transducer is characterized by a sampling aperture size and a sampling time (the *exposure interval*), and the image sensor output measures the number of photons detected per sampling aperture per exposure interval.

Consider an object surface which has a uniform radiance. If its conjugate image surface is to be perceived of as having an exactly uniform irradiance, the number of photons detected in each sampling aperture during each exposure interval must be the same. The quantum statistical nature of light and light detection is such that this does not happen except as a statistical fluke.

If the number of photons detected within a given sampling aperture during each of a large number of exposure intervals is analyzed, it is found that the number detected in each exposure interval varies randomly about an average value $\bar{n}$ with a variance which equals $\bar{n}$ and a standard deviation $\bar{n}^{1/2}$. The same statistics apply for the nominally uniform image irradiance when comparing the number of photons detected in different sampling apertures during the same exposure interval. As a result, when comparing adjacent sampling apertures, there is a statistical uncertainty as to whether a difference in output signal is due to real structure in the object surface or to the random statistical nature of light.

If the number of photons detected in one sampling aperture during an exposure interval is n_s for this nominally uniform image irradiance, $n_s - \bar{n}$ represents the apparent image structure. The ratio of this apparent *signal* to the Poisson statistical standard deviation $\bar{n}^{1/2}$ is termed the *signal-to-noise ratio* (SNR).

$$\text{SNR} = |n_s - \bar{n}|/\bar{n}^{1/2} \tag{128}$$

The SNR is a measure of the probability that $n_s - \bar{n}$ represents a real deviation in the radiance of the object surface.

The above discussion and the nomenclature of Section II allow a qualitative refinement of Overington's terminology: *detection* involves the discrimination of signal from noise; *recognition* involves the discrimination of target from clutter; *identification* involves the recognition of detail internal to the target, which is characteristic of an individual member of the target class. Each upward step requires either an increase in the SNR or a decrease in the size of the sampling aperture at the same SNR. In either case, more photons must be detected per unit area on the image surface.

The optical design parameters playing the most fundamental part in SNR determination are the entrance pupil diameter D_p, the optical transmittance $\tau = \tau_m\tau_p$ (see Section V,B), and the focal ratio F at the image sensor. D_p sets the fundamental lower limit on the sampling aperture diameter as referenced to the object surface. The image irradiance is proportional to τ/F^2, and F also sets the lower limit on the size of the PSF referenced to the image surface. Image quality plays a part, in that image degradation smears out the PSF, enlarging the sampling aperture.

The three-step acquisition scale of Overington can be revised for special applications. The classic paper by Johnson (1958), discussed in more detail shortly, uses four steps: (1) detection, (2) shape orientation, (3) shape recognition, and (4) detail recognition. Johnson's paper deals with a specific military application, the use of night viewing devices to locate and identify military vehicles. It has far broader implications, however, and is worthwhile background reading.

B. Detection of a Point Source Against a
Background Continuum

The simplest object encountered in practice is a point source seen against a uniform background, typified by a star seen through skyglow. This constitutes a pure example of detection, since there is no internal target structure to be identified and no background clutter. The only form of information to be extracted, in the sense of imagery, is the position of the point source. In effect, the question the imaging system must answer is, with that level of confidence can we say that the image structure at (x',y') represents a star and not an artifact of noise in the image background?

Figure 27 represents an imaging system used for this task. It is a television system whose key elements are the optics, photocathode, electron optics, storage target, electron scanning beam, video amplifier and display, each of which can degrade the image viewed by the observer. The task of the observer is to note the location of each stellar image.

Any complete analysis of this imaging system requires consideration of the gain, bandwidth, and noise characteristics of all electronic components within the data processing and display train. The ultimate limits on performance depend on photon statistics, optical system image quality, and photocathode characteristics. For present purposes, which are to determine the effects of degradation in optical system image quality, it is assumed that all components aft of the photocathode have negligible effect. More complete analyses can be found in Biberman (1973) and Bradley (1977).

The target in the camera tube collects photoelectrons over an exposure interval t. The average number of photoelectrons received in any exposure interval is divided into three groups, those attributed to the point source $\bar{n}_s$, those attributed to the background $\bar{n}_b$, and those attributed to dark current from the photocathode $\bar{n}_d$. Here $\bar{n}_s$ represents the total light reaching the image surface from the point source:

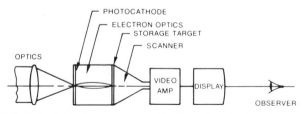

FIG. 27. Schematic of imaging system used in discussing the detection of point sources against a uniform background.

$$\bar{n}_s = \pi D_p^2 \tau S E_a t / 4e \tag{129}$$

where E_a is the entrance pupil irradiance from the point source, S is the photocathode sensitivity in amperes per unit irradiance, and $e = 1.6 \times 10^{-19}$ coulomb per electron.

The photocathode emits an average dark current of $\bar{N}_d$ electrons per unit area per second. If the background radiance is L_b, the average background photoelectron count per unit area per second $\bar{N}_b$ is

$$\bar{N}_b = \pi \tau S L_b / 4 F^2 e \tag{130}$$

The main problem in calculating the SNR is to define the effective sampling area. The actual sampling area is the system PSF which is taken to be the optical system PSF for this analysis. For convenience in calculation, this is replaced by a pillbox-shaped PSF representing an equivalent circular sampling aperture, which is termed the *noise equivalent sampling aperture*. It is based on Schade's equivalent passband concept as modified by Bradley (1977). The definition of the noise equivalent sampling aperture is based on the white noise statistics of photons and photoelectron emission and relates directly to T_L^* as defined by Eq. (120). The area A_s^* of the noise equivalent sampling aperture is

$$A_s^* = 1/T_L^* \tag{131}$$

For a perfect lens with a circular pupil,

$$A_s^*(\text{perf}) = 2.77 \lambda^2 F^2 \tag{132}$$

Therefore, for any lens,

$$A_s^* = 2.77 \lambda^2 F^2 / Q^* \tag{133}$$

and the diameter d_s^* of the noise equivalent sampling aperture is

$$d_s^* = 1.88 \lambda F / Q^{*1/2} \tag{134}$$

The background and dark current contributions may now be written as

$$\bar{n}_b = A_s^* \bar{N}_b t \tag{135}$$

and

$$\bar{n}_d = A_s^* \bar{N}_d t \tag{136}$$

The SNR compares the mean signal $\bar{n}_s$ to the standard deviation in the signal plus background plus dark current:

$$\text{SNR} = \bar{n}_s / (\bar{n}_s + \bar{n}_b + \bar{n}_d)^{1/2} \tag{137}$$

Equations (129) through (137) may now be used to indicate how the optical system, represented by τ and Q^*, affects the SNR.

If the exposure time is fixed, there are three limiting cases:

(1) $\quad$ SNR $\propto \tau, \qquad \bar{n}_s \gg \bar{n}_b + \bar{n}_d$ $\hfill$ (138)

(2) $\quad$ SNR $\propto \tau Q^{*1/2}, \qquad \bar{n}_s \ll \bar{n}_b + \bar{n}_d \quad$ and $\quad \bar{n}_d \gg \bar{n}_b$ $\hfill$ (139)

(3) $\quad$ SNR $\propto (\tau Q^*)^{1/2}, \qquad \bar{n}_s \ll \bar{n}_b + \bar{n}_d \quad$ and $\quad \bar{n}_b \gg \bar{n}_d$ $\hfill$ (140)

If the exposure time is adjusted for uniform background exposure, τ drops out of the equations.

In the first case, for a bright point source, the optical system is only a light bucket, and its image quality is of little significance. The distinction between the second and third cases, for a dim star against either a high dark current or a bright background, is of more interest to the user than to the optical engineer, except in that it points out that subtle changes in user requirements can change the weighting given optical system parameters such as transmittance. For the purposes of this chapter, Eq. (140) can be used as the basis of a one-number merit function relating point source detection to image quality. Since τ_m is usually assumed to be unity, the merit function is $Q^{*1/2}$ or $(\tau_p Q^*)^{1/2}$. The traditional merit function for this application is $\mathcal{D}$ or $\tau_p \mathcal{D}$.

A similar analysis of SNR can be made for a photoelectronic photometer consisting of a lens, a pinhole aperture, a photomultiplier, and a photoelectron counter. In this case, the image quality merit function of most interest is $EE(r)$, where r is the radius of the pinhole aperture. The objective of the optical system designer becomes one of maximizing $EE(r)$ while minimizing r.

C. Limiting Resolution and Threshold Modulation

All object surfaces more complex than arrays of well-separated point sources fall into the category of continuous-tone imagery. With a few exceptions where extreme dynamic range requires special treatment, as in coronagraphs, where diffuse scattering is an extreme problem, optical systems for continuous-tone imagery are judged in terms of their ability to discriminate fine detail. The ability to discriminate fine detail is called *resolving power*, and the size of the finest detail which can be resolved is specified by the *limiting resolution*.

The two-point resolution criteria introduced in Section VI,F illustrate two characteristics common to most forms of limiting resolution:

(1) the structural form of the test object is known in complete detail in advance, so that the observer need only be asked to identify the most closely spaced image structural elements he or she can resolve;

(2) the criteria for determining that the structural elements are resolved is arbitrary to a considerable extent.

Typically, the resolution criterion involves comparison of the brightness midway between the elements to the peak brightness of each element. In the case of two-point resolution, it may be pointed out that the Rayleigh or Sparrow limit set by the entrance pupil diameter can be exceeded greatly if the image sensor has an adequate SNR and a small enough sampling aperture to map irradiance contours of the image structure; a comparison to the known lens PSF reveals the presence of the second point source and allows the separation and orientation of the two to be estimated.

1. Limiting Resolution

Three-bar resolution charts represent the most common form of target used for testing limiting resolution in the United States. The basic pattern consists of three bars of the same size and shape arranged so that the spaces between them are themselves bars of the same size but have the same radiance (or opacity) as the background. Each bar is 1 resel ($\frac{1}{2}$-cycle) wide and 5 resels long, and the pattern is 5×5 resels. The radiance profiles are square waves, and each is specified by its fundamental spatial frequency and target contrast, as defined by Eq. (2). Target contrast can be related to the modulation defined in Eq. (88) by

$$M = (C - 1)/(C + 1), \qquad C = (1 + M)/(1 - M) \qquad (141)$$

Examples of low- and high-contrast three-bar resolution charts are shown in Fig. 28, in the standard U.S. Air Force format. Each spatial frequency is represented by a pair of orthogonal three-bar charts, and the spatial frequencies of adjacent pairs advance by a ratio of the sixth root of 2.

Limiting resolution of an imaging system is tested by visual examination of images of test charts such as those in Fig. 28. The tests are usually done at sufficient magnification and with sufficient average illumination to minimize limitations imposed by the eye of the observer. The limiting resolution spatial frequency is defined as the maximum spatial frequency at which the apparent luminance between the bars is still noticeably different from the apparent luminance of the bars.

Limiting resolution is a subjective measurement and varies considerably from observer to observer for a standardized test procedure. It varies for the same observer as a function of changes in the test procedure, the form of the test target, and the physical condition of the observer. The limiting resolution of any particular imaging system must therefore be

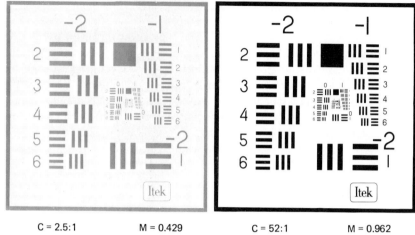

C = 2.5:1   M = 0.429   C = 52:1   M = 0.962

FIG. 28. Typical low- and high-contrast three-bar resolution charts. (Courtesy of Itek Corporation.)

treated statistically, expressing the probability that a given limiting resolution can be achieved. It is possible to estimate these probability levels based on a SNR analysis (Rosell and Willson 1973) but, for purposes of exposition, the concept of threshold modulation is more satisfactory.

2. Threshold Modulation

The concept of the *threshold modulation curve* (TM), or the *aerial image modulation curve* (AIM), as it is also called*, is most widely used by designers of photographic cameras for aerial reconnaissance. The TM curve for a photographic emulsion represents the sinusoidal modulation which must be present in the aerial image of a three-bar target at its fundamental spatial frequency if the target is to be resolvable in the developed photograph. The concept has its origin in a paper by Sandvik (1928). Figure 29, from data published by Lauroesch *et al.* (1970) gives a typical example.

TM curves are derived empirically from experiments with human observers viewing photographic exposures of targets having a fixed array of spatial frequencies and varying contrasts. The observers determine the target contrast for limiting resolution at each spatial frequency, and this information is used to determine the required modulation in the aerial

---

* Purists make the distinction that TM applies only to standard three-bar square-wave targets and that AIM applies only to sinusoidally modulated targets.

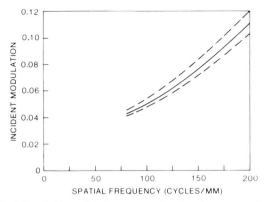

FIG. 29. Typical threshold modulation curve, showing the mean value and bounds for plus or minus one standard deviation. (Redrawn from Fig. 22 of Lauroesch *et al.*, 1970.)

image, based on the known characteristics of the optics used in the experiment. TM curves are derived from statistical analysis of a series of such experiments by the same and different observers.

A single TM curve for an emulsion represents a particular set of circumstances which must be accounted for in any systems analysis. Each TM curve implies a specified emulsion batch receiving a specified exposure, the emulsion being developed for a specified time and temperature in a specified developer. Each measurement of a TM curve is subject to variation, depending on a wide variety of experimental conditions. Lauroesch *et al.* (1970) outlines most of these and should be read by anyone intending to use TM curves. The mean TM curve represents the average of a number of observers and should really be replaced by the band between the curves for plus or minus one standard deviation in the experimental results. In practice, this is seldom done.

The high-contrast limiting resolution for a lens–film combination is determined by plotting the lens MTF and the film TM curves on the same graph and noting the spatial frequency at which they cross. Limiting resolution for low-contrast targets can be determined by either of two techniques: (1) Multiply the lens MTF by the target modulation and cross the result with the TM curve; (2) divide the TM curve by the target modulation and cross the result with the lens MTF. The two techniques are mathematically equivalent, but the latter is preferred by many lens designers who frequently deal with a single emulsion type used with many different lenses.

Figure 30 illustrates the second technique and shows some of the shortcomings of using a single value of limiting resolution as a merit function for choosing between different classes of lens designs. Two lens

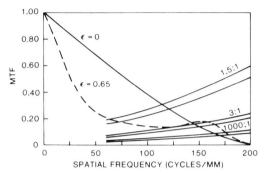

FIG. 30. Determination of the limiting resolution by crossing the lens MTF with the emulsion TM curves.

MTFs are shown, one for a perfect lens with no obstruction and the other for a perfect lens with a centered circular obstruction of diameter ratio $\epsilon = 0.65$. Three target contrasts are considered for the same film, $C = 1.5:1$, $3:1$, and $1000:1$ ($M = 0.20$, 0.50, and 0.998, respectively).

For the highest-contrast target, the obstructed lens gives slightly better resolution (178 versus 168 cycles/mm). For the lowest-contrast target, the unobstructed lens gives far better limiting resolution (118 versus 75 cycles/mm). If limiting resolution is a valid merit function for the application of the lens–film combination, it is imperative that the appropriate target contrast be used. For more general applications, some method of weighting high- and low-contrast limiting resolution is needed.

Figure 30 also illustrates the ambiguity inherent in using TM curves and limiting resolution. The value of the limiting resolution for the highest- and lowest-contrast targets is relatively unambiguous, since the MTF and TM curves cross at relatively steep angles. At 3:1 target contrast, however, the MTF and TM curves nearly coincide over a long-spatial-frequency range (120–160 cycles/mm) for the lens with a central obstruction. In this region, very small shifts in either curve, or even the statistical ambiguity in the TM curve, can lead to a large shift in limiting resolution. However, the general appearance of the recorded image will change very little.

One quantity which has received some attention as a merit function, and which relates more closely to the general appearance of the image, is the area lying between the MTF and the TM curves. This is termed the *modulation transfer function area* (MTFA) (Snyder, 1973) or the *threshold quality factor* (TQF) (Charman and Olin, 1965). Snyder has shown the MTFA to relate closely to the subjective performance of photointerpreters viewing general scenes. Whatever its value may prove to be in the future, the MTFA has not as yet replaced limiting resolution as the principal photographic image quality merit function; MTFA is an abstract concept, not as readily related to by the lay user as limiting resolution.

The MTFA may be thought of as a logical extension of relative edge response, which measures the area under the combined lens–film MTF. The only difference is that the MTFA incorporates the TM curve for the human eye and accounts for some limitations imposed by film granularity. Relative edge response can replace the MTFA when only the optical system is being discussed.

The TM curve used in Figs. 29 and 30 is for Eastman Kodak 3404, a particularly high-resolution emulsion. Figure 31 shows TM curves for a series of Eastman Kodak aerial emulsions measured by Itek (Brock *et al.*, 1966) in the early 1960s. They are included as being representative of a number of different resolution classes of photographic emulsion. It should be noted that none of these curves necessarily represent the performance of current emulsions bearing the same product designation.

The use of the TM curve is largely confined to photographic systems. The reason for this is tied to the fact that a single TM curve represents only one exposure level. With photographic emulsions, there is an optimum exposure level for obtaining the maximum recorded information, and it is customary to vary the exposure time with scene illumination to maintain constant exposure. Thus performance can be adequately specified with one TM curve. In most electrooptical imaging systems, the exposure time is fixed by a frame rate, and variations in scene illumination are compensated for by adjusting gain between the image transducer and the display. A different TM curve would be required for each image irradiance level. For this reason, more traditional forms of television SNR

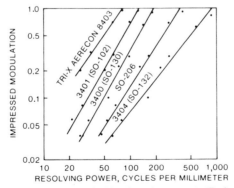

FIG. 31. Aerial film modulation thresholds (three-bar target). (Redrawn from Fig. 3-18 of Brock *et al.*, 1966.)

analyses are retained, even when limiting resolution is the desired end result. More detailed discussions of these techniques can be found in Biberman (1973).

Even with photographic systems, there are limitations on the practical usefulness of TM curves. The most notable of these is the scarcity of published TM curves for commonly available photographic emulsions. Manufacturers publish MTF curves for many of their photographic emulsions, when TM curves are not available, and these may also be used to estimate the limiting resolution of the complete imaging system. But aside from scarcity of data and the basic statistical uncertainty of the performance of human observers, there are technical limitations to both TM and MTF data for photographic emulsions. Brock (1970) gives a good summary of most of the problems.

Photographic emulsions are nonlinear in their response to light, and representing their imaging performance by a linear function such as the MTF is at best an approximation. The TM curve, which is measured empirically, represents an improvement, but it has its own limitations. Square-wave targets contain higher order harmonics of their fundamental frequency which may or may not be reproduced in the aerial image imposed on the film. The credibility of limiting resolutions calculated by crossing the lens MTF with the photographic emulsion TM curve depends to some extent on how that lens compares to the lens used in measuring the TM curve. If there is a substantial difference in focal ratios of the two lenses or in the general shape of their MTFs, due, for example, to a large central obstruction in one, the results may be in error.

In spite of the above philosophical limitations on the process of calculating limiting resolution using the TM curve, it has found extensive use among the designers of aerial reconnaissance cameras. In general, measured limiting resolutions do bear out predictions made using the TM curve, when care is taken in its application. Beyond this, the TM curve is a very useful tool in qualitative examinations of the imaging process, as is shown in Fig. 30.

3. Johnson's Application of Limiting Resolution

The continuing popularity of limiting resolution as an image quality merit function is due in large part to the brilliant work of Johnson (1958) in relating his acquisition scale to limiting resolution. Johnson's interests were in evaluating the usefulness of night viewing devices for military applications, but his results are more generally applicable. Johnson compared observed three-bar limiting resolution with the observed minimum dimension of objects at the threshold of detection, shape orientation,

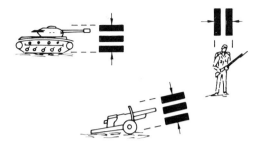

METHOD OF OPTICAL IMAGE TRANSFORMATION

FIG. 32. Johnson's method of relating three-bar limiting resolution to the dimensions of specific targets for minimum detectable target size. (Redrawn from Johnson, 1958.)

shape recognition, and detail recognition in the manner shown in Fig. 32. His results are reproduced in Table III, which shows the transformation between three-bar limiting resolution and level of acquisition for a number of military targets. The entire table has been reproduced to show the degree of general agreement and detailed variation in the limiting resolution required for each level of acquisition. The standard deviation lies within 20–25% of the mean at each level.

Johnson claims that these target transformations hold independent of target contrast and scene SNR, provided only that the three-bar target contrast is the same as that for the target acquisition experiment. He also found that, when the SNR was defined for a sampling aperture $\frac{1}{2}$ cycle (1 resel) in diameter at the limiting spatial frequency, the product of the

TABLE III

OPTICAL IMAGE TRANSFORMATIONS[a]

Target	Detection	Orientation	Recognition	Identification
Truck	0.90	1.25	4.5	8.0
M-48 tank	0.75	1.2	3.5	7.0
Stalin tank	0.75	1.2	3.3	6.0
Centurion tank	0.75	1.2	3.5	6.0
Half-track	1.0	1.5	4.0	5.0
Jeep	1.2	1.5	4.5	5.5
Command car	1.2	1.5	4.3	5.5
Soldier (standing)	1.5	1.8	3.8	8.0
105-mm howitzer	1.0	1.5	4.8	6.0
Average	1.0 ± 0.25	1.4 ± 0.35	4.0 ± 0.8	6.4 ± 1.5

[a] Number of cycles per minimum dimension (broadside view) at limiting resolution required to perform stated level of acquisition (from Johnson, 1958).

target contrast C and the SNR is roughly constant over at least the range $1:1 < C < 10:1$. Based on his Fig. 11,

$$C \times \text{SNR} \simeq 9.1 \pm 1.5 \qquad (142)$$

with the product increasing slightly as C increases. In general, the lower the target contrast, the higher the SNR required for resolution.

Johnson's tests were done with a single lens and several different image sensor–display systems. It seems likely that, if the tests were redone using two lenses having MTFs as disparate as those shown in Fig. 30, the results would differ significantly. If the target contrast were $3:1$, for example, it seems likely that higher magnification would be required for the obstructed lens to perform a given higher level of acquisition than for the unobstructed lens.

D. Raster Processes

In all the previous discussion, no distinction was drawn between image sensors or displays having lateral continuity and those in which the image is sampled at discrete positions. In the former, the transfer characteristics of the image-forming system can be represented totally by convolving component PSFs or cascading component OTFs. If the image is sampled, the process of sampling will in itself add artifacts to the displayed image.

The array of sampling apertures by which the image is scanned is known as a *raster*. In *point rasters*, the sampling apertures are fixed in a two-dimensional matrix of points. In a *line raster*, the image is scanned continuously in one direction and sampled in the orthogonal direction, using one sampling aperture or a linear array of sampling apertures. Halftone screens used in printing, coherent fiber optics, and two-dimensional diode arrays are typical examples of point rasters. Line rasters are commonly found in television systems, facsimile transmission systems, and infrared cameras. Most rasters are rectangular in format. Exceptions include coherent fiber bundles using hexagonal rasters, and halftone screens which sample in the radial direction and are continuous in the tangential direction. For present purposes, attention is confined to the rectangular line raster used in television systems.

Television cameras typically scan in the horizontal (x') direction and sample in the vertical (y') direction. Their MTFs are most conveniently specified in terms of the Cartesian spatial frequencies ν'_x and ν'_y. For analytic purposes, the complete television system is divided up into the camera, the display, and the *raster process* by which spatial frequencies are transformed from camera image to display image. It is the raster process which can introduce artifacts into the displayed image.

The performance of the camera is represented by $T_c(\nu'_x, \nu'_y)$, the cascaded MTF of the optical system plus the image sensor. The performance of the display is represented by its MTF, $T_d(\nu'_x, \nu'_y)$. Since the image is scanned continuously in the x' direction, the modulation transfer charactersitics are fully represented by $T_c(\nu'_x) \times T_d(\nu'_x)$. Discussion of artifacts introduced by the raster process may be confined to the y' dimension. For simplicity, the y subscript and prime have been dropped from the notation.

Mathematically, the raster can be represented by a series of delta functions spaced at intervals d in the y direction. The Fourier transform of this pulse train is a set of spatial frequency harmonics of unit amplitude, with the fundamental frequency $\nu_r = 1/d$ and harmonic frequencies $2\nu_r$, $3\nu_r$, etc. This is the spatial frequency equivalent of a carrier frequency, and the raster process is the equivalent of a carrier modulated by a signal which is the product of $T_c(\nu)$ and the object spectrum $S_o(\nu)$. The display system, in effect, demodulates the carrier by heterodyning it with its own raster spatial frequency. The spatial frequency spectrum of the output image is then determined by the relationship between the bandpass limitations of the MTF and the spatial frequency ν_r.

The maximum frequency which can be reproduced unambiguously in demodulating a carrier is one-half the carrier frequency. In sampling theory, this corresponds to the statement that at least two samples per period are required to reproduce a frequency unambiguously. This frequency, ν_s, is called the *sampling limit* or *Nyquist limit* (Nyquist, 1924, 1928):*

$$\nu_s = \nu_r/2 = 1/(2d) \tag{143}$$

If both $S_o(\nu)T_c(\nu)$ and $T_d(\nu)$ are zero for $\nu > \nu_s$, the displayed image will contain no artifacts. If $S_o(\nu)T_c(\nu)$ and/or $T_d(\nu)$ have substantial residuals at $\nu > \nu_s$, the displayed image will contain artifacts.

The displayed image will contain four types of brightness gratings modulated in the y direction:

(1) a raster grating of the form

$$C[1 + 2 \sum^{p} T_d(p\nu_r) \cos(p2\pi\nu_r y)] \tag{144}$$

(2) object data gratings:

---

* Nyquist studied the maximum rate at which telegraph pulse trains can be transmitted over a line of a given frequency bandpass. The Nyquist limit is an inversion of his results, indicating the maximum sine wave frequency which can be correctly reproduced by a given frequency pulse train.

$$S_o(\nu)T_c(\nu)T_d(\nu) \cos(2\pi\nu y + \phi) \tag{145}$$

(3)   raster plus data frequency sum gratings:

$$S_o(\nu)T_c(\nu) \sum^p \{T_d(p\nu_r + \nu) \cos[2\pi(p\nu_r + \nu)y + \phi]\} \tag{146}$$

(4)   raster minus data frequency difference gratings:

$$S_o(\nu)T_c(\nu) \sum^p \{T_d(p\nu_r - \nu) \cos[2\pi(p\nu_r - \nu)y + \phi]\} \tag{147}$$

where C is a constant related to the average radiance of the output image, p is the raster frequency order number, and ϕ is the phase displacement between the peak of the grating and the raster line origin.

Equation (145) represents the desired image modulation, and the other three equations represent the artifacts. Equation (144) is the raster pattern itself and can be suppressed only by controlling the MTF of the display. The sum frequencies of Eq. (146) are higher than the fundamental raster frequency and are suppressed if the raster frequency is suppressed. Even when the raster fundamental frequency is not suppressed, the sum gratings are usually sufficiently reduced to be ignored. It is the difference gratings of Eq. (147) which are usually the most objectionable and the hardest to suppress.

Figure 33 illustrates a graphical technique for determining the system modulation transfer appropriate to each of these gratings. Given a camera spatial frequency ν, a line is traced through the raster transfer diagram. The intersection point on the solid diagonal line marked ν gives the corresponding display spatial frequency. The intersections with the dashed lines D_1, D_2, etc., give the difference display frequencies $(\nu_r - \nu)$, $(2\nu_r - \nu)$, etc., and the intersections with S_1, S_2, etc., give $(\nu_r + \nu)$, $(2\nu_r + \nu)$, etc. The modulation of the image grating is found by multiplying $S_o(\nu)T_c(\nu)$ by the display MTF at the sum or difference spatial frequency.

The generation of artifacts by sum and difference frequencies is termed *aliasing*. The most objectionable aliasing component is usually the D_1 component $(\nu_r - \nu)$ due to scene detail in the spatial frequency domain $\nu_s < \nu < \nu_r$, particularly when $\nu \rightarrow \nu_r$. It is seen most strongly when there is a prominent periodic structure in the scene, or a sharp edge or line running at a slight angle to the raster lines. In the latter case, the edge or line takes on a saw-tooth appearance. In the former case, if the periodic structure has a spatial frequency in this domain, moiré fringes appear at the difference frequency. Two examples of this are seen every day: color television pictures of persons wearing clothes with stripes or other periodic detail of the right size show strong color moiré patterns; halftone re-

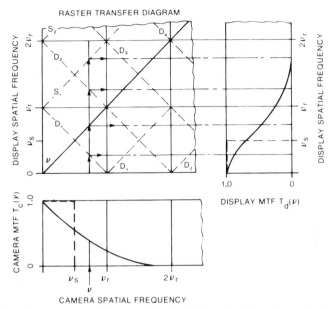

FIG. 33. Schade's diagram for computing the MTF and aliasing frequencies in a display. (Redrawn from Fig. 70 of Schade, 1953.)

productions of objects which are themselves halftone prints show a waffled appearance due to the difference frequency between the two halftone screen raster frequencies.

The optical engineer faced with designing a lens system to go with a sampling image sensor should keep two goals in mind: (1) Any detail which must be reproduced in the output image must have a spatial frequency $\nu < \nu_s$, and (2) if there is substantial object surface detail at spatial frequencies $\nu > \nu_s$, it is desirable to reduce the optical system MTF in this spatial frequency domain. In general it is not possible to do this without reducing the MTF at lower, information-bearing spatial frequencies. It is then necessary to consider which is more important, increasing the information-bearing modulation or decreasing the aliasing. It is possible to do both by apodizing the lens, as discussed in Section X,C. The gains which can be made in this manner are not substantial and cost heavily in terms of pupil transmittance.

Point raster systems show similar forms of aliasing. The basic analysis is similar, and the above conclusions concerning optical image quality also apply. The form of the aliasing is more complex, as a result of the two-dimensional raster, and must be handled by more complex vector analysis as shown by Legault (1973).

IX. SOURCES AND EFFECTS OF WAVEFRONT ERROR

The wavefront aberration function in Section IV,B represents one model of the wavefront error pupil function, and it is a convenient one for discussing wavefront errors inherent in the lens design. In considering the effects of wavefront error in more general terms, other models of the wavefront error pupil function are more useful. For descriptive purposes, treating wavefront deformations as being composed of a series of phase gratings is particularly helpful. For preliminary analysis of high-performance systems, models based on random wavefront error statistics are in wide use at present. Both are introduced here and used in discussing the effects of wavefront error on image quality.

A. THE PHASE GRATING

A *phase grating* is a diffraction grating in which wavefront phase retardation varies instead of transmittance. If the period of the phase grating is d, and if it is assumed for simplicity that the light is normally incident on the grating, light passing through it is diffracted into sidebands at a series of angles α_n measured from the grating normal, where

$$\sin \alpha_n = \pm n\lambda/d \tag{148}$$

and $n = 1, 2, 3, \ldots$, the order number of the sideband. The intensity of each sideband depends upon the phase amplitude of the grating; for very small phase amplitudes, OPD $\ll \lambda$, only the first-order sidebands contain sufficient energy to be seen above the ring structure of the PSF.

The wavefront error pupil function may be thought of as being composed of a spectrum of sinusoidal phase gratings in the same sense that the object surface radiance is modeled by a spectrum of sinusoidal intensity gratings. This concept is useful in qualitatively predicting the types of image degradation associated with different types of wavefront errors. For very simple cases involving a single phase grating, some quantitative analysis can be performed. But the addition of two or more phase gratings of large phase amplitude introduces difficulties, both because of the higher-order sidebands associated with the large phase amplitudes and because sidebands appear corresponding to sum and difference spatial frequencies for each pair of phase gratings.

Barakat and Houston (1966) and Wetherell (1974a) have examined the case of a rotationally symmetric sinusoidal phase grating centered in a circular aperture. The wavefront pupil function for this example is defined as

$$W(\rho) = a \cos(2\pi b\rho + \phi) \tag{149}$$

where a is the phase amplitude in wavelengths, b the spatial frequency in cycles per pupil radius, ρ the normalized radius, as before, and ϕ the phase angle of the sinusoid at $\rho = 0.0$. The PSF and MTF are calculated using the Hankel transforms in Eqs. (67) and (110).

Bakarat and Houston (1966) show that the Strehl definition for this sinusoidal phase grating is

$$\mathcal{D} = [J_0(2\pi a)]^2 \tag{150}$$

Note that $\mathcal{D} = 0.0$ when $a = 0.38274$ wavelength.

Wetherell (1974a) shows that the radius of each (annular) sideband is

$$\begin{aligned}
\alpha_n &\simeq 2nb\lambda/D_p &&\text{(radians in object space)} \\
q'_n &\simeq 2nb\lambda F &&\text{(linear units on the image surface)} \\
\delta'_n &\simeq 1.64\lambda nb &&\text{(Airy radius units)}
\end{aligned} \tag{151}$$

The exact position of the maximum in each sideband varies with ϕ and is within ± 0.6 Airy radius units of the position given by Eq. (151).

The maximum relative intensity in each band is

$$I_n(a,b) \simeq [J_n(2\pi a)]^2/2\pi nb \tag{152}$$

with the exact value varying slightly as a function of ϕ. The energy in each sideband is spread over an annulus roughly 5 Airy radius units wide with the maximum energy concentration at its center. The zonal energy increment

$$\text{ZEI}(\delta'_n \pm 2.5) \simeq 2[J_n(2\pi a)]^2 \tag{153}$$

Figure 34 plots relative intensity and encircled energy versus radius for the first 50 Airy radius units of the PSF for $a = 0.15$, $b = 5.0$, and $\phi = 0.0$. Three sidebands are evident, although the third is too weak to show the normal structure. The first sideband contains 35% of the energy in the PSF. More examples may be found in Wetherell (1974a).

B. Classification of Wavefront Error by Spatial Frequency

The values to be drawn from the phase grating model are more qualitative than quantitative. The model shows the general relationship between phase amplitude and the amount of energy scattered out of the Airy disk. More importantly, it shows the relationship between the characteristic lateral dimensions of the wavefront perturbations and the locations where the diffracted energy reappears in the PSF ring structure. Classifying different forms of wavefront error by spatial frequency is therefore indica-

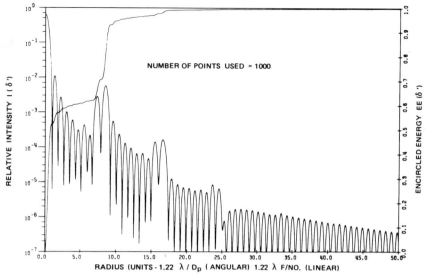

FIG. 34. The PSF and encircled energy for a sinusoidal phase grating of phase amplitude $a = 0.15$ wavelength and spatial frequency $b = 5.0$ cycles per radius ($\phi = 0.0$).

tive of the effects each has on the PSF and OTF. Based on current measurement practices, three categories suggest themselves: (1) figure error and aberrations, (2) ripple, and (3) surface microstructure. The first category is routinely measured during the manufacture of large-aperture, high-performance optical systems. The third category has been given some attention during the last decade, through measurement of its effects (large-angle scattering) and through studies of surface microstructure. Ripple includes everything between the other two domains which is seldom measured and usually ignored.

1. Figure Errors and Aberrations

This class includes most wavefront errors arising from faults in the design, fabrication, and assembly of optical systems. It can be defined as including spatial frequency components in the domain $0 < b < 5$ cycles per pupil radius. The upper limit is arbitrary and was set to approximate current practices in the interferometric testing of large mirrors. The limiting factor is usually the sample grid spacing used in data reduction of interferograms. With a minimum of two samples per cycle required by information theory, $b = 5$ corresponds to a 20×20 matrix of sample points. Sample grids finer than 100×100 have been used but, to minimize data reduction time, a grid of 20×20 points or less is preferred.

Wavefront error components with spatial frequencies below 5 cycles per radius affect energy distribution significantly only within the first 8–10 Airy radius units unless they are of large phase amplitude. They affect image quality for all extremes of object surface radiance dynamic range and for all applications.

2. Ripple

The spatial frequency domain of ripple starts at 5 cycles per radius and extends to a limit where the effects are more commonly thought of as bidirectional scatter. Since the smallest angle at which bidirectional scatter is measured is about $\frac{1}{2}°$ or 100 mrad, the corresponding grating period can be taken as the upper limit on ripple. From Eq. (148), for $n = 1$, the corresponding grating period for visible light is about 100 wavelengths or 50 μm.

Ripple is most evident in large aspherical mirrors where zonal polishing with subaperture-diameter laps tends to introduce fine-scale zonal irregularities in the optical surface. The random polishing technique employing full-aperture laps used for spherical surfaces tends to reduce ripple automatically.

Ripple is of no consequence for very low-contrast imagery such as aerial photography and of little consequence for normal, moderate-contrast imagery unless the ripple phase amplitude is very large. Quantitative techniques for measuring ripple have therefore not been developed. Where ripple can have serious consequences, as in the astronomical search for dim companions to bright stars, qualitative tests such as the Foucault knife-edge test are relied on to indicate the presence of locally steep slopes in the surface during fabrication. Some study of the possible consequences of ripple for visible and ultraviolet imagery in space telescopes has been done by the author, using the phase grating model described above (Wetherell, 1974a).

Where ripple has been a serious factor in image quality, as in solar coronagraphs, the practice has been to use spherical surfaces for the optical elements, relying on the properties of random polishing to reduce both ripple and surface microstructure automatically. Ripple will take on increasing significance with the development of large-aperture space telescopes for use at short wavelengths for long-dynamic-range astronomical imagery. Such telescopes require aspherical surfaces and may force developments in the technology for measuring and reducing ripple.

3. Surface Microstructure

Surface microstructure includes surface irregularities with lateral dimensions of 100 wavelengths down to less than 1 wavelength. It represents the residual roughness left on the optical surface by the polishing process and, for clean mirrors, it is the principal source of bidirectional scattering at angles greater than 0.5°. The standard deviation for the surface height is typically 50–100 Å for normally polished optical surfaces, and it is reduced well below 50 Å only by the use of special "superpolishing" techniques. These techniques tend to be incompatible with maintenance of accurate figures, at least on aspherical mirrors.

Surface microstructure is a problem only where large-angle scattering is a problem. The applications where it is of most concern involve imagery of low-contrast targets within a few degrees of very bright clutter, for example, in photographing the sun's corona. Many infrared applications are sensitive to scattering as well. In the latter case, however, most scattering is caused by other factors such as dust, surface contamination, and scratches.

To the extent that scattering from clean mirrors may be attributed to surface microstructure, the phase grating model can give some insight as to how scattering varies with wavelength. Assume a single spatial frequency phase grating of phase amplitude $a \ll 1.0$, so that energy is scattered into the first-order sideband only. Since a is in wavelengths, $a \propto 1/\lambda$. Expansion of the Bessel function in Eq. (152) shows that the relative intensity $I_1(b) \propto a^2 \propto 1/\lambda^2$. The scatter angle $\alpha_n \propto \lambda$, by Eq. (148), therefore the solid angle into which the light is scattered is proportional to λ^2. In effect, halving the wavelength scatters four times the energy into one-fourth the solid angle, making $E_1(b) \propto 1/\lambda^4$.

In order to determine the change in scattering at a fixed angle as a function of wavelength, the variation in a as a function of spatial frequency b must be known; Eq. (148) may then be used to scale the grating frequency an amount appropriate to the change in wavelength. In practice, this is usually known only indirectly through the monochromatic scatter function $\rho_s(\beta,\lambda)$. If $\rho_s(\beta,\lambda)$ can be represented by a straight line of slope S on a log–log plot, it follows from the above phase grating model that

$$\rho_s(\beta,\lambda_2)/\rho_s(\beta,\lambda_1) = (\lambda_2/\lambda_1)^{-(S+4)} \tag{154}$$

This model is not valid for scattering from dust and similar sources.

C. Random Wavefront Error and the MTF Degradation Function

Random wavefront errors form an important class of factors degrading image quality. The principal sources of random wavefront error are figure error and ripple introduced in fabricating optical surfaces, which are stationary, and atmospheric turbulence, which is time-varying. Both have the property that they cannot be predicted specifically in advance, and both may be treated by the same statistical model, allowing advanced estimation of the average degree of image degradation they will engender.

1. *The MTF Degradation Function Model of O'Neill*

O'Neill (1963) treats the problem of MTF degradation due to random wave-front error statistically. He divides the wavefront error function $W(X,Y)$ (adapting his analysis to our normalized Cartesian pupil coordinates) into a "controlled" part due to design aberrations, $W_a(X,Y)$, and a random part, $W_r(X,Y)$, which is a statistical model based on the ensemble average of a large number of individual examples. $W_r(X,Y)$ is assumed to be Gaussianly distributed, to have a zero mean, a variance ω^2, and a normalized autocorrelation function:

$$\phi_{11}(k,j) = \frac{1}{\omega^2} \int_{-\infty}^{\infty} \int_{-\infty}^{\infty} W_r(X + k, Y + j)W_r(X - k, Y - j)\, dX\, dY \quad (155)$$

The term $\phi_{11}(k,j)$ can also be written as $\phi_{11}(s,\alpha)$ and $\phi_{11}(\nu_n,\alpha)$, since tan $\alpha = k/j$ and $\nu_n^2 = s^2/4 = k^2 + j^2$ (see Fig. 18 and the discussion concerning normalization in Eq. (100) in Section VII,B,1). It is further assumed that the individual realizations of $W_r(X,Y)$ making up the ensemble have the same statistics as the ensemble.

Under these circumstances, the averaged system MTF $T(\nu_n,\alpha)$ may be separated into a rotationally symmetric MTF degradation function due to random wavefront error, $T_\omega(\nu_n)$, and the MTF due to aberrations and diffraction at the pupil boundary, $T_a(\nu_n,\alpha)$. Thus

$$T(\nu_n,\alpha) = T_\omega(\nu_n) \times T_a(\nu_n,\alpha) \quad (156)$$

O'Neill goes on to show that

$$T_\omega(\nu_n) = \exp\{-(2\pi\omega)^2[1 - \phi_{11}(\nu_n)]\} \quad (157)$$

[O'Neill uses an unnormalized form of the autocorrelation function, $\phi_r(\nu_n)$, for which $\phi_r(0) = \sigma^2$, σ^2 being the variance in linear units. In this notation, $T\omega(\nu_n) = \exp\{-(2\pi/\lambda)^2[\sigma^2 - \phi_r(\nu_n)]\}$].

For the most general preliminary analysis, the MTF can be assumed to be rotationally symmetric, and the aberrations lumped together with the random wavefront error. The aberration MTF is then replaced with the perfect lens MTF, and Eq. (157) is written

$$T(\nu_n) \simeq T_\omega(\nu_n) \times T_i(\nu_n) \quad (158)$$

2. *The Autocorrelation Function and Correlation Length*

The properties of the autocorrelation function must be understood to be applied accurately. First, Eq. (155) has the general properties that at $s = 0.0$, $\phi_{11} \equiv 1.0$, its maximum value, and that as $s \to \infty$, $\phi_{11}(s) \to 0.0$.

Second, $\phi_{11}(s)$ first approaches zero at a characteristic displacement $s = l$, and remains in the vicinity of zero for all $s > l$. The displacement l is termed the *correlation length* and is an indication of the spatial frequency content of the wavefront error. The usual inelegant way of stating this is to note that $1/l$ represents the number of "bumps" per pupil diameter in the wavefront.

In statistical analysis, the sample area is infinite and remains constant for all values of function displacement s. Real wavefront error pupil functions are bounded, causing the sample area in an autocorrelation function calculation using Eq. (155) to decrease as s increases, decreasing ϕ_{11}. To offset this effect, ϕ_{11} is usually divided by the normalized sample area which is the perfect lens MTF. This produces spuriously large values of ϕ_{11} as $s \to 2.0$ ($\nu_n \to 1.0$), which can be ignored. Two measured autocorrelation functions are shown in Figs. 35 and 36 (Solomon and Wetherell, 1973). Note that both curves show negative excursions, the magnitude of the negative excursions increasing as l increases. The negative excursion may be thought of as an outgrowth of setting the mean wavefront error to zero, which implies that there is no zero spatial frequency content in the wavefront error pupil function. In this case, Fourier theory shows that

$$\int_0^{2\pi} \int_0^{\infty} \phi_{11}(s,\alpha)s \; ds \; d\alpha = 0.0 \tag{159}$$

Similar autocorrelation functions have been shown by Schwesinger (1972).* These differ from the all-positive autocorrelation functions nor-

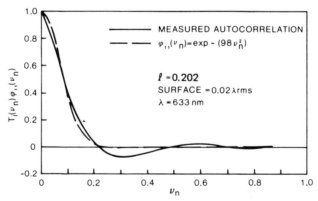

FIG. 35. Measured autocorrelation function for a 48-in.-diameter, $f/2.5$ parabolic primary mirror (zonally averaged). (Courtesy of Itek Corporation.)

---

* An English translation of Schwesinger (1972) is included in Wetherell (1974a).

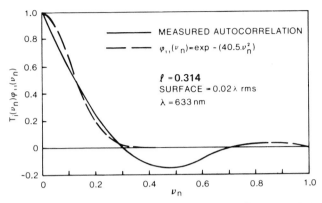

FIG. 36. Measured autocorrelation function for a 12-in.-diameter spherical mirror (zonally averaged). (Courtesy of Itek Corporation.)

mally cited for ensemble averages. The reasons for this have been discussed by O'Neill and Walther (1977) and arise from the fact that the single realizations in an ensemble average need not in themselves have zero means.

A commonly used model (Solomon and Wetherell, 1973; Wetherell, 1974a, Schwesinger, 1972) for the autocorrelation function is the Gaussian model

$$\phi_{11}(s) = \exp(-s^2/l^2) \quad \text{or} \quad \phi_{11}(\nu_n) = \exp(-4\nu_n^2/l^2) \qquad (160)$$

By the ground rules of Eq. (159), this is formally unacceptable, but it may be sufficiently accurate for many uses. Figures 35 and 36 include plots of Eq. (160) and show the degree of mismatch between Eq. (160) and measured autocorrelation functions. Two areas of disagreement occur. The one at lower spatial frequencies is an indication that the Gaussian model is deficient in higher spatial frequency components. The other, at intermediate spatial frequencies, is due to the failure of the Gaussian model to satisfy Eq. (159).

Some justification for using the simple model in Eq. (160) may be taken from the variance in the individual autocorrelation functions. The measured autocorrelation functions in Figs. 35 and 36 are zonally averaged, and their point-by-point values vary considerably over the range of azimuth angles α. The variance in $\phi_{11}(\nu_n,\alpha)$ and the associated MTF degradation function can be quite large, as has been shown by Nicholson (1974).

Figure 37 is one example taken from Nicholson's paper. Nicholson's optical quality factor matches the MTF degradation function in Eq. (157), assuming that $\phi_{11}(\nu_n) = 0.0$ at all spatial frequencies. A Gaussian auto-

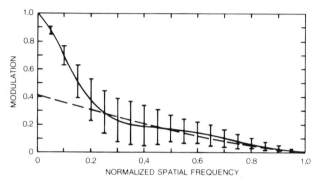

AVERAGE MTF OF CIRCULAR APERTURE WITH RMS WAVEFRONT ERROR OF 0.15λ AND CORRELATION DISTANCE OF 0.3 (16 CASES) ERROR BARS FOR ±2σ

MTF OF CIRCULAR APERTURE MULTIPLIED BY AN OQF OF 0.411 WHERE: OQF = e[-(2πRMS)²]

FIG. 37. Variance in the MTF calculated from 16 synthetic random wavefronts. Each MTF was zonally averaged and then all 16 were averaged. (From Nicholson, 1974, by permission of the author.)

correlation function with a correlation length of about $l = 0.25$ should yield an MTF which stays within the error bars in Fig. 37 over the entire spatial frequency range. Thus the autocorrelation function model in Eq. (160) is of some use, provided that the possible magnitude of the variance from the model is recognized.

A more thorough theoretical analysis of random wavefront error and its variance can be found in papers by Barakat (1971) and Barakat and Blackman (1973).

3. $\mathscr{D}$, $\mathscr{E}$, and Q^* for Random Wavefront Error

Strehl definition, relative edge respone, and Q^* for random wavefront error may be found by substituting Eqs. (156) and (157) into Eq. (112), (122), or (121). If the nature of the autocorrelation function is studied, it will be noted that as $l \to 0$, $T\omega \to \exp(-4\pi^2\omega^2)$ at all spatial frequencies except $\nu = 0$. Equations (112), (122), and (121) then show that

$$\mathscr{D} \to \exp(-4\pi^2\omega^2) \quad \text{as} \quad l \to 0 \tag{161}$$

$$\mathscr{E} \to \exp(-4\pi^2\omega^2) \quad \text{as} \quad l \to 0 \tag{162}$$

and

$$Q^* \to \exp(-8\pi^2\omega^2) \quad \text{as} \quad l \to 0 \tag{163}$$

Equation (163) indicates that for short-correlation-length wavefront error, $Q^{*1/2} = \mathcal{D}$. In this case, the Strehl definition is an adequate indication of point source detectability.

The above equations raise questions as to what values of l will be encountered with real optical systems and whether they will affect the merit functions. The few measurements of real mirror surface autocorrelation functions available indicate that $\frac{1}{10} < l < \frac{1}{3}$ is typical. Equation (159) requires that the real autocorrelation function go negative at $s \ll 2.0$, so there must be an upper limit on l of the order of 0.5. Comparison of Figs. 38 and 26 will give some feel for the effect on the merit functions.

Figure 38 plots the MTF degradation function for $\omega = 0.1$ wavelength rms and various autocorrelation lengths, based on the Gaussian autocorrelation function. Only the lowest portions of the spatial frequency domain are affected by variations in correlation length. The sensitivity curves in Fig. 26 indicate that the Strehl definition and Q^* should be insensitive to variations in correlation length. Thus Eqs. (161) and (163) should be adequate in most circumstances. Figure 26 indicates that relative edge response is sensitive to variations in autocorrelation length. Based on numerical data from synthetic wavefront calculations,

$$\mathcal{E} \simeq \exp[-4\pi^2\omega^2/(1 + l)] \tag{164}$$

should be more accurate than Eq. (162). To the degree that relative edge response relates to MTFA (see Section VIII,B,2), photographic performance as measured by the photointerpreter is dependent on the correlation length as well as the rms wavefront error.

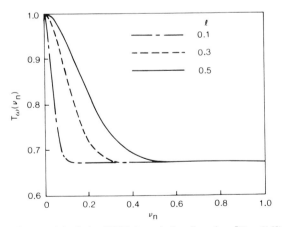

FIG. 38. Gaussian model of the MTF degradation function [Eq. (160)] for $\omega = 0.10$ wavelength rms and various values of the correlation length l.

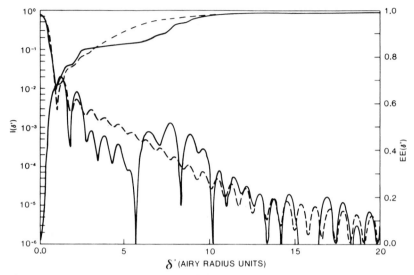

FIG. 39. The PSF and encircled energy for ω = 0.0752 wavelength rms and l = 0.10 calculated from (——) a synthetic wavefront and from (– – –) a Gaussian MTF degradation function model.

Equation (161) matches Eq. (87). Expanding the exponent as a power series leads to Eq. (86). If Eq. (161) is written $\mathscr{D} \simeq [\exp-(2\pi^2\omega^2)]^2$, expansion leads to Eq. (85). Thus O'Neill's random wavefront error model gives a second derivation of Maréchal's equation for Strehl definition.

4. *Computing the* PSF *and the* OTF *for Random Wavefront Error*

Two general approaches can be taken to calculating OTF and PSF data for random wavefront errors. One is to compute a rotationally symmetric MTF using O'Neill's MTF degradation function model. The corresponding PSF is then calculated using Eq. (111). Alternatively, a synthetic wavefront can be created using random number tables and a smoothing function. The PSF is then calculated using Eq. (66) or (67), and the OTF is subsequently calculated using Eq. (108) or (110). The choice of equations depends on whether the synthetic wavefront is random in both directions, or is rotationally symmetric with random variations in the radial direction.

Computation via the MTF degradation function leads to statistically smoothed MTF and PSF curves whose exact form depends on the choice of autocorrelation function model. Computation via the synthetic wavefront technique gives an individual OTF–PSF pair having its own unique features and which can in principle be generated experimentally. To draw statistically valid conclusions from synthetic wavefront calculations it is necessary to generate a large number of wavefronts, using different random number arrays, calculate OTF–PSF pairs, and average them. Figure 37 was generated by such an analysis.

Figure 39 shows two PSFs and their associated encircled energy curves, one generated by each technique. Both are for the same wavefront error, $\omega = 0.07518$ wavelength rms, selected to make $\mathscr{D} = 0.80$ by Eq. (161). Both have a correlation length $l = 0.10$. The PSF drawn with dashed lines was generated using the MTF degradation function and the Gaussian autocorrelation function of Eq. (160). The PSF drawn with solid lines was generated from a rotationally symmetric synthetic wavefront.

The synthetic wavefront is generated as follows. A random number table is used to select OPD values at points spaced $\Delta\rho = 0.006$ apart along a radius. The random numbers are then smoothed by convolution with a Gaussian smoothing function of the form $\exp(-4r^2/l^2)$. [Note: Other shapes of smoothing function can be used, and a square smoothing function has been used for some data presented in Section IX. The shape of the smoothing function is not necessarily related to the shape of the autocorrelation function, but the correlation lengths should be equal.] In the program used to generate Fig. 39, the Hankel transform was integrated by 512-point Gaussian quadrature. The evenly spaced values of OPD were interpolated to the integration points using a parabolic fit, and the entire wavefront scaled to give the specified rms wavefront error.

Comparison of the two PSFs in Fig. 39 shows that there are both qualitative and quantitative differences in the energy distribution. The synthetic wavefront PSF shows regional concentrations of energy indicative of a single dominant spatial frequency component in the wavefront. The PSF generated from the MTF degradation function results from the addition of a perfect lens PSF and a Gaussian scatter PSF and is devoid of the local structural variations to be expected in a real PSF. Some of the qualitative mismatch between PSFs could be eliminated by using an autocorrelation function model more realistic than Eq. (160). No equally convenient model has as yet been defined.

The random surface irregularities in large aspherical mirrors have some zonal structure because of the zonal polishing techniques used, but they are usually dominated by errors random in two dimensions. The PSF in Fig. 39 generated from a zonally averaged model random wavefront is thus not completely realistic, although it may be more realistic than that generated from the Gaussian autocorrelation function model. Figure 40 is a perspective plot of the PSF for a wavefront which is random in two dimensions. Relative intensity is plotted on a linear scale rather than the logarithmic scale in Fig. 39. Deviations from the perfect lens PSF are in

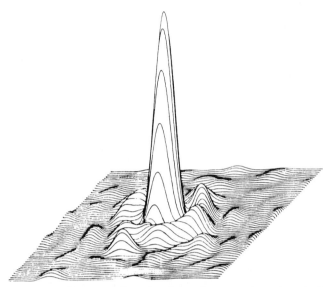

FIG. 40. The PSF for a two-dimensionally random wavefront ($\omega = 0.12$ wavelength rms).

the form of localized peaks of greater relative itensity than the zonal ridges in Fig. 39.

The zonally averaged model is used here for computational convenience, since it allows examination of the effects of fine structure in the pupil function without using gross amounts of computer time. The model is, of course, completely valid for rotationally symmetric pupil functions. There is some question as to the validity of applying results drawn from it to situations involving stationary two-dimensional processes such as random wavefront error.

The author feels that the zonally averaged model is valid for studying zonally averaged merit functions such as encircled energy, and for Strehl definition which measures the volume under the MTF and should be insensitive to variance. There is a legitimate question of quantitative accuracy in the case of relative structural content Q^* and related quantities which are proportional to the volume under the square of the MTF, since the latter is sensitive to variance. Q^* is slightly higher for two-dimensionally random wavefront error than for the corresponding zonally averaged case.

5. The Effects of Varying Correlation Length

There is a tendency to rely on rms wavefront error alone as a measure of image quality, ignoring correlation length or any other means of indicating differences in the spatial frequency distribution of the wavefront error. To illustrate the pitfalls and bounds of acceptability of this practice, several examples are compared, using a fixed value of wavefront error and different correlation lengths. In each case, $\omega = 0.07518$ wavelength rms. The correlation lengths are $l = 0.33, 0.10,$ and 0.03. Both synthetic wavefronts and MTF degradation functions have been used, but the former constitute the primary base for comparisons.

Figure 41 shows the radial profiles for the three wavefronts. The same set of random numbers was used to generate each, with only l changing in the Gaussian smoothing function. The similarity in general features among the three wavefronts is reflected in similarities in their PSFs. The wavefront for which $l = 0.10$ was also used to generate Fig. 39.

Figure 42 shows the MTF for each wavefront. The most significant differences are at spatial frequencies below $\nu_n = 0.20$, where the fine structure of the shorter-correlation-length wavefronts reduces the MTF substantially. The $l = 0.33$ wavefront shows a moderately lower MTF at middle spatial frequencies. This is associated with the larger negative excursion of the autocorrelation function for long correlation lengths noted in Section IX,C,2. The MTF differences at middle spatial frequencies should be small enough to fall within the variance range for the statistical model of MTF degradation.

Figure 43 shows the encircled energy curves for the three PSFs. The principal differences between the curves are a result of the introduction of higher-spatial-frequency components in the wavefront, which diffract en-

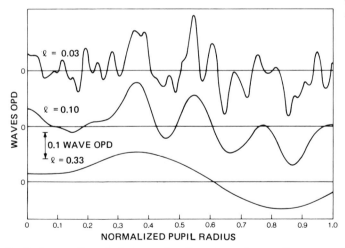

FIG. 41. Radius profiles of three rotationally symmetric wavefront error functions having the same rms phase amplitudes but of different correlation lengths.

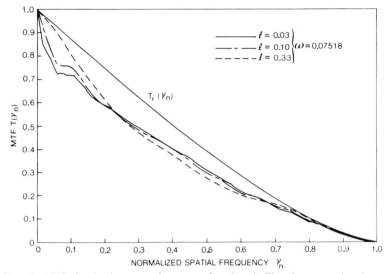

FIG. 42. MTFs for the three wavefront error functions in Fig. 42 compared to the perfect lens MTF.

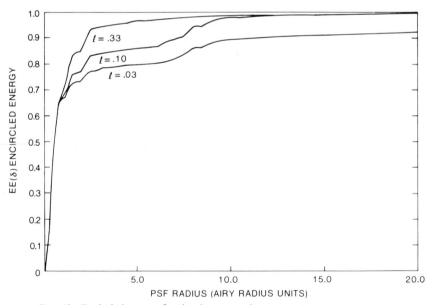

FIG. 43. Encircled energy for the three wavefront error functions in Fig. 41.

TABLE IV

VARIATION IN ONE-NUMBER MERIT FUNCTIONS AND ZEI
AS A FUNCTION OF CORRELATION LENGTH l[a]

	$l = 0.33$		$l = 0.10$		$l = 0.03$	
	W	T	W	T	W	T
Q^*	0.635	0.692	0.634	0.646	0.635	0.641
$Q^{*1/2}$	0.797	0.832	0.796	0.804	0.797	0.801
$\mathcal{D}$	0.796	0.817	0.799	0.802	0.801	0.800
$\mathcal{E}$	0.836	0.859	0.816	0.818	0.808	0.808
EE(1.0)	0.695	0.723	0.674	0.677	0.667	0.671
ZEI for Zone						
0–1[b]	−0.1430	−0.1153	−0.1640	−0.1611	−0.1705	−0.1669
1–3	0.1413	0.1032	0.0590	0.0234	0.0072	−0.0162
3–5	−0.0003	0.0110	−0.0033	0.0545	−0.0042	0.0045
5–10	−0.0004	0.0010	0.1038	0.0740	0.0788	0.0343
10–15	0.0024	0.0001	0.0042	0.0084	0.0116	0.0455
15–20	<0.0001	<0.0001	−0.0002	0.0006	0.0101	0.0401
>20	≪0.0001	≪0.0001	0.0005	0.0001	0.0670	0.0588

[a] For a fixed rms wavefront error $\omega = 0.07518$, comparing data derived from synthetic wavefronts (W) to that derived from the MTF degradation function (T).

[b] Inner and outer radii in Airy radius units.

ergy further outward in the ring structure. For $l = 0.10$, much of the scattered energy appears in the zone between 8 and 10 Airy radius units. For $l = 0.03$, a significant fraction is scattered past 20 Airy radius units.

Table IV shows the energy redistribution in terms of zonal energy increments and lists the values of several one-number merit functions, as well. Data are included from both synthetic wavefront calculations and MTF degradation function calculations to compare their accuracy. One general conclusion which can be drawn from the data is that calculations from the MTF degradation function model are consistently more optimistic than calculations from the synthetic wavefront at longer correlation lengths. This can be attributed directly to the failure of the all-positive Gaussian autocorrelation function model to meet the constraint imposed by Eq. (159). This failure may have adverse consequences when estimating one-number merit functions for optical systems limited by aberrations which have long effective correlation lengths.

More detailed conclusions can be drawn from examination of the synthetic wavefront data: (1) $Q^{*1/2} = \mathcal{D}$ for random wavefront error and is independent of correlation length; (2) $\mathcal{E}$ is consistently higher than $\mathcal{D}$, and the disparity increases as the correlation length increases; (3) the amount of energy diffracted out of the Airy disk increases as the correlation length decreases, as shown by EE(1.0) and ZEI(0–1); (4) at long correlation lengths, virtually all the energy diffracted out of the Airy disk reappears in the first two or three bright rings; and (5) at short correlation lengths, the energy diffracted out of the Airy disk is scattered well out of the central region of the diffraction pattern, with relatively little remaining in the first two or three bright rings.

In general, if $\mathscr{D}$ or Q^* is an appropriate merit function for an optical system degraded only by wavefront error, rms wavefront error by itself will also be appropriate. If $\mathscr{E}$ is the appropriate merit function, or if scattering into the outer rings is important, l must be considered.

D. Sources of Wavefront Error

Aside from design aberrations, sources of wavefront error fall into three categories: (1) static errors due to fabrication and alignment problems, (2) slowly time-varying errors due to environmental interactions with the optical system, (3) rapidly time-varying errors due to atmospheric turbulence. Techniques for calculating these effects are not discussed, only the qualitative nature of their effects. A few references are cited for further enquiry.

1. Fabrication and Alignment

Fabrication effects involve errors in grinding and polishing optical surfaces and errors in thickness or wedging of refracting elements. Alignment errors involve tilt and decentration of elements relative to the optical axis and spacing errors along the axis. Tilt and decentration may also result from wedging in refracting elements. For purposes of computation, each mirror surface or air–glass refracting surface is considered separately, and tilt, decentration, and axial misposition values assigned, irrespective of their causes.

Surface grinding and polishing errors may fall into any of the three spatial frequency domains, ripple usually being found only in large aspherical elements. Their effects have been discussed and can be predicted in advance only in a statistical sense. After the fact, they can be measured and their effects calculated, at least in the case of figure error.

Alignment errors introduce aberrations. The principal aberrations introduced by tilts and decentrations are axial coma and image displacement. Axial coma is similar in appearance and effects to third-order coma, but it is independent of field height. The principal aberrations introduced by axial displacement are defocus and spherical aberration. In refracting lenses, misalignment also introduces chromatic aberrations. All field aberrations can be affected by misalignment.

Many theoretical analyses of misalignment have been published, including those of Ruben (1964), Hopkins and Tiziani (1966), and Rimmer (1970).

2. Environmental Effects

The environmental effects discussed here are confined to those which cause slowly varying changes in the optical system. Rapidly varying effects are more properly reviewed in the context of image motion. The principal changes which vary slowly with time are a result of temperature changes. In very large optical systems, a change in orientation relative to the gravity vector can cause misalignments and mirror distortions through mechanical flexure of optical components and their supporting structures. In long-focal-length refractors, changes in air pressure can cause defocus and other aberrations if not compensated for. But temperature changes cause the most difficulty in maintaining image quality in large, high-performance optical systems.

Thermal effects can be divided into two categories, *thermal soaks,* in which the temperature changes uniformly throughout the optical system, and *thermal gradients,* in which the temperature varies from point to point within the optical system. Both can be serious but, generally speaking, thermal gradients are the most difficult to control. Thermal effects arise because the linear dimensions of most structural and optical materials vary with temperature and because the index of refraction of most refracting materials varies with temperature.

In an all-reflecting optical system, it is possible to eliminate the effects of thermal soaks by exactly matching the thermal expansion coefficients of the mirrors and their supporting structures. The only result of thermal soaks will be a net change in focal length, which can usually be tolerated. Where thermal expansion coefficients do not match, mirror spacing changes at a different rate from mirror surface curvature, and defocus and other aberrations may be introduced. In general, thermal soaks introduce only low-spatial-frequency wavefront errors. An exception to this rule may occur in large mirror substrates if the thermal expansion coefficient varies from point to point. If this is the case, both random figure error and ripple may be introduced if the operating temperature differs substantially from the temperature at which the mirror was built and tested.

Thermal gradients are usually divided into radial and lateral gradients for purposes of exposition. Radial gradients are common in edge-mounted elements, because the heat conduction paths run through the mount structure into the element at its edge. The mount structure is typically metal having a far greater thermal conductivity than the element. Thus the temperature of the entire mount may be stabilized before heat is conducted inward to the axis of the element. Radial gradients introduce surface deformations resembling changes in the radius of curvature, which produce defocus and spherical aberration and can affect field aberrations. Lateral gradients across a lens system tend to cause its structure to "hot dog," tilting, decentering, and despacing elements within the lens system. The principal aberrations introduced are axial coma and image displacement, but other aberrations are introduced as well. Lateral gradients within the optical element can introduce cylindricity, causing axial astigmatism.

If the optical system contains refracting elements, both the thermal expansion coefficient α and the variation of index of refraction with temperature, dn/dT, must be taken into consideration. Both thermal soaks and thermal gradients become complex issues, and both chromatic and monochromatic aberrations must be considered. According to Reitmayer and Schroeder (1975), the effects of thermal soaks are proportional to

$$\gamma = (dn/dT)(n - 1) - \alpha \tag{165}$$

and the effects of thermal gradients are proportional to

$$\Gamma = (dn/dT)(n - 1) + \alpha \tag{166}$$

Selecting a glass type that makes either γ or Γ zero makes the lens element insensitive to either thermal soaks or thermal gradients. So long as both α and dn/dT are finite, both conditions cannot be met in the same element. Thermal gradients can also introduce strain in the glass, if α is too large, which can degrade image quality even when $\Gamma = 0$.

Techniques for calculating thermal and other environmental effects on image quality have been developed by many people involved in designing high-performance optical systems, and a number of papers have been written. Barnes (1966) discusses effects of the aerospace thermal environment. Schwesinger (1954) considers mechanical flexure in large mirrors. Spitzer and Boley (1967), Ogrodnik (1970), and Malvick (1970) discuss thermal deformations of large mirrors. "Optical Telescope Technology" (NASA SP-233) contains a number of papers on large optical system technology, including environmental effects. Barnes (1975) gives substantial data on thermal effects in reconnaisance aircraft windows.

3. *Atmospheric Turbulence*

Large-aperture imaging systems working through all or part of the atmosphere can have their image quality degraded in several ways; of these atmospheric turbulence is the most significant. Other atmospheric effects include reduction in target contrast due to scattered sunlight, and lateral color introduced by refraction in the dispersing medium of air at large zenith angles.

Atmospheric turbulence results in localized variations in the index of refraction, which produce rapidly changing random wavefront errors. If the entrance pupil diameter of the telescope exceeds the characteristic cell size of the air turbulence, the effect is to add a time-varying random wavefront error component to the pupil function. If the cell size is large compared to the entrance pupil diameter, the effect is wavefront tilt, causing image motion. In lateral scale, for the largest-aperture telescopes, the spatial frequency range of the wavefront error component can extend over the domain of figure error and into the domain of ripple.

Hufnagel and Stanley (1964) have analyzed the image quality degradation produced by atmospheric turbulence. Based on this analysis, Hufnagel (1964) has presented a statistical model for the average MTF of a lens of focal length f whose static MTF is $T_1(\nu_x,\nu_y)$:

$$T(\nu_x,\nu_y) = T_1(\nu_x,\nu_y) \exp[-2\pi^2\sigma^2 f^2(\nu_x^2 + \nu_y^2)^{5/6}] \tag{167}$$

where σ is a strength factor for the turbulence.

Coulman (1966) discusses the dependence of the OTF on horizontal range through the atmosphere. Weiner (1967) discusses the effects of atmospheric turbulence on aerial reconnaissance. Wyant (1976) collects a number of papers on imaging through the atmosphere, including several on active optics techniques for compensating for atmospheric turbulence in the optical system.

Air turbulence can arise from sources other than normal atmospheric conditions. Boundary layer turbulence at the skin of high-speed aircraft can degrade image quality for aerial reconnaissance systems. Fisher (1976) gives a brief survey of the effects of boundary layer turbulence on image quality. Lorah and Rubin (1965) survey the effects of atmospheric turbulence, with emphasis on infrared imagery.

X. SOURCES AND EFFECTS OF AMPLITUDE TRANSMITTANCE VARIATION

Variations in the amplitude transmittance pupil function seldom arise inadvertently. Aperture obstructions are usually inherent parts of the optical systems in which they are found. The truncated Gaussian intensity profiles in laser beam projectors are a consequence of the laser emission process. In most other cases, amplitude transmittance variations are deliberately introduced to modify the PSF in a planned manner. Amplitude transmittance variations can arise inadvertently through, for example, damage to mirror coatings caused by environmental factors; but such cases are the exception.

The process of changing the energy distribution in the PSF by deliberate manipulation of the pupil function so as to improve some measure of image quality is known as *apodization*. The term derives from an early application of apodization—suppression of the ring structure. A literal translation of its Greek roots gives "to apodize" the meaning "to remove the feet," the "feet" being the bright rings in the diffraction pattern. The meaning of the term has been extended by usage to include all forms of modifying the PSF through pupil function manipulation. In general, only the amplitude transmittance pupil function $P(\rho,\theta)$ is altered. Two cases are discussed which also involve modifications to $W(\rho,\theta)$.

The principal references for this discussion are Jacquinot and Roizen-Dossier (1964) and Barakat (1962a, 1962b, 1963). Numerical data were generated at Itek Corporation for this chapter.

A. LUNEBERG'S APODIZATION PROBLEMS AND PUPIL CLASSIFICATION

Luneberg, in his 1944 Brown University Notes (Luneberg, 1964), examined four different measures for image quality based on the PSF and determined the boundary conditions for improving them by manipulating the pupil function. Luneberg provided a detailed solution only to the first of what are now called the Luneberg apodization problems. Many workers have provided solutions to the other three (Jacquinot and Roizen-Dossier, 1964). The general forms of the solutions offer a convenient means of classifying amplitude transmittance pupil functions, so the problems are worth reviewing briefly.

If Φ_i is the total flux in the PSF, $E_i(0)$ is the irradiance at the center of the Airy disk, and δ'_0 is the radius of the first dark ring in Airy radius units, the four Luneberg apodization problems may be stated as follows, listing them in the order cited by Barakat (1962a):

(1)   For constant Φ_i, what form of $P(\rho,\theta)$ will maximize $E_i(0)$?

(2)   For constant Φ_i and maximum $E_i(0)$, what form of $P(\rho,\theta)$ will set the radius of the first dark ring to any selected value $\delta'_0 < 1.0$?

(3)   For constant Φ_i, what form of $P(\rho,\theta)$ will maximize encircled energy $EE(\delta')$ for any selected value of δ'?

(4)   For constant Φ_i and maximum $E_i(0)$, what form of $P(\rho,\theta)$ will most improve the Sparrow resolution limit?

The first three problems are specified in terms of one spread function, and their solutions apply equally to coherent and incoherent light (Barakat, 1962a). The fourth problem involves two point sources, and it has separate separate solutions for coherent (Barakat, 1962b) and incoherent (Barakat, 1963) light.

Luneberg solved the first problem explicitly; the solution is a uniform pupil, i.e., the perfect lens. For incoherent light, the solutions to the second and fourth problem are similar, since the second represents an improvement in Rayleigh two-point resolution and both require a decrease in the diameter of the Airy disk. Both involve an amplitude transmittance pupil function which increases monotonically from center to edge, and both cause the transfer of energy from Airy disk to the ring system. Some solutions also involve manipulation of $W(\rho,\theta)$. The third problem is the classic one of suppressing the ring system and involves an amplitude transmittance pupil function which decreases monotonically from center to edge.

Based on the above discussion, a five-level classification of amplitude transmittance pupil functions suggests itself:

Class 1. $P(\rho,\theta) \equiv 1.0$ at all points in the pupil
Class 2. $P(\rho,\theta) = 1.0$ or 0.0 at all points in the pupil
Class 3. $P(\rho,\theta)$ increases monotonically from center to edge
Class 4. $P(\rho,\theta)$ decreases monotonically from center to edge
Class 5. $P(\rho,\theta)$ varies nonmonotonically

A centered circular obstruction may be considered part of class 2 or class 3 and has the same effect on the PSF as solutions to Luneberg's second and fourth apodization problems. Not all forms of obstructions meet the definition for class 3, so a separate class for all forms of obstructions has been defined. Irregular external pupil boundaries are also treated as part of class 2.

B. CLASS-2 PUPIL FUNCTIONS

1. The Centered Circular Obstruction

The PSF for a perfect lens with a centered circular obstruction is defined by Eq. (202) in Appendix B. Its Strehl definition and pupil transmittance are given by Eq. (82). Figure 44 shows the PSF and encircled energy for a centered circular obstruction of diameter ratio $\epsilon = 0.45$ and illustrates the manner in which a central obstruction affects the PSF.

The main effects of the central obstruction are to reduce the Strehl definition and the diameter of the Airy disk and to transfer energy out-

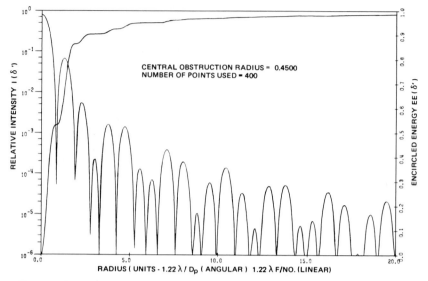

CENTRAL OBSTRUCTION RADIUS = 0.4500
NUMBER OF POINTS USED = 400

RADIUS (UNITS - 1.22 λ / D_D (ANGULAR) 1.22 λ F/NO. (LINEAR)

FIG. 44. The PSF for a perfect lens with a circular pupil and a centered circular obstruction of diameter ratio $\epsilon = 0.45$.

ward into the ring system. The width, location, and height of the bright rings change. The minima between bright rings still go to zero (noting again that the nonzero minima in the PSF plots are computer artifacts), but their locations are shifted. Figure 45 shows the shift in locations for the first four minima as a function of ϵ. Note that the radius of the first minimum is reduced by about one-third for very large obstructions.

The envelope of the ring system shows a periodicity which is a result of the beat between the two oscillating Bessel functions in Eq. (202). This beat produces a series of super-minima whose radii δ'_{sm} are given by

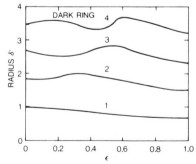

FIG. 45. Radius in Airy radius units of the first four dark rings in the PSF as a function of the central obstruction diameter ratio ϵ.

TABLE V

ENCIRCLED ENERGY WITHIN THE FIRST SIX SUPER-MINIMA OF RADII
$\delta'_{sm} \simeq 1.6n/(1 - \epsilon)$ AS A FUNCTION OF ORDER NUMBER n

n	1	2	3	4	5	6
$EE(\delta'_{sm})$	0.903	0.950	0.966	0.975	0.980	0.983

$$\delta'_{sm} \simeq 1.6n/(1 - \epsilon) \tag{168}$$

where n is an order number, 1,2,3, ... These super-minima have the interesting property that the encircled energy in each is the same for the same order number, regardless of the obstruction size in the absence of wavefront error. Table V gives values for the first six super-minima (Tschunko, 1974).*

There is no exact analytic solution for the encircled energy for a perfect lens with a centered circular obstruction. Some data derived by numerical analysis can be found in the literature (Young, 1970; Goldberg and McCullock, 1969), and more are given here.

Figure 46 shows encircled energy as a function of ϵ for seven different

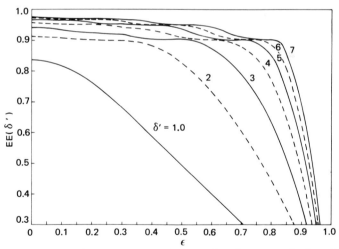

FIG. 46. Encircled energy versus central obstruction diameter ratio for selected PSF radii.

---

* Tschunko (1974) gives an equation for the first super-minimum which implies that the constant in Eq. (168) should be 1.64. This will produce no significant change in the results shown in Table V.

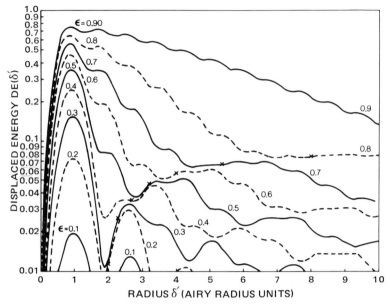

FIG. 47. Displaced energy as a function of radius for different central obstruction diameter ratios.

radii, $\delta' = 1-7$ Airy radius units. Note that there are flat plateaus in the individual curves, notably at encircled energy values of about 90 and 95%. These correspond to the first two superminima. Based on numerical analysis for central obstruction diameter ratios from $\epsilon = 0.0$ to 0.95, the values given in Table V are accurate to 1 in the third decimal.

Figure 47 shows displaced energy as a function of radius for central obstruction diameter ratios from $\epsilon = 0.1$ to 0.9. Here $DE(\delta')$ is plotted on a logarithmic scale and shows values above 1%. Displaced energy represents the fraction of the total energy in the PSF which has been diffracted outward from the given radius when compared with the perfect lens PSF. Positive slopes indicate zones in which there is less energy than for a perfect lens, and negative slopes indicate zones in which there is more energy. For small obstructions, the principal energy transfer is from the Airy disk to the first bright ring. For large obstructions, the energy is spread far into the outer rings. The × marks in Fig. 47 correspond to the first-order super-minimum for each central obstruction diameter ratio. The presence of super-minima is revealed by the confluence of the curves for different diameter ratio obstructions.

The MTF for a perfect lens with a centered circular obstruction has been determined analytically by O'Neill (1956), and his equation can be found in Appendix B. Table VI lists MTF values for central obstruction diameter ratios from 0.0 to 0.95 at normalized spatial frequency incre-

TABLE VI

MTF, $\mathscr{D}$, τ_p, Q^*, AND $\mathscr{E}$, FOR A PERFECT LENS WITH A
CENTERED CIRCULAR OBSTRUCTION OF DIAMETER RATIO ϵ

	ϵ						
v_n	0.00	0.05	0.10	0.15	0.20	0.25	0.30
0.05	0.093636	0.93370	0.92957	0.92531	0.92059	0.91526	0.90918
0.10	0.87289	0.87006	0.86150	0.85199	0.84222	0.83139	0.81913
0.15	0.80973	0.80675	0.79771	0.78234	0.76615	0.74937	0.73068
0.20	0.74706	0.74392	0.73440	0.71822	0.69485	0.67047	0.64481
0.25	0.68504	0.68174	0.67176	0.65477	0.63025	0.59737	0.56286
0.30	0.62384	0.62039	0.60994	0.59216	0.56650	0.53209	0.48773
0.35	0.56364	0.56004	0.54913	0.53058	0.50379	0.46788	0.42158
0.40	0.50463	0.50088	0.48953	0.47021	0.44232	0.41278	0.38149
0.45	0.44701	0.44312	0.43133	0.41664	0.39984	0.38007	0.35686
0.50	0.39100	0.38950	0.38507	0.37772	0.36740	0.35394	0.33708
0.55	0.33683	0.33767	0.34023	0.33982	0.33564	0.32818	0.31735
0.60	0.28476	0.28547	0.28763	0.29131	0.29662	0.29747	0.29362
0.65	0.23507	0.23566	0.23745	0.24049	0.24487	0.25075	0.25832
0.70	0.18812	0.18859	0.19002	0.19245	0.19596	0.20066	0.20673
0.75	0.14429	0.14466	0.14575	0.14761	0.15031	0.15391	0.15856
0.80	0.10409	0.10435	0.10514	0.10648	0.10843	0.11103	0.11438
0.85	0.06815	0.06832	0.06884	0.06972	0.07099	0.07269	0.07489
0.90	0.03739	0.03748	0.03776	0.03825	0.03894	0.03988	0.04108
0.95	0.01332	0.01335	0.01345	0.01363	0.01388	0.01421	0.01464
$\mathscr{D},\tau_\mathrm{p}$	1.0000	0.9975	0.9900	0.9775	0.9600	0.9375	0.9100
Q^*	1.0000	0.9922	0.9696	0.9332	0.8852	0.8276	0.7633
$\mathscr{E}$	1.0000	0.9967	0.9874	0.9724	0.9524	0.9275	0.8982

	ϵ						
v_n	0.35	0.40	0.45	0.50	0.55	0.60	0.65
0.05	0.90218	0.89401	0.88436	0.87278	0.85864	0.84096	0.81823
0.10	0.80506	0.78868	0.76936	0.74620	0.71792	0.68258	0.63715
0.15	0.70940	0.68472	0.65568	0.62092	0.57851	0.52554	0.45749
0.20	0.61602	0.58288	0.54403	0.49762	0.44108	0.37055	0.30967
0.25	0.52593	0.48402	0.43518	0.37705	0.32653	0.28327	0.24333
0.30	0.44058	0.38920	0.34454	0.30589	0.27019	0.23647	0.20422
0.35	0.37382	0.33355	0.29852	0.26619	0.23570	0.20660	0.17858
0.40	0.34653	0.30672	0.27182	0.24144	0.21327	0.18659	0.16105
0.45	0.32964	0.29764	0.25980	0.22707	0.19910	0.17334	0.14907
0.50	0.31639	0.29124	0.26074	0.22360	0.19197	0.16550	0.14139
0.55	0.30285	0.28421	0.26070	0.23122	0.19416	0.16306	0.13763
0.60	0.28564	0.27369	0.25919	0.23529	0.20669	0.16931	0.13837
0.65	0.26020	0.25621	0.24738	0.23340	0.21337	0.18569	0.14769
0.70	0.21438	0.22395	0.22659	0.22205	0.21137	0.19373	0.16724
0.75	0.16444	0.17178	0.18093	0.19239	0.19549	0.18958	0.17532
0.80	0.11862	0.12391	0.13052	0.13878	0.14923	0.16264	0.16573
0.85	0.07766	0.08113	0.08545	0.09086	0.09770	0.10648	0.11800

(Continued)

TABLE VI (Continued)

				ϵ			
v_n	0.35	0.40	0.45	0.50	0.55	0.60	0.65
0.90	0.04261	0.04451	0.04688	0.04985	0.05360	0.05842	0.06474
0.95	0.01518	0.01586	0.01670	0.01776	0.01910	0.02081	0.02306
$\mathcal{D},\tau_p$	0.8775	0.8400	0.7975	0.7500	0.6975	0.6400	0.5775
Q^*	0.6948	0.6238	0.5516	0.4791	0.4080	0.3393	0.2733
$\mathcal{E}$	0.8646	0.8271	0.7858	0.7399	0.6905	0.6375	0.5830

			ϵ			
v_n	0.70	0.75	0.80	0.85	0.90	0.95
0.05	0.78792	0.74549	0.68186	0.57519	0.36368	0.16295
0.10	0.57660	0.49184	0.36471	0.25370	0.16358	0.08042
0.15	0.36681	0.28869	0.22405	0.16475	0.10843	0.05381
0.20	0.25838	0.21122	0.16658	0.12362	0.08180	0.04070
0.25	0.20553	0.16927	0.13413	0.09984	0.06618	0.03296
0.30	0.17310	0.14289	0.11340	0.08449	0.05602	0.02789
0.35	0.15145	0.12504	0.09923	0.07390	0.04898	0.02437
0.40	0.13641	0.11248	0.08916	0.06633	0.04391	0.02182
0.45	0.12589	0.10356	0.08191	0.06082	0.04019	0.01994
0.50	0.11881	0.09735	0.07675	0.05683	0.03746	0.01854
0.55	0.11471	0.09342	0.07330	0.05405	0.03551	0.01752
0.60	0.11366	0.09167	0.07140	0.05236	0.03423	0.01683
0.65	0.11661	0.09241	0.07115	0.05173	0.03360	0.01643
0.70	0.12837	0.09686	0.07301	0.05236	0.03367	0.01634
0.75	0.15059	0.11062	0.07838	0.05478	0.03465	0.01661
0.80	0.15681	0.13504	0.09383	0.06051	0.03706	0.01742
0.85	0.13362	0.13578	0.11962	0.07738	0.04247	0.01915
0.90	0.07331	0.08545	0.10385	0.10208	0.06037	0.02313
0.95	0.02612	0.03045	0.03700	0.04800	0.07011	0.04091
$\mathcal{D},\tau_p$	0.5100	0.4375	0.3600	0.2775	0.1900	0.0975
Q^*	0.2121	0.1563	0.1070	0.0638	0.0298	0.0072
$\mathcal{E}$	0.5179	0.4519	0.3812	0.3042	0.2149	0.1235

ments of 0.05. Values are also listed for the Strehl definition, relative edge response, and Q^*.

The principal effects of the central obstruction are the suppression of low- and middle-spatial-frequency MTFs, and enhancement of the MTF at very high spatial frequencies. The latter is a consequence of reducing the Airy disk diameter, and the former is the consequence of transferring energy from the Airy disk to the ring system. All the one-number merit functions decrease as ϵ increases.

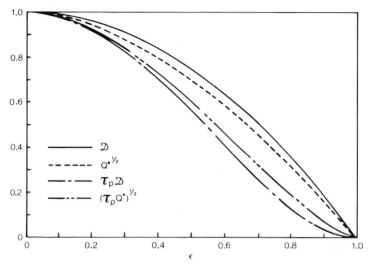

FIG. 48. Merit functions for point source detection for a perfect lens with a centered circular obstruction of diameter ratio ϵ.

Figure 48 plots $\mathscr{D}$, $Q^{*1/2}$, $\tau_p \mathscr{D}$, and $(\tau_p Q^*)^{1/2}$ as a function of ϵ to compare the traditional merit function of peak image irradiance to the more exact $Q^{*1/2}$ and $(\tau_p Q^*)^{1/2}$ for point source detection. For fixed exposure times, where SNR is proportional to $(\tau_p Q^*)^{1/2}$ (see Section VII,B), using the peak irradiance $\tau_p \mathscr{D}$ is pessimistic, since the latter ignores reduction in noise due to the reduction in background exposure. If the exposure time is adjusted to give a fixed background exposure, where SNR is proportional to $Q^{*1/2}$, $\tau_p \mathscr{D}$ is highly pessimistic and $\mathscr{D}$ is optimistic. Strehl definition is a useful merit function for pure wavefront error but must be used with circumspection when aperture obstructions are present.

Strehl definition, relative edge response, and Q^* may be fit by exponential equations for values of $\epsilon < 0.60$. These approximations are convenient for purposes of preliminary image quality analysis.

$$\mathscr{D}_\epsilon \equiv 1 - \epsilon^2 \simeq \exp(-\epsilon^2), \qquad \epsilon < 0.60 \tag{169}$$

$$Q_\epsilon^* \simeq \exp(-3\epsilon^2), \qquad \epsilon < 0.60 \tag{170}$$

$$\mathscr{E}_\epsilon \simeq \exp(-1.2\epsilon^2), \qquad \epsilon < 0.60 \tag{171}$$

2. Other Forms of Pupil Obstruction

Most obstructed pupils show departures from circular symmetry, if only because spiders are used to support the devices obstructing the pupil. There are seldom exact analytic solutions for the PSF and MTF in these cases. It is sometimes inconvenient to solve these problems numerically as well because of the very fine grid of sample points needed to adequately represent variations in the amplitude transmittance pupil function. The PSF and MTF can be determined experimentally using a near-perfect lens with an appropriate mask, with the MTF being determined by using two masks, a uniform light source, and a photometer. Measuring transmission through the two masks as a function of mask displacement gives values for the MTF.

Two noteworthy papers giving the PSF and MTF for irregular pupils and obstructions have been published by Mahan *et al.* (1965) and Tschunko and Sheehan (1971).

C. Class-3 Pupil Functions

Class-3 pupil functions include solutions to Luneberg's second and fourth apodization problems. Barakat's (1962a) solution to the second problem is used to illustrate the class, since the solution to the fourth problem is very similar (Barakat, 1963).

Barakat's solution has the general form

$$P(\rho) = \frac{1 - C(\beta)J_0(\beta,\rho)}{1 - C(\beta)J_0(\beta)} \tag{172}$$

where $\beta = 3.83171\delta_0'$, $J_0(\ )$ and $J_1(\ )$ are Bessel functions of the first kind, and

$$C(\beta) = 2J_1(\beta)/\beta[J_0^2(\beta) + J_1^2(\beta)] \tag{173}$$

The radius $\delta_0' \leq 1.0$ Airy radius units may be selected arbitrarily. It will be found that $P(\rho)$ is negative over part of the pupil if δ_0' is small. A negative value of $P(\rho)$ is interpreted as being equal to the corresponding positive value of $P(\rho)$ coupled with $W(\rho) = \lambda/2$.

As an example, take the minimum value of δ_0' for which $P(\rho)$ is all-positive. This is approximately $\delta_0' = 0.8276$ and leads to the function

$$P(\rho) = 0.76193[1 - J_0(3.171\rho)] \tag{174}$$

A plot of Eq. (174) has been included in Fig. 50 where it can be contrasted to class-4 pupil functions.

Barakat's solution is not a pure amplitude transmittance pupil function in all cases. It is interesting to compare his solution to an all-wavefront-error pupil function solution found by Wilkins (1950). The latter solution is of the form

$$W(\rho) = \left\{ \begin{array}{ll} \lambda/4, & 0 \leqq \rho < \rho_0 \\ 0, & \rho = \rho_0 \\ -\lambda/4, & \rho_0 < \rho \leqq 1.0 \end{array} \right\} \tag{175}$$

where ρ_0 is determined by solving

$$2\rho_0 J_1(3.8317\delta'_0\rho_0) = J_1(3.8317\delta'_0) \tag{176}$$

For $\delta'_0 = 0.8276$ Airy radius units, Eq. (176) gives $\rho_0 = 0.3124$.

The narrowing of the central maximum in these two examples approximates that produced by a central obstruction diameter ratio $\epsilon = 0.50$, and the image quality of all three should be compared. Figure 49 shows MTFs for all three examples. Table VII lists one-number merit function values for each. In general, Wilkins' phase apodization is superior where a fixed exposure time dictates maximizing transmittance. Barakat's apodization function is superior where the exposure time can be extended to offset the loss in transmittance. The central obstruction produces image quality values between the other two examples. Note that there is a disparity between the Strehl definition and $Q^{*1/2}$ for Barakat's apodization, but not for Wilkins' phase apodization which is after all pure wavefront error.

On balance, apodization produced by class-3 pupil functions is of academic rather than practical interest. The only measure of image quality

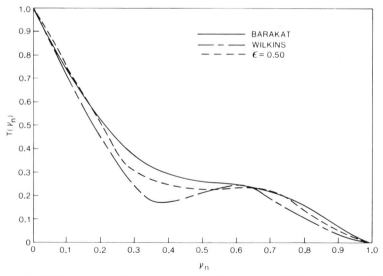

FIG. 49. MTFs for the apodization functions of Barakat and of Wilkins compared to that for a central obstruction producing the same radius for the first minimum in the PSF.

TABLE VII

SINGLE-NUMBER MERIT FUNCTIONS FOR PUPIL FUNCTIONS
WHICH REDUCE THE DIAMETER OF THE AIRY DISK BY THE SAME AMOUNT

	Eq. (174)	Eq. (175)	$\epsilon = 0.50$
Q^*	0.560	0.387	0.479
τ_p	0.481	1.000	0.750
$\mathcal{D}$	0.828	0.649	0.750
$Q^{*1/2}$	0.748	0.622	0.692
$\mathcal{E}$	0.792	0.669	0.740
$\tau_0\mathcal{D}$	0.398	0.649	0.563
$(\tau_p Q^*)^{1/2}$	0.519	0.622	0.599
EE(0.83)	0.53	0.42	0.48

which is improved over an unapodized lens is that of two-point resolution. While the resolution of close binaries is of interest in astronomy, they are usually of very different brightness. This form of apodization reduces the radius of the first minimum but increases the brightness of the first bright ring. To take advantage of this form of apodization, it would be necessary to use narrow bandpass wavelength filtration and to eliminate any forms of wavefront error which might fill in the first minimum. Wilkins' phase apodization has no application where the MTF is an important merit function, since it always reduces the MTF below that of a perfect lens, as will any pure wavefront error.

D. CLASS-4 PUPIL FUNCTIONS

Amplitude transmittance pupil functions which decrease monotonically toward the edge of the pupil suppress the outer ring system to varying degrees, increasing the encircled energy at radii closer to the center of the PSF. The cost of obtaining a significant increase in the fraction of the PSF's total energy lying within a given radius is a reduction in the image irradiance at all points in the PSF. In effect, apodization reduces image irradiance selectively, the greatest suppression occurring in the outer rings for class-4 pupil functions. Thus apodization of this form is useful only where the lower pupil transmittance can be tolerated. An exception to this statement occurs in the case of laser beam optics, where the nature of stimulated emission imposes a Gaussian intensity profile on the laser beam. Any further modification of $P(\rho)$ involves relatively small losses in effective pupil transmittance.

Class-4 amplitude transmittance pupil functions have several potential applications. Maximizing encircled energy may be useful in some forms of stellar photometry. Suppressing the ring structure may be useful where long-dynamic-range imagery is involved, as in the search for dim companions to very bright stars. Apodization of this form alters the MTF in a manner which can reduce aliasing in sampled imagery. Apodization is also being considered by some investigators for optimizing the PSF profile for laser fusion experiments (Vanderhoff, 1977).

1. Forms of $P(\rho)$ to Be Discussed

The most commonly analyzed apodization problem is Luneberg's third problem, maximizing the encircled energy within a given radius. Since only one point in the encircled energy curve is specified, it may be expected that there are different solutions to the problem which have different energy distributions inside and outside the given radius. There is no exact analytic solution to Luneberg's third apodization problem, and published solutions are the result of numerical analyses of preselected forms for $P(\rho)$.

To illustrate the degree of variability in the PSFs produced by different class-4 pupil functions, four different forms are examined:

(a) $P(\rho) = 1 - 1.563\rho^2 + 0.6565\rho^4$ (177)

(b) $P(\rho) = 0.181 + 0.426(1 - \rho^2) + 0.257(1 - \rho^2)^2$
$$+ 0.136(1 - \rho^2)^3 \tag{178}$$

(c) $P(\rho) = \exp(-\rho^2/c^2)$ (179)

(d) $P(\rho) = 0.181 + 0.426(1 - \rho^4) + 0.257(1 - \rho^4)^2$
$$+ 0.136(1 - \rho^4)^3 \tag{180}$$

Functions (a) and (b) were published as solutions to the problem of maximizing EE(1.0), by Barakat (1962a) and Straubel (1935), respectively. (Straubel's solution predates Luneberg's classification of the four apodization problems by 9 years.) Function (c) is a truncated Gaussian function and is used to show the effects of varying its parameter c on the PSF. The parameter c represents the normalized radius at which the amplitude transmittance is $1/e = 0.3679$ and the intensity transmittance is $1/e^2 = 0.1353$.

Functions (a), (b), and (c) are similar, all being power series of the form $1 - \Sigma(-1)^n A_n \rho^{2n}$, $n = 1, 2, 3, \ldots$. Function (d) is an arbitrary variation of Straubel's solution and shows the effects of changing the basic shape of $P(\rho)$ to one having the same value at $\rho = 1.0$ but a higher total pupil transmittance.

Figure 50 plots $P(\rho)$ versus ρ for all four class-4 functions plus the class-3 pupil function in Eq. (174). Three examples of form (c) are included. These examples are used in illustrating the effects of class-4 apo-

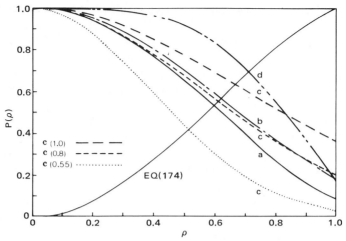

FIG. 50. Amplitude transmittance pupil functions from Eq. (174) and Eqs. (177)–(180).

dization on the inner PSF structure, the outer PSF structure, and several one-number merit functions. Form (c) is used to show the effects of class-4 apodization on the MTF.

2. *Inner Structure of the* PSF

Table VIII lists several values of encircled energy, excluded energy, and pupil transmittance used in discussing the effects of these apodization functions on the structure of the PSF. In function (c), $c = 0.80$, which is close to its optimum value for maximizing EE(1.0). Functions (b) and (c) produce very similar values for the quantities listed in Table VIII, except for XE(9.0) which is discussed in Section X,D,3. These similarities might

TABLE VIII

$P(1.0)$, ENCIRCLED ENERGY, PUPIL TRANSMITTANCE, AND
EXCLUDED ENERGY FOR APODIZATION FUNCTIONS (a)–(d) AND A PERFECT LENS

	(a)	(b)	(c)^a	(d)	Perfect
$P(1.0)$	0.0935	0.1810	0.2096	0.1810	1.0000
EE(1.0)	0.9567	0.9672	0.9667	0.9540	0.8378
EE(1.3)	0.9941	0.9846	0.9833	0.9625	0.8604
τ_p	0.262	0.319	0.306	0.511	1.000
XE(1.0)	0.0433	0.0328	0.0333	0.0460	0.1622
XE(9.0)	0.000594	0.001874	0.002611	0.001188	0.018192

$^a c = 0.80.$

be expected from a comparison of the two amplitude transmittance curves shown in Fig. 50, and it is probable that a refinement in the choice of c would bring the results even closer. For function (a), EE(1.0) and τ_p are significantly lower. This suboptimal performance is probably the result of using only the ρ^2 and ρ^4 terms during the optimization analysis. Note, however, that EE(1.3) is one percentage point higher for (a) than for either (b) or (c).

Pupil function (d) produces lower values for EE(1.0) and EE(1.3) but has a substantially higher pupil transmittance. When a comparison is made to the perfect lens PSF in terms of excluded energy, function (d) reduces XE(1.0) by 72%, and function (b) reduces XE(1.0) by 80%. The pupil transmittance of (d) is 1.6 times that of (b), which may be sufficient justification to prefer (d) for some applications. The comparison between (a) and (d) is even more favorable to (d) unless the greater suppression of the first bright ring by (a) is an overriding consideration.

Figure 51 shows the effects on EE(1.0) and EE(1.3) of varying the parameter c in pupil function (c). Encircled energy has been plotted as a function of τ_p to allow insertion of data points for pupil functions (a), (b), and (d). Figure 51 emphasizes the familial relationship between (a) (b), and (c), and the degree to which (d) departs from these characteristics.

Figure 52 plots displaced energy versus radius for pupil function (c), with values of c ranging from 0.5 to 2.0. The ZEI plots have been inserted to compare the energy redistribution within 2 Airy radius units for pupil functions (a) and (c). The displaced energy curves show the main effect on

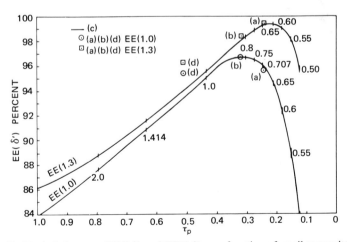

FIG. 51. Encircled energy EE(1.0) and EE(1.3) as a function of pupil transmittance for apodization functions (a)–(d) in Eqs. (177)–(180).

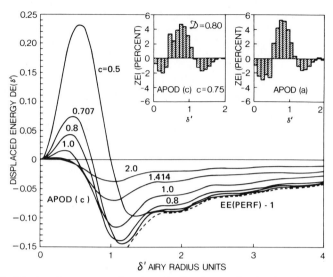

FIG. 52. Displaced energy as a function of radius for apodization function (c). The insets show ZEIs for apodization functions (a) and (c).

the inner structure of the PSF of varying c; as c decreases, the inward displacement of energy increases until overridden by the outward displacement caused by the reduction in the effective entrance pupil diameter.

3. Outer Structure of the PSF

Figure 53 plots excluded energy as a function of radius out to 9 Airy radius units for the class-4 pupil functions in Fig. 50, adding the perfect lens excluded energy curve for comparison. In the region well away from the center of the PSF, class-4 apodization uniformly reduces the PSF, retaining the same envelope slope as for the perfect lens. Based on the data for XE(9.0), $P(1.0)$, and τ_p listed in table VIII,

$$XE(\delta') \simeq XE_i(\delta')P^2(1.0)/\tau_p \qquad (181)$$

where $XE_i(\delta')$ is the excluded energy for the perfect lens. Thus the degree of suppression in the outer ring structure is directly proportional to $P^2(1.0)$ which is the energy transmittance at the edge of the pupil.

If the exponent of ρ in Eqs. (178) and (180) is made a variable, so that the terms become $1 - \rho^a$, then as a increases, $P(\rho)$ at small values of ρ will approach 1.0 and the radius at which the curve rolls off toward the fixed value of $P(1.0)$ will increase. Equation (181) indicates that $XE(\delta')$ decreases slightly as τ_p increases but remains essentially the same at very

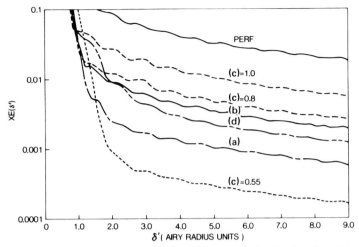

FIG. 53. Excluded energy as a function of radius for apodization functions (a)–(d); the perfect lens excluded energy curve is shown for comparison.

long radii, dependent solely on the value of $P(1.0)$. As a increases, the radius at which the equilibrium slope in $XE(\delta')$ is reached increases, and the inner portions of the excluded energy curve lift toward that of the perfect lens. Curves (b) and (d) in Fig. 53 illustrate the effect.

This discussion suggests that an apodization problem of more interest than Luneberg's third problem is that of maximizing $EE(1.0)$ for arbitrarily chosen values of τ_p while minimizing $P(1.0)$.

4. Effects of Class-4 Pupil Functions on the MTF

Class-4 amplitude transmittance functions enhance the MTF at low spatial frequencies and suppress it at high spatial frequencies. Figure 54 illustrates the effect by comparing the perfect lens MTF to the MTFs for three different form (c) pupil functions with $c = 0.50, 0.80$, and 1.414. The degree of low-spatial-frequency MTF enhancement increases as c deeases, until it is offset by a reduction in the effective cutoff spatial freency.

Figure 55 illustrates how the effect might be used to reduce aliasing in an optical system using a sampling image sensor. A sampling aperture spacing of 25 μm is assumed, coupled with an $F = 38$ optical system operating in visible light. The dashed line represents a perfect lens MTF, and the solid line is the MTF for a form (c) pupil function with $c = 0.80$. The area under the MTF increases for data-bearing spatial frequencies below ν_s, and the MTF is reduced in the region between ν_s and the raster

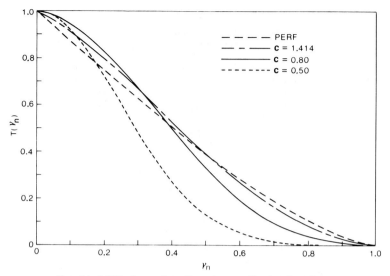

FIG. 54. MTFs for various Gaussian apodization functions.

frequency ν_r, where the most objectionable forms of aliasing can arise. There is a practical question as to whether the degree of aliasing suppression achieved is worth the loss in pupil transmittance ($\tau_p = 0.306$). A pupil function similar to (d) might offer a more acceptable balance of transmission and MTF modification.

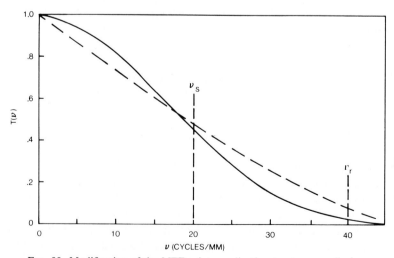

FIG. 55. Modification of the MTF using apodization to suppress aliasing.

TABLE IX

One-Number Merit Functions for Pupil Functions (a)–(d)

	(a)	(b)	(c)[a]	(d)
Q^*	0.837	0.946	0.955	0.977
τ_p	0.262	0.319	0.306	0.511
$\mathcal{D}$	0.730	0.827	0.836	0.864
$Q^{*1/2}$	0.915	0.973	0.977	0.989
$\mathcal{E}$	0.903	0.959	0.972	0.976
$\tau_p\mathcal{D}$	0.191	0.264	0.256	0.441
$(\tau_p Q^*)^{1/2}$	0.469	0.550	0.541	0.706

[a] $c = 0.80$.

5. One-Number Merit Functions

Table IX lists single-number merit functions for the four apodization functions in Table VIII. As before, the individual merit functions match fairly closely for (b) and (c), with those not involving pupil transmittance slightly favoring (c). Although nominally optimized in the same manner, function (a) shows noticeably lower performance in all cases. Function (d) shows marginally better performance for all merit functions not involving pupil transmittance, and substantially better performance where pupil transmittance is involved. For point source detection, Strehl definition gives a substantially pessimistic indication of image quality, and $Q^{*1/2}$ or $(\tau_p Q^*)^{1/2}$ should be used in its place.

Figure 56 shows the variation of these merit functions with c for pupil function (c). Here $Q^{*1/2}$ and $\mathcal{E}$ are sufficiently close in value (1–2% maximum difference) to be plotted as the same curve. Note that both Q^* and $\mathcal{E}$ exceed 1.0 for $c > 0.90$. Thus an apodized lens can exceed a perfect lens in performance by two user-related merit functions if the loss in τ_p can be offset by increased exposure time. Maximizing Q^* represents another problem in apodization analysis.

E. Apodization in the Presence of Wavefront Error

Apodization can suppress the effects of diffraction at the edge of the pupil, but it does not necessarily suppress diffraction due to wavefront error. The amount of suppression of wavefront diffraction depends upon the form of the wavefront error. Figure 57 shows three different wavefronts all having about the same rms value, $\omega \simeq 0.075$ wavelength rms, but substantially different spatial frequency characteristics. Low-spatial-frequency categories are represented by pure third-order spherical aber-

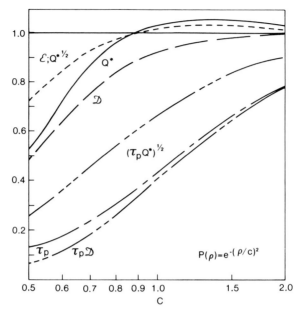

FIG. 56. Merit functions for Gaussian apodization.

ration, $W_{40} = 0.252$ wavelength OPD, and by a combination of third
and fifth-order spherical aberration, $W_{40} = 1.48$ and $W_{60} = -1.48$ wavelengths OPD. The third wavefront is a rotationally symmetric wavefront
generated from a random number table using a square smoothing function
which leaves a high-spatial-frequency residual. The PSFs were calculated

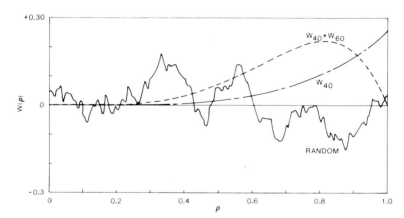

FIG. 57. Wavefront error radius profiles for wavefronts used to show the effects of combining apodization with wavefront error.

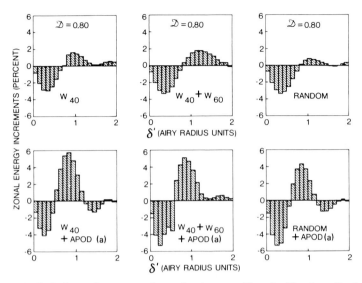

FIG. 58. ZEIs for various types of wavefront error with and without apodization.

for all three wavefronts with and without apodization of form (a) [Eq. (177)].

Figure 58 and Table X show ZEI data for each of the six cases. Data on the effects of apodization by itself are included in Table X, and the corresponding ZEI plot is in Fig. 52. The figures show ZEI for zones 0.1 Airy radius units wide to indicate detailed variations in how the energy is redistributed by wavefront error with and without apodization. The ZEI data in Table X can be used for more quantitative comparisons.

Wavefront error displaces energy outward from the Airy disk, the distance the energy is displaced being dependent on the spatial frequency

TABLE X

ZEI DATA FOR APODIZATION COMBINED WITH WAVEFRONT ERROR

$W(\rho)^a$	—	W_{40}		$W_{40} + W_{60}$		Random	
$P(\rho)^b$	(a)	—	(a)	—	(a)	—	(a)
ω	—	0.7513	0.7513	0.7517	0.7517	0.7518	0.7518
ZEI(0–1)	0.1189	−0.0953	0.0750	−0.1547	−0.0253	−0.1632	−0.0589
ZEI(1–1.8)	−0.0321	0.0473	0.0029	0.1009	0.0880	0.0227	−0.0063
ZEI(1.8–)	−0.0868	0.0480	−0.0780	0.0537	−0.0628	0.1404	0.0574

[a] See Fig. 57.
[b] See Eq. (177).

characteristics of the wavefront error. Apodization displaces energy outward from the zone $\delta' = 0$–0.5 Airy radius units and inward from the zone $\delta' = 1.2$ to infinity (see Fig. 52). Apodization plus wavefront error combines these effects but not necessarily in the manner predicted by adding the ZEI data for apodization and wavefront error taken separately. In the case of pure third-order spherical aberration, the inward transfer of energy is greater than that predicted by the sum. For $W_{40} + W_{60}$, apodization is not effective at suppressing the first bright ring ($\delta' = 1$–1.8) but is more effective than indicated by the sum in suppressing the outer parts of the PSF. In both cases, the portion of the wavefront associated with the transfer of energy which is best suppressed is that near the edge of the pupil where transmittance is the lowest. For the random wavefront error, the sum of individual ZEIs agrees more closely with the combined data.

Table XI illustrates the ineffectiveness of apodization at suppressing diffraction by higher spatial frequency wavefront error components by looking at excluded energy data. In the absence of wavefront error, the form (a) pupil function suppresses XE(20.0) by more than a factor of 30. With the random wavefront model in Fig. 57, apodization suppresses XE(20.0) by a factor of 1.5, leaving it higher than for the perfect lens value. Apodization cannot be relied on to reduce the outer ring system of the PSF unless it is accompanied by a reduction in optical surface ripple causing higher-spatial-frequency ripple on the wavefront.

Table XII lists one-number merit functions for the six examples in Fig. 58, plus the pupil function (a) by itself. In general, conclusions about one-number merit functions drawn earlier concerning apodization by itself also apply when wavefront error is present. The interaction of wavefront error effects and apodization effects can be predicted in most cases by multiplying the merit functions for wavefront error by the corresponding merit functions for apodization. The actual values will be a few percent higher than the product for Q^* and $\mathscr{E}$, and slightly lower than the product for $\mathscr{D}$. Third-order spherical aberration represents an exception to this rule, the actual merit function for the combination being substantially higher than predicted by the product. Q^* is higher for W_{40} plus apo-

TABLE XI

EXCLUDED ENERGY AT $\delta' = 20$ AIRY RADIUS UNITS FOR A
PERFECT LENS AND AN ABERRATED LENS WITH AND WITHOUT APODIZATION

	Perfect	Perfect + (a)	Random[a]	Random + (a)
XE(20.0)	0.0083	0.00027	0.020	0.0133

[a] $\omega = 0.07518$, $l = 0.125$, square smoothing function.

TABLE XII

ONE-NUMBER MERIT FUNCTIONS FOR APODIZATION COMBINED
WITH WAVEFRONT ERROR

$W(\rho)^a$	—		W_{40}		$W_{40} + W_{60}$		Random	
$P(\rho)^b$	(a)	—	(a)	—	(a)	—	(a)
ω	—	0.7513	0.7513	0.7517	0.7517	0.7518	0.7518
Q^*	0.838	0.680	0.719	0.636	0.568	0.607	0.538
τ_p	0.262	1.000	0.262	1.000	0.262	1.000	0.262
$\mathcal{D}$	0.730	0.797	0.649	0.795	0.578	0.799	0.575
$Q^{*1/2}$	0.915	0.824	0.848	0.798	0.753	0.779	0.733
$\mathcal{E}$	0.903	0.854	0.844	0.832	0.771	0.817	0.739
$\tau_p\mathcal{D}$	0.191	0.797	0.170	0.795	0.151	0.799	0.151
$(\tau_p Q^*)^{1/2}$	0.469	0.824	0.434	0.798	0.386	0.779	0.375

^{a} See Fig. 57.
^{b} See Eq. (177).

dization than for W_{40} alone. Similar results should be found with pure defocus.

F. CLASS-5 PUPIL FUNCTIONS

Class 5 is a catchall which accounts for all other forms of variation in amplitude transmittance which do not fit into the first four classes. Two examples are discussed.

1. Apodization of Obstructed Pupils

Central obstructions cannot always be avoided on occasions when suppression of the outer ring structure is necessary. Oliver (1975) has suggested the following sonine function as an apodization form:

$$P(\rho) = \left[1 - \frac{(1 + \epsilon - 2\rho)^2}{(1 - \epsilon)^2} \right]^{(n-1)}, \quad \epsilon \leq \rho \leq 1.0$$
$$= 0.0, \quad \rho < \epsilon \quad (182)$$

where n is an order number, $n = 1$ is the unapodized obstruction, and higher-order numbers produce greater apodization. Figure 59 shows $P(\rho)$ for $\epsilon = 0.325$, $n = 1, 2, 3$. Order numbers higher than 3 produce a narrower transmission band and are not of great interest.

Figure 60 shows displaced energy within 3 Airy radius units of the center of the PSF. It indicates the price paid in transfer of energy to the first bright ring for suppression of the outer ring structure. Table XIII lists

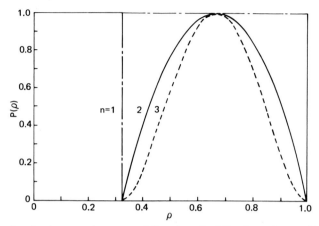

FIG. 59. Amplitude transmittance pupil functions for Oliver's sonine apodization function for a central obstruction of diameter ratio $\epsilon = 0.325$.

the one-number merit functions and shows the degree of suppression in the outer ring structure by giving excluded energy values at 10 and 20 Airy radius units. The degree of suppression is impressive, in the absence of wavefront error. In view of the above comments about wavefront error, there is nothing to be gained by going beyond $n = 2$.

Figure 61 shows the MTFs for the obstructed lens and the two apodization functions in Fig. 59. The two apodization functions reduce the

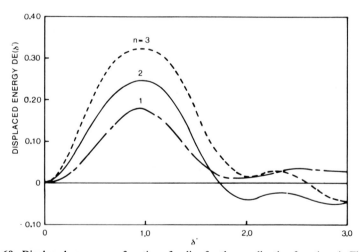

FIG. 60. Displaced energy as a function of radius for the apodization functions in Fig. 59.

TABLE XIII

ONE-NUMBER MERIT FUNCTIONS AND
EXCLUDED ENERGY FOR AN OBSTRUCTED PUPIL ($\epsilon = 0.325$)
WITH SONINE APODIZATION

n	1^a	2	3
$Q*$	0.733	0.569	0.430
τ_p	0.894	0.477	0.363
$\mathcal{D}$	0.894	0.745	0.626
$Q*^{1/2}$	0.856	0.793	0.699
$\mathcal{E}$	0.884	0.793	0.699
$\tau_p\mathcal{D}$	0.800	0.356	0.228
$(\tau_p Q*)^{1/2}$	0.810	0.521	0.395
XE(10.0)	0.0251	0.00034	0.00003
XE(20.0)	0.0126	0.00004	b

^{a} $n = 1$ corresponds to the central obstruction alone.
^{b} Less than computer roundoff.

MTF substantially, but the losses for $n = 2$ should be tolerable if the concomitant loss in pupil transmittance is acceptable.

2. Inadvertent Apodization

Inadvertent variations in the amplitude transmittance pupil function can arise in some optical systems as a result of environmental effects

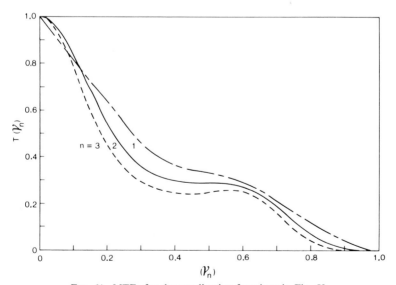

FIG. 61. MTFs for the apodization functions in Fig. 59.

damaging mirror and antireflection coatings. To show the magnitude of
the effects which might occur in extreme cases, consider a sinusoidal amplitude transmittance function of the form

$$P(\rho) = a_1 + a_2 \cos(2\pi b\rho) \qquad (183)$$

where $a_1 + a_2 \equiv 1.0$ and b is the spatial frequency in cycles per radius.
The modulation of this function ranges from zero when $a_1 = 1.0$ and $a_2 =$
0.0, to unity when $a_1 = a_2 = 0.50$. It can thus represent the most extreme
range of situations encountered.

Figure 62 shows the PSF and encircled energy for $a_1 = a_2 = 0.50$ and
$b = 5$ cycles per radius. The effect produced is similar to that produced
by a phase grating, but with no higher-order sidebands. Thus the effects
produced by random variations in the amplitude transmittance should be
similar to those produced by random wavefront error; Fig. 62 represents
the worst possible image degradation encountered.

Table XIV lists one-number merit functions for three examples covering the full range of modulations possible. For $a_2 < 0.25$, the most significant loss is still that of transmittance.

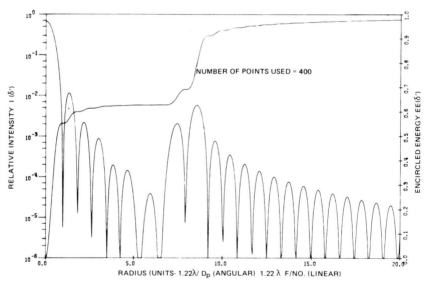

FIG. 62. The PSF for sinusoidal amplitude transmittance variations, where $a_1 = 0.50$,
$a_2 = 0.50$, and $b = 5.0$ cycles per radius.

TABLE XIV

One-Number Merit Functions for the
Apodization Function of Eq. (183) with
$b = 5$ Cycles per Radius

a_1	0.95	0.75	0.50
a_2	0.05	0.25	0.50
Q^*	0.996	0.889	0.420
τ_p	0.904	0.594	0.375
$\mathscr{D}$	0.999	0.947	0.667
$Q^{*1/2}$	0.998	0.943	0.648
$\mathscr{E}$	0.998	0.943	0.652
$\tau_p \mathscr{D}$	0.903	0.563	0.250
$(\tau_p Q^*)^{1/2}$	0.949	0.726	0.397
DE(1.0)	0.0016	0.047	0.285

XI. SOURCES AND EFFECTS OF IMAGE MOTION

Image motion is not a product of the optical system itself but involves motion of the image relative to the image sensor induced by outside influences. It is still a factor which must be considered in image quality analysis, since it can affect the choice of optical configuration. The analysis of image motion and image motion compensation techniques is a major topic and is not dealt with here. Our course is to comment on the forms and sources of image motion most commonly encountered and to model the effects of image motion on image quality.

A. Forms and Sources of Image Motion

The forms of image motion encountered depend upon the type of imaging system and its application. The range of image motion forms is well represented by the two extremes encountered in aerial reconnaissance and astronomy.

In aerial reconnaissance, the camera and object surface are moving rapidly with respect to each other. High-performance reconnaissance cameras usually have elaborate image motion compensation systems built into their film transports, which reduce the effective image motion to a small residual within the exposure interval. The result is an image motion residual which varies from point to point in the image surface but which is similar in character at all points. In general, the exposure interval is far shorter than the period of any random image motions, and the image moves in a single direction, or at least along a segment of a definable path. Thus the effects on image quality of a specifically definable form of image motion must be analyzed.

In astronomy, the relative motions of object surface and optical system are slow, but they are still small fractions of the exposure interval. Residual motion arises primarily from atmospheric turbulence for ground-based telescopes, and from pointing instabilities for both ground-based and satellite-borne telescopes. Motion of the image is typically two-dimensional, and the exposure interval is many times the characteristic periods of most of the image motion components. The effects of this type of image motion are most conveniently analyzed statistically.

Techniques for analyzing the specific forms of image motion encountered with each type of aerial reconnaissance camera are beyond the scope of this chapter. The interested reader is directed to Jensen (1968) who gives a good introductory review, and to Brown (1969) who defines calculation techniques for computing image motion from the camera's motion relative to the ground. For present purposes, the residual image motion is assumed to be either linear or sinusoidal.

For statistical analysis of two-dimensional image motion, the Gaussian model for random walk is used, assuming rotational symmetry is approached over the long exposure interval. The discussion presented here is an adaptation and updating of material from Solomon and Wetherell (1973).

B. Effects on the PSF, MTF, and One-Number Merit Functions

Image motion does not interact with other factors degrading image quality in a coherent manner and thus can never compensate for other forms of image degradation; it can only add to the loss in image quality.

For purposes of calculating image quality, image motion can be represented by a PSF, which is a record of the motion of a point source object during the exposure interval, weighted according to its velocity at each point along the trajectory of motion. Convolving the image motion PSF with the optical system PSF gives the combined effective PSF for the recorded image. Alternatively, taking the Fourier transform of the image motion PSF gives an image motion OTF or MTF degradation function which can be multiplied by the optical system OTF. The procedure is identical to that used in treating sampling apertures. In general, it is assumed that the intensity of the point source remains fixed during the exposure interval. If the intensity of the point source fluctuates rapidly enough, it will affect the velocity weighting function used in generating the image motion PSF.

The image motion PSF can be quite irregular in a real situation, affecting both the MTF and phase transfer function. The examples considered here have symmetry which allows them to be treated solely in terms of their MTF degradation functions.

1. MTF Degradation Functions for Image Motion

If the image motion lies totally along a straight line, it will cause a loss in modulation for spatial frequency components modulated along that line. At right angles to that line, their will be no image motion MTF degradation. At angles between these extremes, the image degradation varies according to the cosine of the angle away from the image motion line.

Consider two forms of image motion along a straight line, linear and sinusoidal. In linear image motion, the image point moves a distance $a = vt_e$, where v is the image velocity and t_e is the exposure interval. In sinusoidal image motion, the image point oscillates a distance $\pm a'$ about a fixed point. In both cases, the motion is assumed to lie along a line at an angle γ from the y' axis. For linear image motion, the MTF degradation function is

$$T_{lm}(v',\alpha) = \frac{\sin[\pi a v' \cos(\alpha - \gamma)]}{\pi a v' \cos(\alpha - \gamma)} \qquad (184)$$

Treatment of the sinusoidal image motion depends upon the relationship between the period of the motion and t_e. The MTF degradation function given here is valid if the period is an exact divisor of t_e, or if the period is so much smaller than t_e that assuming it is an exact divisor will produce tolerable errors. For sinusoidal motion,

$$T_{sm}(v',\alpha) = J_0[2\pi a' v' \cos(\alpha - \gamma)] \qquad (185)$$

where $J_0[\ \ ]$ is the zero-order Bessel function of the first kind.

Two-dimensional image motion is modeled on the assumption of random-walk statistics, with σ_m being the standard deviation of the image point motion. The image motion PSF is taken to be rotationally symmetric and of Gaussian form:

$$I_m(r) = \exp(-r^2/2\sigma_m^2) \qquad (186)$$

The units for σ_m may be radians if referenced to object space, or units of length if referenced to the image surface. In writing the image motion MTF degradation function, it is convenient to normalize σ_m by multiplying by the cutoff spatial frequency v_0:

$$\begin{aligned}
\sigma &= \sigma_m \nu_0 \\
&= \sigma_m / \lambda F, \quad (\sigma_m \text{ in linear units}) \\
&= \sigma_m D_p / \lambda, \quad (\sigma_m \text{ in radians})
\end{aligned} \right\} \qquad (187)$$

The MTF degradation function for Gaussian image motion is then

$$T_\sigma(\nu_n) = \exp(-2\pi^2 \sigma^2 \nu_n^2) \qquad (188)$$

Equation (188) is used to model image motion in all the numerical examples which follow.

Examination of Eqs. (184), (185), and (188) shows the primary effect of image motion to be a suppression of modulation, which grows progressively worse as spatial frequency increases. Note that for both linear and sinusoidal image motion, the MTF degradation function can go negative, indicating spurious resolution. The same property was noted with square and circular sampling apertures in Section VII,E. Figure 63 shows the effect of Gaussian image motion on the optical system MTF for values of σ from 0.1 to 0.5.

2. The Effects of Image Motion on the PSF

Image motion by its very nature cannot displace energy far without leveling the Airy disk completely. Small amounts of image motion fill in the minima in the ring system by smearing out the bright rings, as shown by the ZEI plot in Fig. 64. Most of the energy transfer occurs between the center and edge of the Airy disk. The periodic form of the ZEI outside the

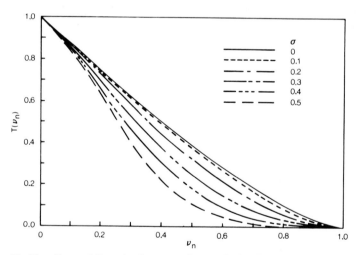

Fig. 63. The effects of Gaussian image motion on the MTF, where σ is the normalized image motion as defined by Eqs. (186) and (187).

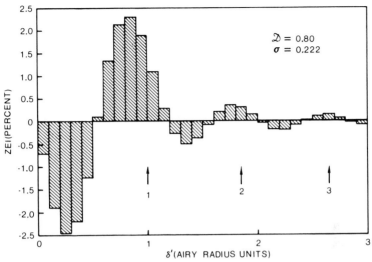

FIG. 64. ZEIs for an amount of Gaussian image motion reducing the Strehl definition to 80%. The arrows indicate the first three minima in the perfect lens PSF.

Airy disk shows the filling in of the minima (arrows) by broadening of the adjacent bright rings.

3. The Effects of Image Motion on One-Number Merit Functions

Image motion has a strong effect on MTF-based merit functions but little effect on merit functions related to encircled energy where $\delta' > 0.8$ Airy radius units. Figure 65 illustrates these differences by plotting several merit functions against σ. Note that, when $\mathcal{D} = 0.80$, the displaced energy DE(1.0) is less than 1%. Even when $\mathcal{D} = 0.50$, DE(1.0) < 8%. Figure 65 also shows a strong discrepancy between $\mathcal{D}$ and $Q^{*1/2}$, indicating that the former is a poor measure of point source detectability when image motion is present.

As with wavefront error and central obstruction diameter ratio, it is convenient to model the effects of image motion on the merit functions with simple equations. The following expressions approximate the data presented in Fig. 65. For Strehl definition (see Appendix C),

$$\begin{aligned} \mathcal{D}_\sigma &\simeq 1/(1 + 5.5\sigma^2), \qquad \sigma < 1.0 \\ &\simeq \exp(-4.3\sigma^2), \qquad \sigma < 0.3 \end{aligned} \qquad (189)$$

For relative edge response,

$$\mathcal{E}_\sigma \simeq \exp(-2.2\sigma^2), \qquad \sigma < 0.35 \qquad (190)$$

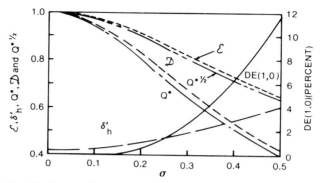

FIG. 65. One-number merit functions for Gaussian image motion.

And for normalized noise equivalent structural content,

$$Q_\sigma^* \simeq 1/(1 + 6\sigma^2), \qquad \sigma < 0.5$$
$$\simeq \exp(-5.5\sigma^2), \qquad \sigma < 0.3 \qquad (191)$$

Image motion is one form of image degradation which can be well represented by the half-power diameter of the PSF. The half-power diameter, when expressed as a fraction of the diameter of the first minimum in the perfect lens PSF, is numerically equal to the radius δ_h' in Airy radius units at which the relative intensity

$$I(\delta_h') = \mathscr{D}/2.0 \qquad (192)$$

Figure 65 includes a curve of δ_h' versus σ, which may be approximated by the equation

$$\delta_h' \simeq 0.96(0.19 + \sigma^2)^{1/2} \qquad (193)$$

The actual half-power diameter d_h' is then

$$d_h' \simeq 2.35\lambda F(0.19 + \sigma^2)^{1/2} \text{ (linear units)}$$
$$\approx 2.35\lambda(0.19 + \sigma^2)^{1/2}/D_p \text{ (radians)} \qquad (194)$$

For $\sigma \geqslant 1.0$, this reduces to $2.35\sigma_m$.

XII. MERIT FUNCTION MODELS FOR
PRELIMINARY ANALYSES

When making performance expectation calculations during the early stages of choosing parameters for high-quality optical systems, it is useful to have approximate models for estimating image quality. Rough estimates of the MTF can be made using the MTF degradation function models given in this chapter, coupled with the equation for the MTF of a lens with a centered circular obstruction given in Appendix B. Models for Strehl definition and other merit functions have been developed by the author (Solomon and Wetherell, 1973) and are presented in this section in somewhat modified form.

These models present the lens system in normalized form, representing it by its exit pupil and defining it with four *image quality parameters*, the rms wavefront error ω, the correlation length l, the central obstruction diameter ratio ϵ, and the normalized Gaussian image motion σ. To relate these models to actual optical system design parameters, it is necessary to know the wavelength λ, the focal ratio F, and the entrance pupil diameter D_p. The relationships between these three design parameters and the image quality parameters have been given earlier.

There is one basic reservation which should be kept in mind in using these models. All assume that the effects of the image quality parameters are separable and do not interact, so that it is valid to state, for example, that $Q^* = Q_\omega^* \times Q_\epsilon^* \times Q_\sigma^*$. This assumption is only approximately true in general and can be wrong in specific cases, as when long-correlation-length wavefront error is present. The models should be considered valid only for small amounts of each image quality parameter and at best are accurate only to within a few percent.

For more accurate modeling, it is preferable to construct synthetic pupil functions and compute the PSF or OTF by Fourier transformation or autocorrelation. Statistical models relevant to the particular application can then be generated by using a number of different synthetic pupils and averaging the resultant PSFs or OTFs.

A. STREHL DEFINITION $\mathscr{D}$

Strehl definition models for the individual image quality parameters were given in Eqs. (161), (169), and (189). Correlation length is of less significance unless very long and has not been included in the model. Two different Strehl definition models can be written. An entirely nonexponential model can be written using Eq. (86) in place of Eq. (161):

$$\mathscr{D} \simeq (1 - \epsilon^2)(1 - 4\pi^2\omega^2)/(1 + 5.5\sigma^2)$$
$$\epsilon < 0.7, \qquad \sigma < 0.3, \qquad \omega < 0.12, \qquad \mathscr{D} > 0.40 \qquad (195)$$

For use with scientific calculators, an all-exponential form is convenient:

$$\mathscr{D} \simeq \exp[-(4\pi^2\omega^2 + \epsilon^2 + 4.3\sigma^2)]$$
$$\epsilon < 0.6, \qquad \sigma < 0.3, \qquad \omega < 0.12, \qquad \mathscr{D} > 0.40 \qquad (196)$$

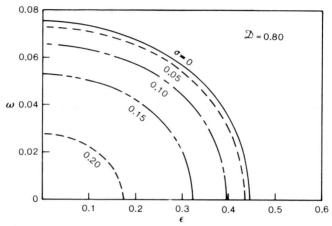

FIG. 66. Combinations of wavefront error ω, image motion σ, and central obstruction diameter ratio ϵ which just meet Maréchal's criterion for Strehl definition [generated with Eq. (196)].

Interactions between image quality parameters restrict the range of $\mathscr{D}$ over which the equations can apply. (See Appendix C for further comments.)

The performance criterion most commonly associated with the Strehl ratio is Maréchal's criterion for diffraction-limited performance, $\mathscr{D} \geq 0.80$. Figure 66 plots the combinations of image quality parameters which just meet the criterion. Table XV lists image quality parameters which meet the criterion individually or in balanced combinations.

B. EE(1.0) AND ITS RELATIONSHIP TO STREHL DEFINITION

The Strehl ratio is often taken as an indication of the energy distribution in the PSF. Some of the shortcomings of this assumption have

TABLE XV

VALUES OF ω, ϵ, AND σ WHICH PRODUCE A STREHL DEFINITION
OF 80% INDIVIDUALLY OR IN EQUALLY WEIGHTED COMBINATIONS

	ω	ϵ	σ	$\omega + \epsilon^a$	$\omega + \epsilon + \sigma^b$
ω	0.0752	—	—	0.0532	0.0434
ϵ	—	0.447	—	0.325	0.268
σ	—	—	0.22	—	0.12

[a] $\mathscr{D}_\omega = \mathscr{D}_\epsilon = (0.80)^{1/2} = 0.894$.
[b] $\mathscr{D}_\omega = \mathscr{D}_\epsilon = \mathscr{D}_\sigma = (0.80)^{1/3} = 0.928$.

already been illustrated by plotting ZEIs for $\mathcal{D} \simeq 0.80$ for a central obstruction (Fig. 13), for Gaussian apodization (Fig. 52), for various forms of wavefront error (Fig. 58), and for image motion (Fig. 64). One particular measure of energy distribution commonly associated with Strehl definition is EE(1.0). Since EE(1.0) is sometimes used as a one-number merit function, this relationship is worth closer examination.

Biberman (1973, p. 25) states that EE(1.0) is about 0.84 times the Strehl ratio. He appears to be referring to the actual irradiance ratio, so that in our notation his statement becomes

$$\text{EE}(1.0) \simeq 0.84\tau_\text{p}\mathcal{D} \tag{197}$$

While Eq. (197) is approximately correct in some instances and almost always represents the minimum energy in the Airy disk, it can be very pessimistic in many other instances. Figure 67 illustrates the scale of the discrepancies.

Equation (197) holds well for random wavefront error when $l < 0.10$, and for central obstructions when $\epsilon < 0.50$. It begins to break down for long-correlation-length wavefront error such as defocus (W_{20}) or third-order spherical aberration (W_{40}.) There is little correlation between $\mathcal{D}$ and EE(1.0) for image motion and a negative correlation for apodization. Substituting $\mathcal{D}$ for $\tau_\text{p}\mathcal{D}$ makes things worse. Thus Eq. (197) must be used with considerable circumspection.

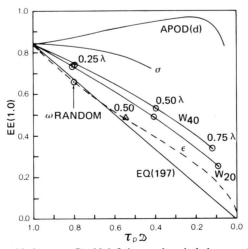

FIG. 67. Relationship between Strehl definiton and encircled energy EE(1.0) for various forms of image quality degradation.

C. Normalized Noise Equivalent Structural Content Q^*

Models of Q^* for the individual image quality parameters were given in Eqs. (163), (170), and (191). As with the Strehl definition, correlation length is not a critical factor and has been ignored in generating this model. Based on the limits of accuracy for the Strehl definition models of Eqs. (196) and (197), only an all-exponential model is suggested here:

$$Q^* \simeq \exp[-(8\pi^2\omega^2 + 3\epsilon^2 + 5.5\sigma^2)]$$
$$\epsilon < 0.6, \qquad \sigma < 0.3, \qquad \omega < 0.12, \qquad Q^* > 0.20 \qquad (198)$$

The value of Q^* most closely corresponding to Maréchal's criterion is $Q^* = 0.64$. Figure 68 and Table XVI show the combinations of ω, ϵ, and σ which just satisfy this criterion. Comparison to Fig. 67 and Table XV shows Q^* to be more sensitive to ϵ and less sensitive to σ than the Strehl definition.

Some reservations must be mentioned concerning the accuracy of this model. A general reservation has been stated concerning the assumption that the effects of individual image quality parameters are separable. In addition, Q^* is sensitive to the variance in the MTF, where Strehl definition is not, because Q^* is computed from the volume under the square of the MTF. Equation (198) may still be used for preliminary estimates, but modeling with synthetic wavefronts to compute actual MTF functions may be preferred when accuracy is desired.

D. Relative Edge Response $\mathscr{E}$

Relative edge response models for the individual image quality parameters were given in Eqs. (164), (171), and (190). Relative edge response is sensitive to correlation length which has been incorporated into Eq. (164). The combined model for relative edge response is

TABLE XVI

Values of ω, ϵ, and σ Which Produce $Q^* = 0.64$
Individually or in Equally Weighted Combinations

	ω	ϵ	σ	$\omega + \epsilon^a$	$\omega + \epsilon + \sigma^b$
ω	0.0752	—	—	0.0532	0.0434
ϵ	—	0.386	—	0.273	0.223
σ	—	—	0.306	—	0.164

[a] $Q_\omega^* = Q_\epsilon^* = (0.64)^{1/2} = 0.800$.
[b] $Q_\omega^* = Q_\epsilon^* = Q_\sigma^* = (0.64)^{1/3} = 0.862$.

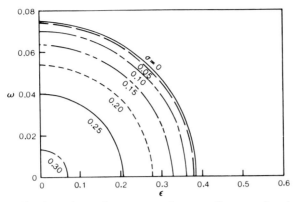

Fig. 68. Combinations of wavefront error ω, image motion σ, and central obstruction diameter ratio ϵ producing $Q^* = 0.64$ [generated with Eq. (198)].

$$\mathcal{E} \simeq \exp\{-[4\pi^2\omega^2/(1 + l) + 1.2\epsilon^2 + 2.2\sigma^2]\}$$
$$\epsilon < 0.6, \qquad \sigma < 0.35, \qquad \omega < 0.12 \tag{199}$$

On the basis of wavefront error, $\mathcal{E} \geqq 0.80$ may be considered the equivalent of Maréchal's criterion. No figure has drawn showing combinations of image quality parameters producing this value of relative edge response, because of the added complexity of the fourth parameter. A comparison of Eqs. (196) and (199) shows relative edge response slightly more sensitive to ϵ than Strehl definition, and considerably less sensitive to σ.

APPENDIX A. LERMAN'S EQUATION FOR THE VARIANCE OF AN OBSTRUCTED PUPIL

Lerman (1969a) has derived an equation for the variance V in the aberration polynomial in Eq. (35) when there is a centered circular obstruction of diameter ratio ϵ. Lerman's equation is reproduced here using the wave aberration coefficients in Table II.

$$\omega^2 = V = \frac{W_{20}^2}{12} (1 - 2\epsilon^2 + \epsilon^4) + \frac{W_{20}W_{40}}{6} (1 - \epsilon^2 - \epsilon^4 + \epsilon^6)$$

$$+ \frac{W_{20}W_{60}}{20} (3 - 2\epsilon^2 - 2\epsilon^4 - 2\epsilon^6 + 3\epsilon^8)$$

$$+ \frac{W_{20}W_{80}}{15} (2 - \epsilon^2 - \epsilon^4 - \epsilon^6 - \epsilon^8 + 2\epsilon^{10})$$

(*Equation continues*)

$$+ \frac{W_{20}W_{22}}{12}(1 - 2\epsilon^2 + \epsilon^4) + \frac{W_{20}W_{42}}{15}(1 - \epsilon^2 - \epsilon^4 + \epsilon^6)$$

$$+ \frac{W_{40}^2}{45}(4 - \epsilon^2 - 6\epsilon^4 - \epsilon^6 + 4\epsilon^8)$$

$$+ \frac{W_{40}W_{60}}{6}(1 - \epsilon^4 - \epsilon^6 + \epsilon^{10})$$

$$+ \frac{W_{40}W_{80}}{105}(16 + 2\epsilon^2 - 12\epsilon^4 - 12\epsilon^6 - 12\epsilon^8 + 2\epsilon^{10} + 16\epsilon^{12})$$

$$+ \frac{W_{40}W_{22}}{12}(1 - \epsilon^2 - \epsilon^4 + \epsilon^6)$$

$$+ \frac{W_{40}W_{42}}{45}(4 - \epsilon^2 - 6\epsilon^4 - \epsilon^6 + \epsilon^8)$$

$$+ \frac{W_{60}^2}{112}(9 + 2\epsilon^2 - 5\epsilon^4 - 12\epsilon^6 - 5\epsilon^8 + 2\epsilon^{10} + 9\epsilon^{12})$$

$$+ \frac{W_{60}W_{80}}{20}(3 + \epsilon^2 - \epsilon^4 - 3\epsilon^6 - 3\epsilon^8 - \epsilon^{10} + \epsilon^{12} + 3\epsilon^{14})$$

$$+ \frac{W_{60}W_{22}}{40}(3 - 2\epsilon^2 - 2\epsilon^4 - 2\epsilon^6 + 3\epsilon^8)$$

$$+ \frac{W_{60}W_{42}}{12}(1 - \epsilon^4 - \epsilon^6 + \epsilon^{10})$$

$$+ \frac{W_{80}^2}{225}(16 + 7\epsilon^2 - 2\epsilon^4 - 11\epsilon^6 - 20\epsilon^8 - 11\epsilon^{10} - 2\epsilon^{12} + 7\epsilon^{14} + 16\epsilon^{16})$$

$$+ \frac{W_{80}W_{22}}{30}(2 - \epsilon^2 - \epsilon^4 - \epsilon^6 - \epsilon^8 + 2\epsilon^{10})$$

$$+ \frac{W_{80}W_{42}}{105}(8 + \epsilon^2 - 6\epsilon^4 - 6\epsilon^6 - 6\epsilon^8 + \epsilon^{10} + 8\epsilon^{12})$$

$$+ \frac{W_{11}^2}{4}(1 + \epsilon^2) + \frac{W_{11}W_{33}}{4}(1 + \epsilon^2 + \epsilon^4)$$

$$+ \frac{W_{11}W_{31}}{3}(1 + \epsilon^2 + \epsilon^4) + \frac{W_{11}W_{51}}{4}(1 + \epsilon^2 + \epsilon^4 + \epsilon^6)$$

$$+ \frac{W_{22}^2}{16}(1 + \epsilon^4) + \frac{W_{22}W_{42}}{48}(5 + \epsilon^2 + \epsilon^4 + 5\epsilon^6)$$

$$+ \frac{5W_{33}^2}{64}(1 + \epsilon^2 + \epsilon^4 + \epsilon^6) + \frac{3W_{33}W_{31}}{16}(1 + \epsilon^2 + \epsilon^4 + \epsilon^6)$$

$$+ \frac{3W_{33}W_{51}}{20}(1 + \epsilon^2 + \epsilon^4 + \epsilon^6 + \epsilon^8) + \frac{W_{31}^2}{8}(1 + \epsilon^2 + \epsilon^4 + \epsilon^6)$$

$$+ \frac{W_{31}W_{51}}{5}(1 + \epsilon^2 + \epsilon^4 + \epsilon^6 + \epsilon^8)$$

(Equation continues)

$$+ \frac{W_{42}^2}{360} (17 + 7\epsilon^2 - 3\epsilon^4 + 7\epsilon^6 + 17\epsilon^8)$$

$$+ \frac{W_{51}^2}{12} (1 + \epsilon^2 + \epsilon^4 + \epsilon^6 + \epsilon^8 + \epsilon^{10}) \tag{200}$$

APPENDIX B. THE MTF AND PSF FOR A PERFECT LENS WITH A CENTRAL OBSTRUCTION

O'Neill (1956) has derived an equation for the MTF of a perfect lens with a circular pupil and a centered circular obstruction of diameter ratio ϵ. The form given here has been adapted to the present notation, and O'Neill's erratum has been incorporated.

$$T(\nu_n) = (A + B + C)/(1 - \epsilon^2) \tag{201}$$

where

$$A = \frac{2}{\pi} [\arccos \nu_n - \nu_n (1 - \nu_n^2)^{1/2}], \qquad 0 \le \nu_n \le 1.0$$

$$= 0, \qquad \nu_n > 1.0$$

$$B = \frac{2\epsilon^2}{\pi} \left\{ \arccos \left(\frac{\nu_n}{\epsilon} \right) - \left(\frac{\nu_n}{\epsilon} \right) \left[1 - \left(\frac{\nu_n}{\epsilon} \right)^2 \right]^{1/2} \right\}, \qquad 0 \le \frac{\nu_n}{\epsilon} \le 1.0$$

$$= 0, \qquad \frac{\nu_n}{\epsilon} > 1.0$$

$$C = -2\epsilon^2, \qquad 0 \le \nu_n \le (1 - \epsilon)/2$$

$$= \frac{2}{\pi} \left\{ \epsilon \sin \phi + \frac{\phi}{2} (1 + \epsilon^2) - (1 - \epsilon^2) \arctan \left(\frac{1 + \epsilon}{1 - \epsilon} \tan \frac{\phi}{2} \right) \right\} - 2\epsilon^2,$$

$$\frac{(1 - \epsilon)}{2} < \nu_n < \frac{(1 + \epsilon)}{2}$$

$$= 0, \qquad \nu_n \ge \frac{(1 + \epsilon)}{2}$$

and

$$\phi = \arccos \left[\frac{(1 + \epsilon^2 - 4\nu_n^2)}{2\epsilon} \right]$$

The PSF for this lens is also rotationally symmetric, having a relative intensity $I_\epsilon(q')$ as a function of radius q' given by

$$I_\epsilon(q') = \frac{1}{1 - \epsilon^2} \left[\frac{2J_1(\pi q'/\lambda F)}{\pi q'/\lambda F} - \frac{2\epsilon^2 J_1(\pi \epsilon q'/\lambda F)}{\pi \epsilon q'/\lambda F} \right]^2 \tag{202}$$

TABLE XVII

THE STREHL DEFINITION FOR COMBINATIONS OF CENTRAL OBSTRUCTION DIAMETER RATIO ϵ AND GAUSSIAN IMAGE MOTION σ AS DEFINED IN EQS. (186) AND (187)[a]

	ϵ									
σ	0.0	0.05	0.10	0.15	0.20	0.25	0.30	0.35	0.40	0.45
0.00	1.000	1.000	1.000	1.000	1.000	1.000	1.000	1.000	1.000	1.000
0.05	0.988	0.988	0.988	0.987	0.987	0.987	0.987	0.986	0.986	0.985
0.10	0.953	0.952	0.952	0.952	0.951	0.950	0.948	0.947	0.945	0.943
0.15	0.898	0.898	0.898	0.896	0.895	0.892	0.890	0.887	0.883	0.879
0.20	0.831	0.831	0.830	0.828	0.825	0.822	0.817	0.813	0.807	0.801
0.25	0.756	0.756	0.754	0.752	0.748	0.744	0.738	0.732	0.724	0.716
0.30	0.680	0.679	0.678	0.674	0.670	0.665	0.658	0.650	0.642	0.632
0.35	0.606	0.606	0.603	0.600	0.595	0.589	0.582	0.573	0.564	0.554
0.40	0.538	0.537	0.535	0.531	0.526	0.520	0.512	0.503	0.494	0.483
0.45	0.476	0.476	0.473	0.469	0.464	0.458	0.450	0.442	0.433	0.423
0.50	0.422	0.421	0.419	0.415	0.410	0.404	0.397	0.389	0.380	0.371
0.55	0.375	0.374	0.372	0.368	0.364	0.358	0.351	0.344	0.336	0.328
0.60	0.334	0.333	0.331	0.328	0.323	0.318	0.312	0.306	0.299	0.293
0.65	0.298	0.298	0.296	0.293	0.289	0.285	0.279	0.274	0.268	0.263
0.70	0.268	0.267	0.265	0.263	0.260	0.256	0.251	0.247	0.242	0.237
0.75	0.241	0.241	0.239	0.237	0.234	0.231	0.227	0.223	0.220	0.216
0.80	0.218	0.218	0.216	0.215	0.212	0.209	0.206	0.203	0.201	0.198
0.85	0.198	0.198	0.197	0.195	0.193	0.191	0.188	0.186	0.184	0.182
0.90	0.181	0.180	0.179	0.178	0.176	0.175	0.173	0.171	0.169	0.168
0.95	0.165	0.165	0.164	0.163	0.162	0.160	0.159	0.158	0.157	0.156
1.00	0.152	0.152	0.151	0.150	0.149	0.148	0.147	0.146	0.145	0.145

	ϵ									
σ	0.50	0.55	0.60	0.65	0.70	0.75	0.80	0.85	0.90	0.95
0.00	1.000	1.000	1.000	1.000	1.000	1.000	1.000	1.000	1.000	1.000
0.05	0.985	0.984	0.983	0.983	0.982	0.981	0.980	0.979	0.979	0.981
0.10	0.941	0.939	0.936	0.933	0.930	0.927	0.924	0.920	0.917	0.916
0.15	0.875	0.870	0.865	0.859	0.853	0.847	0.841	0.834	0.827	0.823
0.20	0.794	0.787	0.779	0.771	0.762	0.752	0.743	0.733	0.723	0.715
0.25	0.708	0.698	0.688	0.677	0.666	0.655	0.643	0.630	0.618	0.608
0.30	0.622	0.611	0.599	0.587	0.575	0.562	0.549	0.536	0.524	0.513
0.35	0.543	0.531	0.519	0.507	0.494	0.481	0.468	0.456	0.443	0.433
0.40	0.472	0.461	0.449	0.437	0.425	0.413	0.402	0.390	0.379	0.370
0.45	0.412	0.402	0.391	0.380	0.369	0.359	0.348	0.338	0.328	0.321
0.50	0.362	0.352	0.343	0.333	0.324	0.315	0.306	0.297	0.289	0.283
0.55	0.320	0.312	0.304	0.296	0.288	0.280	0.273	0.265	0.258	0.253
0.60	0.286	0.279	0.272	0.265	0.259	0.252	0.246	0.240	0.234	0.229
0.65	0.257	0.251	0.246	0.240	0.235	0.229	0.224	0.218	0.212	0.203
0.70	0.233	0.228	0.224	0.219	0.215	0.210	0.206	0.201	0.196	0.187
0.75	0.212	0.209	0.205	0.202	0.198	0.194	0.190	0.186	0.181	0.174

(*Continued*)

TABLE XVII (*Continued*)

σ	0.50	0.55	0.60	0.65	0.70	0.75	0.80	0.85	0.90	0.95
					ε					
0.80	0.195	0.192	0.189	0.187	0.184	0.181	0.177	0.174	0.169	0.163
0.85	0.180	0.178	0.176	0.173	0.171	0.169	0.166	0.163	0.159	0.153
0.90	0.166	0.165	0.164	0.162	0.160	0.158	0.156	0.153	0.150	0.144
0.95	0.155	0.154	0.153	0.152	0.150	0.149	0.147	0.144	0.141	0.136
1.00	0.144	0.144	0.143	0.143	0.142	0.140	0.139	0.137	0.134	0.131

[a] From Mahajan, 1978.

Equation (202) is normalized so that $I_\epsilon(0)$ is the Strehl definition.

APPENDIX C. IMAGE MOTION IN OBSTRUCTED LENSES

A study of image motion for lenses with central obstructions, by Mahajan (1978), came to the present author's attention while completing the final draft for this chapter. Mahajan's results give quantitative illustrations of the interaction between image motion and aperture obstructions and throw some light on the validity of the Strehl definition model in Eqs. (195) and (196). Mahajan's Table 1 is reproduced here as Table XVII with his gracious consent.

Table XVII lists the Strehl definition for various combinations of central obstruction diameter ratio ϵ and Gaussian image motion σ, where σ is defined as in Eq. (187). The Strehl definition has been normalized to 1.0 at $\sigma = 0.0$ for each value of ϵ, which simplifies the task of comparing Strehl definition losses at different obstruction diameter ratios. Examination of any row in Table XVII shows that, for fixed σ, the larger the value of ϵ, the smaller the Strehl definition. Thus for applications where Strehl definition is an adequate measure of image quality, lenses with large aperture obstructions are more sensitive to image motion than lenses with small obstructions.

The source of the above effect seems to lie in the fact that a large central obstruction narrows the diameter of the central maximum in the PSF, so that lateral smearing can reduce the peak intensity more than with an unobstructed lens. If so, there should be less interaction between image motion and wavefront error, since the latter does not normally reduce the diameter of the central maximum (Wilkin's apodization being the exception).

ACKNOWLEDGMENTS

The author would like to extend his thanks to three individuals and two institutions whose aid and support were instrumental in completing this chapter. Ira M. Egdall wrote and operated the computer programs used to generate much of the data presented in this chapter and deserves special thanks. A careful review and criticism of the first draft by John M. Vanderhoff has been very helpful. The chapter could not have been written at all without the support of Richard J. Wollensak and Itek Corporation. Finally, this report had its beginnings in two study contracts from the National Aeronautics and Space Administration related to the space telescope program, contracts NASw-2313 and NAS8-30639. The final reports from these studies (Solomon and Wetherell, 1973; Wetherell, 1974a) and a paper derived from the earliest study (Wetherell, 1974b) form the foundation of this chapter.

REFERENCES

Abbe, E. (1873). *Schultzes. Arch. mikrrosk. Anat.* **9**, 413.
Barakat, R. (1962a). *J. Opt. Soc. Am.* **52**, 264.
Barakat, R. (1962b). *J. Opt. Soc. Am.* **52**, 276.
Barakat, R. (1963). *J. Opt. Soc. Am.* **53**, 274.
Barakat, R. (1971). *Opt. Acta* **18**, 683.
Barakat, R., and Blackman, E. (1973). *Opt. Acta* **20**, 901.
Barakat, R., and Houston, A. (1966). *Appl. Opt.* **5**, 1850.
Barnes, W. P., Jr., (1966). *Appl. Opt.* **5**, 701.
Barnes, W. P., Jr., ed. (1975). "Reconnaissance and Surveillance Window Design Handbook," AFAL-TR-75-200. Air Force Avionics Laboratory, AFAL/RWI, Wright-Patterson AFB, Ohio.
Biberman, L. M., ed. (1973). "Perception of Displayed Information." Plenum, New York.
Biberman, L. M., and Nudelman, S., eds. (1971). "Photoelectronic Imaging Devices." Plenum, New York.
Boivin, A., Dow, J., and Wolf, E. (1967). *J. Opt. Soc. Am.* **57**, 1171.
Born, M., and Wolf, E. (1959). "Principles of Optics," 1st ed. Pergammon, Oxford.
Boutry, G. A. (1962). "Instrumental Optics" (R. Auerbach, transl.) p. 51. Wiley (Interscience), New York.
Bradley, W. C. (1977). *Opt. Eng.* **16**, 249.
Brock, G. C. (1970). "Image Evaluation for Aerial Photography." The Focal Press, London.
Brock, G. C., Harvey, D. I., Kohler, R. J., and Myskowski, E. P. (1966). "Photographic Considerations for Aerospace," 2nd ed. Itek Corporation, Lexington, Massachusetts.
Brown, E. B. (1969). *In* "Evaluation of Motion Degraded Images," p. 45, NASA SP-193. Superintendent of Documents, GPO, Washington, D.C.
Buchdahl, H. A. (1968). "Optical Aberration Coefficients." Dover, New York.
Charman, W. N., and Olin, A. (1965). *Photgr. Sci. Eng.* **9**, 385.
Coulman, C. E. (1966). *J. Opt. Soc. Am.* **56**, 1232.
Cox, A. (1964). "A System of Optical Design." Focal Press, London.

Cruickshank, F. D., and Hill, G. A. (1960). *J. Opt. Soc. Am.* **50**, 379.

Dainty, J. C., and Shaw, R. (1974). "Image Science." Academic Press, New York.

Duffieux, P. M. (1946). "L'Integrale de Fourier et ses Applications a l'optique." Société Anonyme des Imprimeries, Oberthur; English translation, "The Fourier Integral and Its Application to Optics," Defense Documentation Center #459473, Ameron Station, Alexandria, Virginia (1965).

Fisher, R. W. (1976). Aerial reconnaissance systems (E. Shea, ed.), *Proc. Soc. Photo-Opt. Instrum. Eng.* **79**, 44.

Frieser, H. (1935). *C. R., Proc. IX Congr. Int. Photogr., Paris,* p. 207.

Goldberg, J. L., and McCullock, A. W. (1969). *Appl. Opt.* **8**, 1451.

Goodman, J. W. (1968). "Introduction to Fourier Optics," McGraw-Hill, New York.

Hecht, E., and Zajac, A. (1974)."Optics," p. 354. Addison-Wesley, Reading, Massachusetts.

Herzberger, M. (1947). *J. Opt. Soc. Am.* **37**, 485.

Hopkins, H. H. (1950). "Wave Theory of Aberrations." Oxford Univ. Press, London and New York.

Hopkins, H. H. (1974). Image assessment and specification (D. Dutton, ed.), *Proc. Soc. Photo-Opt. Instrum. Eng.* **46**, 2.

Hopkins, H. H., and Tiziani, H. J. (1966). *Brit. J. Appl. Phys.* **17**, 33.

Houston, J. B., Jr., ed. (1974). "Optical Shop Notebook." Optical Society of America, Washington, D.C.

Hufnagel, R. E. (1963). *In* Image evaluation for reconnaissance, Proceedings of a symposium held April 3–4, Itek Report 9048-6, Itek Corporation, Lexington, Massachusetts.

Hufnagel, R. E., and Stanley, N. R. (1964). *J. Opt. Soc. Am.* **54**, 52.

Inglestam, E. (1961). *Opt. Acta* **8**, 359.

Jacquinot, P., Roizen-Dossier, B., (1964). *In* "Progress in Optics," Vol. III (E. Wolf, ed.), p. 31. North Holland Publ., Amsterdam.

Jensen, N. (1968). "Optical and Photographic Reconaissance Systems," Chap. 8, Wiley, New York.

Johnson, J. (1958). *In* "Image Intensifier Symposium, Oct. 1958, p. 249. "U.S. Army E.R.D.L., Corps of Engineers. Fort Belvoir, Virginia PB151813.

Kingslake, R. (1965a). *In* "Applied Optics and Optical Engineering," Vol. I (R. Kingslake, ed.), Chap. 6. Academic Press, New York.

Kingslake, R. (1965b). *In* "Applied Optics and Optical Engineering," Vol. III (R. Kingslake, ed.), Chap. 1. Academic Press, New York.

Lauroesch, T. J., Fulmer, G. G., Edinger, J. R., Keene, G. T., and Kerwick, T. F. (1970). *Appl. Opt.* **9**, 875.

Legault, R. (1973). *In* "Perception of Displayed Information" (L. M. Biberman, Ed.), Chap. 7. Plenum, New York.

Lerman, S. H. (1969a). Effects of surface irregularities on image quality, Itek Research and Development Report 5742, 1 Jan.–30 June, Itek Corporation, Lexington, Massachusetts.

Lerman, S. H. (1969b). Effects of surface irregularities on image quality, Itek Research and Development Report 5742, 1 July–31 December, Itek Corporation, Lexington, Massachusetts.

Levi, L. (1969). *Appl. Opt.* **8**, 607.

Linfoot, E. H. (1956). *J. Opt. Soc. Am.* **46**, 740.

Lorah, L. D., and Rubin, E. (1965). *In* "Handbook of Military Infrared Technology" (W. L. Wolfe, ed.), p. 825. Office of Naval Research, Superintendent of Documents, GPO, Washington, D.C.

Luneberg, R. K. (1964). "Mathematical Theory of Optics," pp. 344–353. Univ. of California Press, Berkeley.

MacKenzie, D. (1928). *Trans. Soc. Motion Pict. Eng.* **12**, 730.

Mahajan, V. N. (1978). *Appl. Opt.* **17**, 3329.

Mahan, A. I., Bitterli, C. V., and Cannon, S. M. (1965). *In* Photographic and spectroscopic optics, *J. Appl. Phys. Japan,* **4**, Suppl. 1, p. 276.

Malacara, D., ed. (1978). "Optical Shop Testing," Wiley, New York.

Malvick, A. J. (1970). *Appl. Opt.* **9**, 2481.

Maréchal, A. (1947). *Rev. d'Optique* **26**, 257 (cited in Born and Wolf 1959, p. 467).

Martin, L. C. (1966). "The Theory of the Microscope," Amer. Elsevier, New York.

Mertz, P., and Gray, F. (1934). *Bell Syst. Tech. J.* **8**, 464.

Muray, J. J., Nicodemus, F. E., and Wunderman, J. (1971). *Appl. Opt.* **10**, 1465.

Nicholson, D. S. (1974). Effective systems integration and optical design (G. W. Wilkerson and R. W. Poindexter, eds.), *Proc. Soc. Photo-Opt. Instrum. Eng.* **54**, 163.

Nyquist, H. (1924). *Bell Syst. Tech. J.* **3**, 324.

Nyquist, H. (1928). *AIEE Trans.,* **47**, 617.

Ogrodnik, R. F. (1970). *Appl. Opt.* **9**, 2028.

Oliver, B. M. (1975). *In* "Imaging in Astronomy" (W. Wetherell, ed.), p. WB-7. Optical Society of America, Washington, D.C.

O'Neill, E. L. (1956). *J. Opt. Soc. Am.* **46**, 285, 1096.

O'Neill, E. L. (1963). "Introduction to Statistical Optics," pp. 99–101. Addison-Wesley, Reading, Massachusetts.

O'Neill, E. L., and Walther, A. (1977). *J. Opt. Soc. Am.* **67**, 1125.

Overington, J., (1976). "Vision and Acquisition." Pentech Press (Crane, Russak and Company), London.

Lord Rayleigh (J. W. Strutt) (1896). *Phil. Mag.* **XLII**, 167.

Lord Rayleigh (J. W. Strutt). (1964a). "Scientific Papers of Lord Rayleigh," Vol. IV, p. 235. Dover, New York.

Lord Rayleigh (J. W. Strutt) (1964b). "Scientific Papers of Lord Rayleigh," Vol. I, Dover, New York.

Reitmayer, F., and Schroeder, H. (1975). *Appl. Opt.* **14**, 716.

Richards, B., and Wolf, E. (1959). *Proc. Phys. Soc. London, Sect. A* **253**, 358.

Rimmer, M. P. (1962). *In* "Summer School Notes, 1962," Vol. 2. Institute of Optics, University of Rochester, Rochester, New York.

Rimmer, M. P. (1970). *Appl. Opt.* **9**, 533.

Rose, A. (1973). "Vision: Human and Electronic," Plenum, New York.

Rosell, F. A., and Willson, R. H. (1973). *In* "Perception of Displayed Information" (L. M. Biberman, ed.), Chap. 5, Plenum, New York.

Ruben, P. L. (1964). *J. Opt. Soc. Am.* **54**, 45.

Sandvik, O. (1928). *J. Opt. Soc. Am.* **16**, 244.

Sayanagi, K. (1966). "Image Structure and Transfer." Institute of Optics, University of Rochester, Rochester, New York.

Schade, O. H. (1951). *J. Soc. Motion Pict. Telev. Eng.* **56**, 137.

Schade, O. H. (1952). *J. Soc. Motion Pict. Telev. Eng.* **58**, 181.

Schade, O. H. (1953). *J. Soc. Motion Pict. Telev. Eng.* **61**, 97.

Schade, O. H. (1955). *J. Soc. Motion Pict. Telev. Eng.* **64**, 593.

Schwesinger, G. (1954). *J. Opt. Soc. Am.* **44**, 417.

Schwesinger, G. (1972). *Optik* (Stuttgart) **34**, 533.

Scott, F., Scott, R. M., Shack, R. V. (1963). *Photogr. Sci. Eng.* **7**, 345.

Shack, R. V. (1974). *In* Image assessment and specification (D. Dutton, ed.), *Proc. Soc. Photo-Opt. Instrum. Eng.* **46**, 39.

Shaw, R., ed. (1976). "Selected Readings in Image Evaluation." Society of Photographic Scientists and Engineers, Washington, D.C.

Smith, W. J. (1966). "Modern Optical Engineering," McGraw-Hill, New York.

Snyder, H. L. (1973). *In* "Perception of Displayed Information" (L. M. Biberman, ed.), Chap. 3. Plenum, New York.

Solomon, L., and Wetherell, W. B. (1979). Large space telescope image quality analyses, Itek Report 72-9486-1a (revised ed.), Itek Corporation, Lexington, Mass (final report on NASA contract NASw-2313, 4 Dec. 1973).

Spitzer, L., Jr., and Boley, B. A. (1967). *J. Opt. Soc. Am.* **57**, 901.

Straubel, R. (1935). "Pieter Zeeman Verhandelingen," p. 302. Nijhoff, The Hague (cited in Jacquinot and Roizen-Dossier, 1964).

Strehl, K. (1902). *Z. Instrumentenkd.* **22**, 213 (cited in Born and Wolf, 1959).

Tschunko, H. F. A. (1974). *Appl. Opt.* **13**, 22.

Tschunko, H. F. A., and Sheehan, P. J. (1971). *Appl. Opt.* **10**, 1432.

Vanderhoff, J. M. (1977). Systems integration and optical design II (S. Refermat, ed.), *Proc. Soc. Photo-Opt. Instrum. Eng.* **103**, 48.

Walther, A. (1965). *In* "Applied Optics and Optical Engineering," Vol. I (R. Kingslake, ed.), Chap. 7. Academic Press, New York.

Weiner, M. M. (1967). *Appl. Opt.,* **6**, 1984.

Wetherell, W. B. (1974a). Ultraviolet and visible scattered light effects on the performance of the large space telescope (LST), Itek Report 74-9507-1, Itek Corporation, Lexington, Massachusetts (final Report on NASA contract NAS8-30639, 12 Dec. 1974).

Wetherell, W. B. (1974b). *In* "Space Optics" (B. J. Thompson and R. R. Shannon, eds.), p. 55. National Academy of Sciences, Washington, D.C.

Wilkins, J. E., Jr., (1950). *J. Opt. Soc. Am.* **40**, 222.

Wyant, J. C., ed. (1976). Imaging through the atmosphere, *Proc. Soc. Photo-Opt. Instrum. Eng.* **75**.

Young, A. T. (1970). *Appl. Opt.* **9**, 1874.

"Military Standardization Handbook Optical Design," MIL-HDBK-141, Defence Supply Agency, Washington, D.C. 1962.

"Optical Telescope Technology," NASA SP-233, 1970, Superintendent of Documents, GPO, Washington, D.C., 1970.

APPLIED OPTICS AND OPTICAL ENGINEERING, VOL. VIII

CHAPTER 7

Circuits for Detectors of Visible Radiation

WILLIAM SWINDELL

Optical Sciences Center, University of Arizona, and Department of Radiology, Arizona Health Sciences Center, Tucson, Arizona

I. INTRODUCTION

This chapter deals with designing circuits for use with silicon photodiodes and photomultipliers (PMTs). We are not concerned with detailed physical descriptions of the various devices. The plan is to investigate how external circuit parameters affect the linearity, dynamic range, signal-to-noise ratio, and frequency response of the detector system, and to show how to choose component values that optimize performance in the required manner. This is done for photodiodes in Section II and for PMTs in Section III.

Only basic circuits are considered. Thus, the reader must consult other sources for a discussion of such topics as avalanche photodiodes and extremely fast circuits.

It is not a primary aim of this chapter to make comparisons between photodiodes and PMTs. However, it seems worthwhile to provide a short review of the relative merits of the two devices. In Table I it is seen that a silicon photodiode is normally the preferred detector and that a PMT is superior for very low light levels and for detection over large areas.

317

TABLE I
Comparison between the PMT and the Silicon Photodiode

PMT	Silicon photodiode
Spectral range	
120–1100 nm in several steps; Cs–Te photocathode with LiF window has a response that is useful from 120 to 360 nm; upper wavelength limit about 1080 nm with Ga–In–As photocathodes.	Most single devices cover 350–1050 nm; devices can be optimized to respond down to 200 nm in the ultraviolet and up to 1150 nm in the near infrared.
Noise equivalent power	
Dark current noise equivalent to 10^{-14}–10^{-16} W of incident energy (1 Hz bandwidth); can count single photons in specialized applications.	For 1-Hz bandwidth typically in the range 10^{-12}–10^{-14} W; can reach quantum noise limit with coherent detection.
Response time	
Pulse rise time <5 ns; special-purpose tubes ≤ 0.5 ns; also transit time delay of several nanoseconds may be important.	Intrinsic speed limited by junction design; commercial systems available with a 10–90% rise time of $<10^{-9}$ s and a bandwidth of 350 MHz; unbiased mode upper limit is typically 10^6 Hz.
Power supply requirements	
500–5000 V; requires high stability; tube gain depends strongly on voltage.	Bias voltage 0–100 V; associated amplifiers require ± 15 V; high stability not necessary.
Size limitations	
Photocathode diameter 0.5–14 in.; glass envelope can compromise optics; small detector arrays impractical.	Wide range, 0.003–25 cm linear dimension; good access to active area; detector arrays and special shapes available.
Stability	
Sensitive to temperature changes; subject to fatigue and drift.	Sensitive to temperature changes; essentially stable with time, i.e., 0.02% per hour, 1% per year.
Effect of magnetic field	
Requires shielding from ac fields or if used on moving platform, from any static field present.	No significant effect.

TABLE I (*Continued*)

PMT	Silicon photodiode
Output signal level	
Responsivity up to 10^6 A/W (internal amplification up to 5×10^7).	Less than 1 A/W; requires external amplifier in many applications.
Ruggedness	
Glass envelope; microphonic; ruggedized versions available.	Inherently rugged.
Spatial uniformity of response	
Often uniform to within 20% but sometimes may vary several hundred percent over photocathode.	Normally uniform to ≈5% fluctuation over active area.

The optical heterodyne receiver is discussed only briefly. The main advantage of this system is its ability to detect very low-level signals. In fact, for a bandwidth Δf and quantum efficiency η, the noise equivalent power (NEP) for a well-designed system is given by NEP = $\Delta f(h\nu/\eta)$, which is the theoretical quantum noise limit. However, heterodyne detection requires the presence of an optical local oscillator, and this restricts the technique to coherent systems and to systems where the shift in frequency for the local oscillator can be produced conveniently. The need to prevent phase cancellations over the incident wavefront makes the system more convenient for use in the infrared where alignment considerations are less critical. Thus the system has applications in infrared tracking and communications systems where the problems associated with monochromaticity, angular alignment, small-angle field of view, and local oscillators are accommodated with relative ease.

Finally, it should be realized that many of the manufacturers of detection equipment produce excellent detector–amplifier packages, ready to plug in and use. It is frequently better to buy the package, since even if it costs more, it is often worth it in the long run.

II. THE PHOTODIODE

We start by representing the diode with its equivalent circuit (Section II,A). From this description it is possible to derive expressions for the linearity and frequency response of the system. The important distinction between the biased and unbiased mode of operation is also made. Section II,B introduces the operational amplifier (op-amp). We show how to choose components that optimize the circuit for the desired application.

A. Basic Considerations

1. *Equivalent Circuit*

The principal components in the diode equivalent circuit are shown in Fig. 1. They are the junction shunt resistance R_J, the series resistance R_S, the junction capacitance C_J, the case capacitance C_c, the current generator representing the photoelectrically generated current I_p, the shot noise generator $\overline{i_s}$, and the Johnson noise generator $\overline{i_j}$.

The junction shunt resistance depends strongly on the active area of the device, varying approximately as the inverse square of the area. Since $R_J \gg R_S$, R_J is often identified with the dc resistance of the device. It varies with the level of illumination and with the bias conditions.

The value of R_J at zero bias and zero illumination is of special interest, since this parameter figures prominently in calculating the linearity. To distinguish it from R_J, which is a function of the operating point, it is given a special symbol, R_J^*, and is called the characteristic resistance. Being inversely proportional to the reverse leakage current [see Eq. (10)] R_J^* is strongly temperature-dependent.

Series resistance R_S is the sum of the lead, contact, and undepleted bulk silicon resistance. The latter is given by $\rho d/A$, where ρ is the intrinsic resistivity, A is the junction area, and d is the thickness of the undepleted region. The amount of contact resistance is determined by a design compromise. The contacts should be thick in order to keep the resistance small. However, in the case of a Schottky-barrier photodiode, the gold layer should be thin enough to be transparent and, for a planar-diffused diode, the diffused contact layer should be thin enough to ensure that most of the photons are absorbed in the junction. Lead resistance is

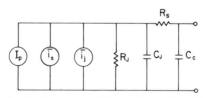

FIG. 1. Equivalent circuit for a silicon photodiode.

usually negligible. Generally speaking, the series resistance increases as the active area of the diode decreases.

In the unbiased mode, the series resistance is relatively independent of operating conditions, but its magnitude is important in determining the linearity and response time of the device. In the biased mode of operation, the series resistance is a strong function of bias, decreasing as the bias voltage is increased, but it does not determine system linearity as it does in the photovoltaic mode.

The shunt capacitance C_J depends on the area of the junction and on the applied bias voltage. Without bias, the capacitance is typically a few thousand (~ 3000) picofarads per square centimeter of area. This value may be reduced by at least an order of magnitude with the application of a reverse-bias voltage.

2. *The Load Line*

The (conventional) current I that flows in the external circuit is the algebraic sum of two components:

$$I = I_D + I_p \tag{1}$$

The diode current I_D is related to the voltage across the junction V_J:

$$I_D = I_0[1 - \exp(-qV_J/\beta kT)] \tag{2}$$

where I_0 is the reverse leakage current, q is the electronic charge, β is a constant close to unity (and is itself weakly dependent upon V_J), k is Boltzmann's constant, and T is the absolute temperature of the junction. We follow the usual practice of making $\beta = 1$ and writing Eq. (2) in the form

$$I_D = I_0[1 - \exp(-V_J/0.026)] \quad (T = 300 \text{ K}) \tag{3}$$

for V_J measured in volts. A positive V_J represents a reverse bias on the diode.

The photogenerated current I_p is strictly proportional to the incident radiant flux Φ except at very high flux levels when saturation and ultimately destruction take place; i.e.,

$$I_p = \mathcal{R}\Phi \tag{4}$$

where $\mathcal{R}$ is the responsivity (amperes/watt) of the photodiode.

The external circuit is represented by a bias voltage supply V_B and a load resistor R_L across which the load voltage V_L is developed. The low-frequency equivalent circuit thus appears as shown in Fig. 2.

V_J and I are related through the circuit parameters

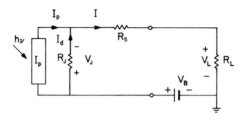

FIG. 2. Low-frequency equivalent circuit for a reverse-biased photodiode.

$$V_J = V_B - I(R_S + R_L) \qquad (5)$$

V_J and I are also related through Eqs. (1), (2), and (4); thus the actual operating point of the diode is given by simultaneously satisfying both sets of equations. Figure 3 shows Eq. (1) plotted for several values of I_p. Superimposed on this graph are three typical load lines defined in Eq. (5). The intersection of the load line with the diode curves determines how the current I in the external circuit varies as a function of the irradiance.

3. Unbiased versus Biased Operation

The presence of an external reverse bias defines the biased mode of operation. It is exemplified by load line 1 in Fig. 3. The reverse bias increases the electric field in the junction. This reduces the electron and hole transit times, thus decreasing the intrinsic rise time of the detector. Rise and fall times of less than 10^{-8} s are readily achievable in this mode.

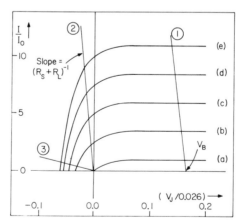

FIG. 3. Characteristic curves for a silicon photodiode with various levels of incident irradiation. Curve (a) is the response of the diode in the absence of incident radiation. Curves (b), (c), (d), and (e) show the effect of increasing the flux level. Three load lines, ①, ②, and ③, are shown.

A second advantage is that the biased mode gives extremely linear operation provided that the dark current is accounted for. However, the dark current, i.e., the reverse leakage current I_0, is also the source of most of the drawbacks of this mode of operation. I_0 is temperature-dependent, approximately doubling for each 10°C rise. It becomes difficult to make reliable measurements of steady photocurrents whose magnitudes are significantly smaller than I_0. In addition, there is both shot and $1/f$ noise associated with I_0. This further increases the difficulty of measuring small signals, especially at low frequencies. With some devices, however, especially those with small active areas, a significant part of I_0 is due to surface leakage and the manufacturer may use guard rings to eliminate this component.

Load lines 2 and 3 typify the unbiased mode of operation. The lack of external reverse bias means that the junction is now self-biased in the forward direction. This gives rise to a more nonlinear performance (see load line 2). However, with a low total load resistance, $R' = R_S + R_L$, linear performance can be achieved over a usefully wide range of photocurrent. High values of R_L, as shown with load line 3, should be avoided unless an almost logarithmic performance of the open-circuited detector is sought. In this case, the junction voltage V_J appears at the device terminals and we have, by combining Eqs. (1), (3), and (4), and setting $I = 0$,

$$V_L = -V_J = 0.026 \ln[(R\Phi/I_0) + 1] \qquad (6)$$

With the absence of dark current and its concomitant $1/f$ noise, performance at low frequencies is enhanced. In the $1-100$ Hz region, it is possible to obtain noise level reductions of a few orders of magnitude in some cases, and it is relatively easy to measure photocurrents that are orders of magnitude lower than the reverse leakage current of the diode. The intrinsic upper-frequency response is considerably reduced but can still be as high as 10^6 Hz so, unless the additional speed of the reverse-biased diode is needed, it is usually advantageous to use the unbiased mode of operation.

4. Nonlinearity

The relationship between the output current I and the incident flux Φ is obtained using Eqs. (1), (2), (4), and (5):

$$I = I_0(1 - \exp\{-q[V_B - I(R_S + R_L)]/\beta kT\}) + \mathcal{R}\Phi \qquad (7)$$

and it is this equation that determines the theoretical linearity of the system.

For the biased mode of operation V_B is at least several volts and often several tens of volts. Thus for practical purposes, Eq. (7) may be rewritten

$$I = I_0 + \mathscr{R}\Phi \tag{8}$$

Apart from the dc offset term, the biased system is as linear as the intrinsic linearity of the photogeneration process itself. In practice this often amounts to better than $\pm 1\%$ over seven decades of incident flux level.

If the dark current I_0 is objectionable, it should be nulled with an appropriate offset current I_0'. Preferably this nulling current should track the temperature dependence of I_0. (One way to do this is to use an auxiliary matched photodiode that is shielded from the radiation and maintained at the same temperature as the active detector.)

If ac coupling is used, the dark current no longer enters into consideration, and the system is intrinsically linear.

In the unbiased mode, the exponential term in Eq. (7) cannot be neglected, and we have, setting $V_B = 0$,

$$I = I_0\{1 - \exp[qI(R_S + R_L)/\beta kT]\} + I_p \tag{9}$$

We now define the characteristic resistance R_J^* of the junction as the dynamic resistance of the junction at zero bias; i.e.,

$$R_J^* = \left.\frac{\overline{dV_J}}{dI_d}\right|_{V_J=0} = \frac{\beta kT}{qI_0} \tag{10}$$

where we have made use of Eq. (2). Equation (9) now becomes

$$I = I_0\{1 - \exp[I(R_S + R_L)/I_0 R_J^*]\} + I_p \tag{11}$$

This is the expression that determines the (nonlinear) relationship between the observed current I and the incident flux Φ (since $I_p = \mathscr{R}\Phi$).

As we will shortly see, the system becomes grossly nonlinear when the exponent in Eq. (11) is greater than (approximately) 0.1. For values less than 0.1, where the expansion $e^x = 1 + x + \frac{1}{2}x^2$ is a valid approximation, we can write

$$I = \frac{I_p}{1 + r}\left[1 - \frac{I_p}{I_0}\left(\frac{r}{1+r}\right)^2\right] \tag{12}$$

where we have set $r = (R_S + R_L)/R_J^*$.

The output current depends parabolically on I_p (i.e., with Φ), and the magnitude of the nonlinearity can be described by the ratio of the nonlinear coefficient to the linear coefficient in Eq. (12).

The nonlinearity NL is thus given by

$$NL = (I_p/I_0)[r/(1 + r)]^2 \tag{13}$$

where Eq. (13) is also subject to the limitation $rI/I_0 < 0.1$. (The reader may wish to include a numerical factor on the order of unity to account for the various ways in which a straight-line fit may be made to a parabola.)

In practice $r \ll 1$, so it is reasonable, using Eq. (12), to write the limits over which Eq. (13) is valid as $rI_p/I_0 < 0.1$. Thus the maximum nonlinearity that can occur in this region is given by

$$NL_{max} = r/10 \qquad (14)$$

We shall see in Section II,B that the effective load resistor R'_L can be made arbitrarily small and that practical values for r can easily be 10^{-3} and less. Usually r can be made sufficiently small such that the corresponding nonlinearity (whose maximum value is given by $r/10$) is acceptably small.

For the situation where the small-quantity expansion is not valid the following approach is helpful.

We define quantities P_1 and P_2 as follows:

$$I = I_p/(1 + P_1) \qquad (15)$$

and

$$dI/dI_p = 1/(1 - P_2) \qquad (16)$$

and nonlinearity can be displayed by the manner in which P_1 and P_2 vary as a function of I. P_1 refers to large-scale signals, while P_2 is valuable for describing signals fluctuating about a nominally steady component. The reason for defining P_1 and P_2 as shown by Eqs. (15) and (16) is that P_1 and P_2 are readily determined, analytically, from Eq. (11):

$$P_1 = \left[\exp\left(\frac{I}{I_0} \frac{R_S + R_L}{R_J^*}\right) - 1 \right] \Big/ \frac{I}{I_0} \qquad (17)$$

$$P_2 = \frac{R_S + R_L}{R_J^*} \exp\left(\frac{I}{I_0} \frac{R_S + R_L}{R_J^*}\right) \qquad (18)$$

These expressions are shown graphically in Fig. 4. Here it is seen that the behavior is divided into two regions. The first region, called the *essentially* linear region, is given by

$$\frac{I}{I_0} \frac{R_S + R_L}{R_J^*} < 0.1 \qquad (19)$$

In this range of load currents, both P_1 and P_2 are essentially independent of current and are given by

$$P_1 = P_2 = (R_S + R_L)/R_J^* \qquad (20)$$

A more practical way to express the range of validity of Eq. (20) is found from Eq. (19). By making the appropriate substitutions we obtain

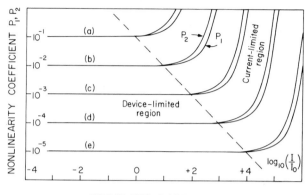

NORMALIZED DIODE CURRENT

FIG. 4. Nonlinearity coefficients P_1 and P_2 depend on the ratio of effective load resist-ance to characteristic resistance $[(R_S + R_L)/R_J^*]$ and photocurrent I. For curves (a) through (e) the value of $[(R_S + R_L)/R_J^*]$ is 10^{-1}, 10^{-2}, 10^{-3}, 10^{-4}, and 10^{-5}, respectively. To the left of the dashed line, P_1 and P_2 are essentially independent of output current. To the right of the line the system rapidly becomes nonlinear.

$$- V_J = I(R_S + R_L) \leq 2.5 \quad \text{mV} \tag{21}$$

The second region is defined by

$$- V_J = I(R_S + R_L) > 2.5 \quad \text{mV} \tag{22}$$

In this range the system rapidly becomes increasingly nonlinear with increasing photocurrent. In any event, $P_1 < P_2$.

There are three important results that stem from this discussion. First, provided that the inequality of Eq. (21) is met, the system should be es-sentially linear as evidenced by the fact that P_1 and P_2 are independent of I/I_0. The precise degree of departure from linearity can be determined by examining Eq. (11), (12), or (14), as appropriate. Second, when Eq. (22) is valid, the system rapidly becomes nonlinear. Equation (11) should be used to determine the amount of nonlinearity. Third, from the linearity point of view, the load resistance R_L should be small, but there is not much point in making R_L significantly smaller than R_S. Lowering R_L simultaneously reduces the amount of nonlinearity and extends the range over which the system is "essentially" linear.

If necessary, the values of R_S and R_J can be determined by experiment (Hamstra and Wendland, 1972).

5. Frequency Response

When a photodiode is loaded by a purely resistive load, its response time is governed by one of two factors. They are (1) the photoelectronic processes within the junction itself such as carrier drift velocity and carrier lifetime, and (2) the electrical characteristics such as junction resistance and capacitance. We now study the frequency response of the system as determined by (2).

We consider the incident radiation to consist of a steady (dc) component upon which is superimposed a sinusoidally time-varying (ac) component. It is convenient to consider the ac component small compared to the dc component, so that the junction resistance can be modeled by a fixed resistor R_D of value $R_D = \Delta V_J / \Delta I_D$. R_D is calculated at the operating point as determined by the dc irradiance level and the external circuit values. For biased operation $R_D > R_J$; for unbiased short-circuit operation $R_D < R_J$.

For a resistive load R_L, and neglecting case capacitance, we have the signal current generator supplying the parallel combination of R_D, C_J, and $R_S + R_L$, thus giving

$$f_c = \frac{1}{2\pi} \frac{R_D + R_S + R_L}{R_D(R_S + R_L)C_J} \qquad (23)$$

where f_c is the frequency at which the output voltage has fallen to $1/\sqrt{2}$ of its low-frequency value. Normally, $R_D \gg (R_S + R_L)$, which gives

$$f_c = \frac{1}{2\pi} \frac{1}{(R_S + R_L)C_J} \qquad (24)$$

The maximum upper frequency response, $f_{c\,max}$, is thus reached by having $R_L \ll R_S$, which gives

$$f_{c\,max} = (1/2\pi)(1/R_S C_J) \qquad (25)$$

f_c is maximized by using small-area diodes and a reverse bias, both of which reduce C_J.

B. The Transconductance Amplifier

1. The Basic Circuit

The circuit shown in Fig. 5 involving an op-amp converts input current I into an output voltage V according to the expression

$$V = -I \left[\frac{AR_{in}R_F R_L - R_o R_{in}R_L}{(A + 1) R_{in}R_L + R_F R_L + R_o(R_L + R_{in} + R_F)} \right] \qquad (26)$$

where R_F is the feedback resistor, R_{in} the amplifier dc input resistance, A the open-loop voltage gain of the amplifier, R_o the amplifier output resist-

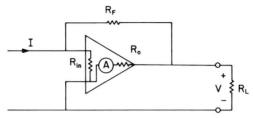

FIG. 5. Equivalent low-frequency circuit for an op-amp operating in transimpedance mode.

ance, and R_L the load resistance. This expression neglects various offset levels that may be present.

Under usual conditions we have $R_{in} \gtrsim R_F$, $R_o \ll R_F$, $R_L \ll R_F$, and $A \simeq 10^4 - 10^7$. Thus we can write

$$V = -IR_F \qquad (27)$$

Note that this voltage is derived at the low-impedance output of the op-amp which makes it easy to drive further stages of amplification or other output devices.

The high-voltage gain of the amplifier requires that the input voltage v be correspondingly small compared to the maximum output (which is usually ≈ 10 V) and that the photodiode feeding the circuit "see" an effective load resistor R_E given by v/I, which to a high degree of accuracy is given by

$$R_E = \frac{v}{I} = \frac{V/A}{V/R_F} = \frac{R_F}{A} \qquad (28)$$

2. Linearity

One virtue of the op-amp configuration is that the output V is almost totally insensitive to changes in the amplifier gain, as is seen by examining Eq. (26); such changes may result from an inadequately stabilized power supply or aging of components. In the biased mode, the linearity is thus determined by the intrinsic linearity of the diode itself. There is a dc offset component of the output voltage caused by the diode reverse leakage current and by offset arising in the amplifier. If it is necessary to null this offset, there are several ways of doing so (see Fig. 6). If ac coupling is used, there is no need to consider nulling the dark current offset.

For the unbiased mode, the equations of Section II,A are applicable. However, the value of R_L to be used in these equations is R_F/A, as given by Eq. (28). This is one of the reasons for using an op-amp in this configuration. As far as the signal is concerned, the system acts like a load re-

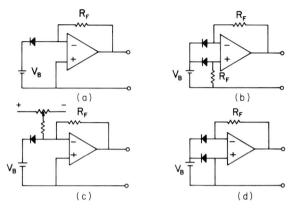

FIG. 6. (a) Straightforward circuit for applying bias to the photodiode. There is an output voltage even when no radiation is incident on the detector because of the reverse-biased leakage current. This may be nulled as shown in (b), where a matched, light-shielded detector is used, or in (c), where the leakage current is canceled by an external current. (d) A large component of the dark current can be eliminated from the signal path by using guard ring devices as shown.

sistor of value R_F, generating an output voltage $-IR_F$. For linearity considerations, however, the effective load is only R_F/A [Eq. (28)], and this is typically in the range of 1–1000 Ω. If it is required to extend the linearity over the widest range of detector currents, we must choose R_F or A such that $R_F/A \ll R_S$, in which case the maximum current for which the system is essentially linear is given by Eq. (19):

$$I = I_0(R_S + R_L)/10R_j^* \tag{29}$$

Equation (7) or Eqs. (17) and (18), with R_F/A replacing R_L, provide general expressions for the linearity for any operating condition.

3. Frequency Response

The parameters that contribute significantly to frequency response of the diode–op-amp combination are shown in Fig. 7. It has been assumed that $R_S = 0$, so that C_J and C_c and the input capacitance of the amplifier can be lumped together as a single input capacitance C_i. Also, the parallel combination of R_J and the amplifier input resistance is denoted here by R_i. The feedback resistor R_F is shunted by capacitor C_F which is composed, in general, of stray capacity plus an additional compensating capacitor. To simplify the analysis, it is assumed that the open-loop (radian) frequency response of the amplifiers $A(\omega)$ has but a single pole and is determined by a dc gain $A(0)$ and a gain bandwidth product G (hertz), so that the voltage gain $A(\omega)$ may be written

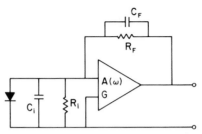

Fig. 7. The ac equivalent circuit for determining the frequency response of the diode–amplifier combination. G is the amplifier gain–bandwidth product.

$$A(\omega)/A(0) = \{1 + [i\omega A(0)/2\pi G]\}^{-1}. \tag{30}$$

See Fig. 8.

If we make the usual assumptions that the output impedance is negligibly small and that terms containing $1/A(0)$ are negligibly small, it is straightforward to show that overall transimpedance $\hat{R}(\omega)$ is given by

$$\hat{R}(\omega) = -R_F[1 + i\gamma 2(\omega/\omega_n) - (\omega/\omega_n)^2]^{-1} \tag{31}$$

where the resonance frequency ω_n is given by

$$\omega_n = [2\pi G/R_F(C_I + C_F)]^{1/2} \tag{32}$$

and the damping factor γ is given by

$$\gamma = \tfrac{1}{2}[C_F R_F + R_F(C_I + C_F)/A + (1 + R_F/R_I)/2\pi G]\omega_n \tag{33}$$

This behavior is summarized in Fig. 9. Note that, for $\gamma < 1.0$, there is gain peaking in the frequency domain (which corresponds to overshoot in the time domain). Also, for $\gamma \gg 1$, the upper band limit shifts from ω_n to $\omega_n/2\gamma$ followed by a 6-dB octave falloff. In many practical applications,

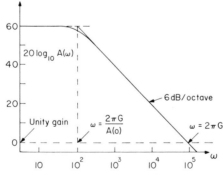

Fig. 8. Frequency dependence of open-loop voltage gain for an op-amp whose low-frequency voltage gain $A(0) = 10^3$ and gain–bandwidth product $G = 10^5/2\pi$.

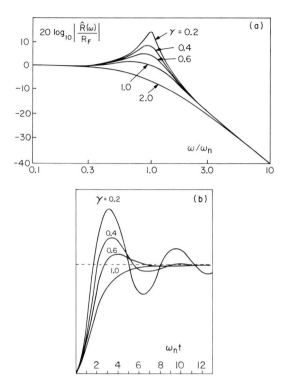

FIG. 9. (a) If the circuit parameters are chosen such that the damping coefficient is less than unity, gain peaking will occur. (b) In the temporal domain this corresponds to the output voltage displaying overshoot phenomena. The ordinate represents system output voltage following a step input.

the first term of Eq. (33) dominates the others, and $C_I \gg C_F$, so we can write

$$\omega_n \simeq (2\pi G/R_F C_I)^{1/2} \qquad (34)$$

and

$$\gamma \simeq \tfrac{1}{2} C_F R_F \omega_n \qquad (35)$$

Thus the order of design is usually:

(1) Choose R_F for desired low-frequency transimpedance, $|\hat{R}(0)| = R_F$.

(2) Select the required bandwidth using Eq. (34) by suitably choosing G and C_I.

(3) Select γ (usually in the range of $\gamma = 1$) with suitable values of C_F using Eq. (35).

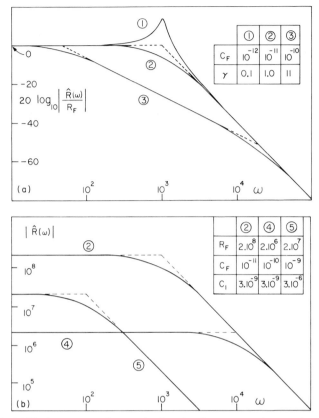

FIG. 10. (a) The importance of correctly choosing C_F to control the damping ratio. (b) By suitably choosing C_F, R_F, and C_i (by adding more capacitance) a variety of frequency responses may be obtained by using a given op-am–diode combination ($G = 10^5$, $C_i = 3 \times 10^{-9}$ F, $R_i = 5 \times 10^7$ Ω).

(4) Check the values ω_n and γ using the more accurate Eqs. (32) and (33).

Some iteration between steps (2) and (3) may be necessary. The final determination of C_F for critical applications is best done experimentally by observing the system step response and adjusting C_F until the required transient output is observed.

Figure 10 illustrates how the frequency response of a system may be varied starting with an op-amp with $G = 10^5$ Hz and a diode specified by $R_i = 5 \times 10^7$ Ω and $C_i = 3 \times 10^{-9}$ F.

4. Noise

For low-level, low-frequency signals, the noise associated with the biased mode of operation may be several orders of magnitude greater than that of the same device operated in the unbiased mode. This is because of the excess noise [often called "flicker" or "one over f" ($1/f$) noise] associated with the dark current. The following argument is based on the unbiased mode for which the noise spectrum at low frequencies is essentially flat; however, extension to a biased operation is straightforward if the noise spectrum is known. We consider all possible sources of noise (see Fig. 11).

The Johnson noise current $\overline{i}_j$, arising in the junction, is given by

$$\overline{i}_j = (4kTB/R_D)^{1/2} \tag{36}$$

where B is the noise equivalent bandwidth and R_D is the diode dynamic resistance at the operating point.

Shot noise $\overline{i}_s$ is associated with the signal current I and is given by

$$\overline{i}_s = (2qIB)^{1/2} \tag{37}$$

The contribution to the noise from the amplifier comes from its current input noise $\overline{i}_n$ and voltage input noise $\overline{e}_n$. These quantities are frequency-dependent and can vary considerably from unit to unit. The manufacturer's literature can be used as a guide, but for critical applications units should be hand-selected for best performance. The input resistance to the amplifier is assumed noiseless, its Johnson noise being contained in $\overline{i}_n$.

Finally, there is the thermal noise of the feedback resistor:

$$\overline{i}_f = (4kTB/R_F)^{1/2} \tag{38}$$

Based on this model, the total (rms) noise voltage $\overline{V}_n$ appearing at the amplifier output is given by

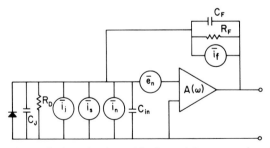

FIG. 11. The noise equivalent circuit used in determining expressions for the noise appearing at the output of the op-amp.

$$\overline{V}_n = Z_F \left[\frac{\overline{e_n^2}}{Z_e^2} + \overline{i_n^2} + \overline{i_J^2} + \overline{i_f^2} + \overline{i_s^2} \right]^{1/2} \tag{39}$$

where

$$Z_F = R_F/(1 + \omega^2 C_F^2 R_F^2)^{1/2}, \qquad Z_E = R_E/(1 + \omega^2 C_e^2 R_e^2)^{1/2},$$
$$C_e = C_J + C_{in} + C_F, \qquad R_e = (R_D^{-1} + R_{in}^{-1} + R_F^{-1})^{-1}$$

It is usually assumed, as it is here, that all noise sources are uncorrelated. In fact, e_n and i_n may be slightly correlated. The signal voltage is given by

$$V_s = -IZ_F \tag{40}$$

and so the full expression for the signal-to-noise ratio is

$$\text{SNR} = \frac{V_S}{V_N} = -I \left[\frac{e_n^2}{Z_e^2} + 4kTB \left(\frac{1}{R_F} + \frac{1}{R_D} \right) + i_n^2 + 2eIB \right]^{-1/2} \tag{41}$$

Equation (41) must be used with caution, because the signal has been defined as a steady quantity V_S. A more thorough treatment would define the signal as the rms fluctuation of the modulated part of the signal voltage; also, the frequency dependence of the responsivity of the detector and of the amplifier would be considered.

III. THE PHOTOMULTIPLIER

A. BASIC CONSIDERATIONS

1. Principles of Operation

Incident radiation is converted into an output electric current in three steps. First, the radiation is absorbed by the photocathode. When the photon energy $h\nu$ is greater than the work function of the cathode material, a photoelectron may be emitted. The efficiency of the process, the quantum efficiency, is typically 0.1 photoelectrons per photon may be as high as 0.25, but sometimes is as low as 0.001. This lack of efficiency is compensated for by the second step, an essentially noise-free multiplication process. The photocurrent emitted from the photocathode is multiplied by secondary emission processes in a series of electrodes called dynodes. Current amplification factors of 10^6 are typical. For example, a gain of 5 at each of nine electrodes yields a multiplication factor of $5^9 \simeq 2 \times 10^6$. Finally, this amplified photocurrent is collected by a final anode and delivered to the external circuit. See Fig. 12.

The cathode photocurrent i_k is given by

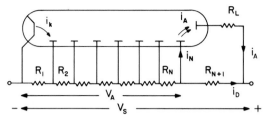

FIG. 12. Basic circuit configuration for PMTs. Arrows show direction of electron current.

$$i_k = (e/hc) \int P(\lambda)\eta(\lambda)\lambda \, d\lambda \tag{42}$$

where $\eta(\lambda)$ may be regarded as the quantum efficiency of the photocathode–window combination and $P(\lambda)$ is the power spectrum of the incident radiation.

The anode current i_A is given by

$$i_A = fMi_k \tag{43}$$

where M is the overall electron multiplication ratio for electrons arriving at the first dynode and f is a factor ($f < 1$) describing the collection efficiency of the first dynode.

The overall responsivity $\mathcal{R}$ is given by

$$\mathcal{R} = \frac{i_A}{\int P(\lambda) \, d\lambda} = \frac{fMe}{hc} \frac{\int P(\lambda)\eta(\lambda)\lambda \, d\lambda}{\int P(\lambda) \, d\lambda} \tag{44}$$

For monochromatic radiation the spectral responsivity $\mathcal{R}_\lambda$ is given by

$$\mathcal{R}_\lambda = \frac{fMe\eta(\lambda)\lambda}{hc} \tag{45}$$

The long-wavelength limit of the spectral range is determined by the work function of the cathode material. Infrared photons lack the energy to eject photoelectrons from the photocathode. The present limit is about 1.3 μm. At the low end, the wavelength limit is really determined by the window material, since the photoelectric effect is operative out to high-energy x rays. The practical limit is between 0.1 and 0.2 μm. Wide coverage such as that provided by the S1 photocathode is usually accompanied by low quantum efficiency.

The noise in a suitably designed and operated tube is just that due to statistical fluctuation of the photon flux (multiplied by a factor which takes into account the quantum efficiency of the cathode and collection efficiency and secondary multiplication yield of the first stage). It is considerably lower than that achieved with photodiodes. Individual photon counting is possible.

PMTs are used extensively in astronomy and in nuclear physics. In astronomical work, the major use is to measure very weak but normally continuous photon fluxes. In nuclear studies, the photomultiplier is used to detect weak flashes of light generated by scintillation events. The flash of light may contain less than 100 photons. Regardless of the application, the output of the PMT is always the same—a series of current pulses. Before discussing such topics as signal-to-noise ratio and linearity, we first investigate the nature of this pulse train and, in particular, the origin of the pulses it contains.

We are not concerned with the temporal structure of each individual pulse, only with the amount of charge it contains. Indeed, we presume that each pulse has been processed through a system whose impulse response is long compared to the actual pulse width. Thus, the measured pulse height can be taken to be an accurate measure of the number of electrons contained in the anode pulse, and thus pulse height and pulse charge are directly proportional to each other.

First, consider the pulse height distribution $s(h)$ of the signal pulses. The number of electrons in the anode pulse is a random variable. It is the result of a sequence of random events that start with the collection efficiency of the first dynode and include the statistical fluctuations inherent in the secondary multiplication process. These latter statistics are still imperfectly understood. Most of the variation in pulse height is created at the first dynode. It is important to keep the cathode-to-first-dynode voltage at the recommended value to reduce this variation. Experimental results have shown that in a good PMT there is only a small fraction of signal pulses larger than twice the average.

The dark pulse height distribution $d(h)$ is quite complicated because many different factors contribute to it. Contributing mechanisms can include ion pulses, radioactive decay of ^{40}K in the glass envelope, cosmic rays, thermally generated electrons, and light produced by electroluminescence and corona discharge. There is usually a preponderance of small pulses with the spectrum often having a shape approximately given by $d(h) \propto h^{-2}$.

Because $s(h)$ and $d(h)$ have substantially different forms, it follows that the optimum detection process involves a weighting scheme which favors pulses most likely to be signal pulses. The most general scheme allows arbitrary weighting factors to be applied. It requires a pulse height analyzer in order to measure $d(h)$, and the total pulse height spectrum $t(h) = d(h) + s(h)$. This is the most flexible, most expensive, and slowest way to count signal photons.

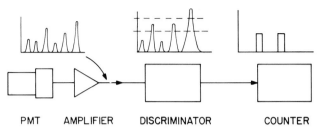

PMT   AMPLIFIER   DISCRIMINATOR   COUNTER

FIG. 13. Block diagram of photon counter using a simple two-level discriminator. Only pulses whose height falls between two preset levels are counted.

For low light levels, straightforward pulse counting with a discriminator set to reject abnormally large and small pulses gives almost optimal results. The weighting function in this case is given by $w(h)$ = constant for h values within the discriminator window and $w(h) = 0$ elsewhere. It is a relatively inexpensive method and is much faster than a pulse height analyzer. A block diagram of the system is shown in Fig. 13. Commercial systems are available that extend the flexibility of this basic arrangement. For example, if the source is monitored with a separate detector, any variations in source strength can be measured and ratio techniques used to compensate for these fluctuations. Also, the signal can be modulated with a chopper or similar device. This allows the background count rate to be continuously updated and subtracted in real time from the signal count rate.

As the intensity of the light source is increased, the probability that the charge pulses will overlap, giving rise to spuriously high and wide pulses, also increases. There is also a pulse pile-up error related to the dead time of the discriminator; this also increases with light intensity for high levels of irradiance. As a result, pulse counting becomes inappropriate and dc (analog) methods should be used. In this case all the pulses are collected and smoothed (integrated) by an appropriate circuit. The weighting function $w(h)$ in this case is simply directly proportional to h. The noise is then measured in terms of the signal current fluctuations in exactly the same way as discussed in Section II.

It is beyond the scope of this chapter to discuss the circuits for discriminators, pulse amplifiers, and the like. Nor can we discuss in detail such subjects as optimum weighting filters, threshold levels, and theoretical expressions for the signal-to-noise ratio. These topics are admirably dealt with elsewhere (see the references at the end of this chapter). Rather, we look at the circuit in the immediate neighborhood of a PMT and show how it may be optimized for these two rather different modes of operation (i.e., pulse counting and analog detection).

2. Gain Control

The overall electron multiplication ratio M for a system of N dynodes is given by

$$M = \prod_{n=1}^{N} \delta_n \tag{46}$$

where the δ_n's are the individual dynode multiplication factors and include losses due to lost electrons, etc. It is assumed that the anode collection efficiency is unity, a condition that is nearly always met.

The gain of each stage depends on the accelerating voltage V between the dynode and the previous electrode in a manner reasonably well approximated by the empirical relationship

$$\delta = AV^P \tag{47}$$

where P and A are constants. Thus if we assume for simplicity that the dynodes are held at equally spaced voltages with a total accelerating voltage of V_A, we have from the preceding two equations

$$M = [A(V_A/N)^P]^N \tag{48}$$

Note that V_A differs from the supply voltage V_S by the voltage drop across the last resistor in the dynode chain. If we insert typical values of $A = 0.2$ and $P = 0.65$ for Cs–Sb dynodes, we will have gain-versus-voltage curves as shown in Fig. 14.

This highly nonlinear dependence is useful, since it allows the overall responsivity of the PMT to be varied over extremely wide limits simply by varying the supply voltage. This is a popular way of controlling the system gain.

If the power supply is fixed, it is possible to achieve modest changes in overall gain simply by varying the voltage on one dynode relative to that of its two nearest neighbors, as shown in Fig. 15.

The precise shape of the gain-versus-dynode-voltage curve depends on the voltage difference between the two outer dynodes and upon the dynode geometry. However, the method may be useful for controlling the gain when only fixed voltage power supplies are available, or when more than one PMT is operated from a common power supply.

From Eq. (48) we obtain

$$dM/M = PN \, dV_A/V_A \tag{49}$$

which states that gain fluctuations will be approximately N times greater than fluctuations in the overall accelerating voltage. In the absence of dynode loading, discussed in Section III,B, and with a totally resistive

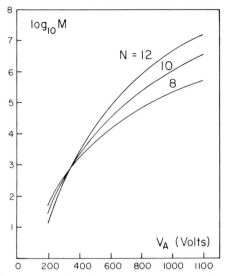

FIG. 14. Typical dependence of overall electron multiplication M as a function of accelerating voltage V_A.

dynode chain, $dV_A/V_A = dV_S/V_S$, which gives the following rule of thumb relating the power supply stability to the required gain stability:

$$dM/M = N\, dV_S/V_S \tag{50}$$

B. NONLINEARITIES

1. *Dynode Voltage Stability*

With reference to Fig. 16, it can be seen that, when anode current i_A is present, dynode currents i_N, i_{N-1}, etc., must also be flowing. These dynode currents flow through the dynode resistor chain and change the operating voltages of the dynodes, thus affecting the gain of the system in a signal-dependent way. The final dynode current dominates the others, and the magnitude of the gain change is readily estimated by neglecting all dynode currents except the largest. We have

$$i_N = i_A[1 - (1/\delta)] \tag{51}$$

In the absence of any anode current i_A the dynode chain (electron) current i_D flows without diversion through resistors R_1 through R_{N+1}. With the signal present, a fraction of i_A is diverted at the last dynode, thus decreasing the current flow in R_{N+1} by the same amount. This gives rise to an increase in voltage at the last dynode of

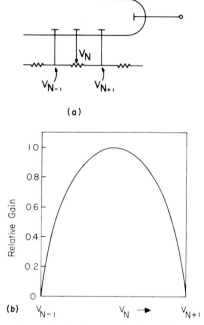

Fig. 15. PMT gain can be controlled by varying the voltage applied to a single dynode (a). Gain variation as predicted by Eq. (47) is shown in (b). In practice, the actual shape of the curve is strongly influenced by the design of the dynode system.

$$\Delta V_N = i_A R_{N+1}$$

V_N, however, is just the V_A of Eq. (49) and, by combining Eqs. (49) and (51) and setting $V_N = V_A$, we obtain

$$\frac{dM}{M} = \frac{i_A(1 - 1/\delta)PNR_D}{V_N} \qquad (52)$$

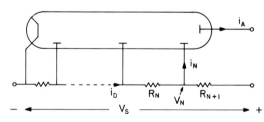

Fig. 16. Current flowing from the last dynode changes the dynode operating voltage in a signal-dependent manner, giving rise to nonlinear amplification. Arrows show direction of electron current.

The dynode chain current i_D is given by

$$i_D = V_N/NR_D \tag{53}$$

and thus, using the last two equations, we can write approximately

$$dM/M = i_A/i_D$$

A rigorous derivation gives a result that differs only slightly. This means, for example, that in order to maintain a gain stability of better than 1 part in 100, the anode current must not be allowed to exceed $\frac{1}{100}$ of the dynode chain current.

The practical problems are that either the anode current must often be kept inconveniently low or the dynode chain current must be made inconveniently high, with the attendant problems of heat dissipation in the dynode resistor chain and the need for additional current capacity in the power supply.

This gain variation is reduced by an order of magnitude by stabilizing the voltage of the last dynode (or dynodes). For the analog mode of operation it is convenient to use a zener diode as shown in Fig. 17a. For pulse counting it is more usual to employ capacitors as shown in Fig. 17b.

2. Space Charge Effects

The characteristic curves for a typical PMT are shown in Fig. 18.

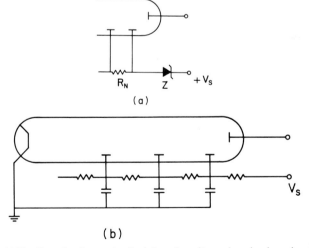

(a)

(b)

Fig. 17. (a) Nonlinearity due to the final dynode voltage changing in a signal-dependent manner is largely eliminated by using a zener diode (Z) to stabilize the operating point. (b) For pulse-counting methods, capacitor stabilization is used.

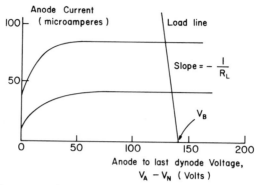

FIG. 18. Typical current–voltage output curves for two levels of illumination for a PMT.

Provided the anode-to-last-dynode potential, $V_A - V_N$, is maintained above a minimum level, usually about 70 V, then all the electrons emitted by the last dynode will be collected by the anode and the measured anode current will be linearly related to the radiant intensity at the cathode. If this voltage gets too low, space charge effects will be present and the output will become nonlinear. Linear operation is obtained by ensuring that the load line does not intersect curved portions of the characteristic curves. This means that the effective source voltage V_B (see Fig. 18), which is simply the anode-to-last-dynode potential, must be maintained at a high enough level. For fixed-voltage power supplies, this simply requires the appropriate choice of dynode chain resistor values. With some tube types, higher voltages between the last two or three dynodes are also recommended by manufacturers to prevent space charge problems. When variable power supplies are used, it is good practice to replace R_{N+1} with a zener diode, thus ensuring that the required voltage is present. A zener diode in this position thus does double duty.

In addition, the effective anode load resistance R_L must be sufficiently small. This can be achieved in practice by using a transconductance amplifier (see Section II,B and Fig. 19). This also gives a convenient way of controlling system gain by using a variable feedback resistor R_F. As with the photodiode, the signal output voltage V_o is given by

$$V_o = i_A R_F \tag{54}$$

with R_F typically in the range of 10^4–10^7 Ω. However, where nonlinearities as determined by the load line are concerned, the effective anode load resistor R_L is given by

$$R_L = R_F/A \tag{55}$$

where A is the open-loop gain of the amplifier. R_L is normally less than 1000 Ω.

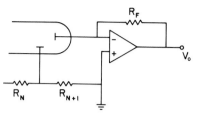

FIG. 19. The use of an op-amp in the output circuit has the advantages of eliminating the effect of space charge buildup, extending the frequency response of the system, and providing a convenient wide-range gain control.

3. Semitransparent Cathode Limitations

End-on semitransparent photocathodes necessarily have a high internal resistance. For high enough cathode photocurrents electric fields are generated that are parallel to the photosensitive surface. These fields may be sufficiently strong to distort the focusing field provided by the first dynode or the focusing electrode. In this way the collection efficiency of the first dynode varies in a manner that is signal-dependent, giving rise to nonlinear operation. Large-area tubes are worse in this respect, and the manufacturer's guidelines should not be exceeded. Cooling the tube, as is sometimes needed to reduce thermal noise, requires a substantial derating of the maximum allowable cathode photocurrent. A 1-in.-diameter bialkali photocathode may have a maximum current rating of 10^{-8} A at room temperature and only 10^{-10} A at $-100°C$. The effect is not present in opaque photocathode systems which can operate at cathode current densities that are higher by several orders of magnitude.

4. Other Nonlinearities

In addition to the previously described sources of nonlinearity there are a number of somewhat unpredictable effects. These include fatigue, hysteresis, and nonuniform sensitivity over the active area. Since these effects differ widely with different tube types, and even among tubes of the same type, and since the behavior in many cases depends upon the recent operating history and environmental conditions, it is not appropriate to discuss them in detail. Rather, in order to make the reader aware of possible problems, we summarize some of the results reported in the literature (Sauerbrey, 1972; Youngbluth, 1970).

From measurements taken on five different types of Cs–Sb (S11) end-on PMTs it was found that, over a 3-h period of constant illumination, the output current varied between a minimum of 10% for the most stable tube and a maximum of 400%. The variation in gain depends on the dynode structure and surface condition, and all parts of the dynode system contribute significantly to the gain instabilities. These tests were run at 20% of the maximum rating specified by the manufacturer. Some tubes showed a more-or-less steady drift over the 3-h period. With others, a major portion of the observed changes took place in the first few minutes. It was concluded that different internal recovery conditions, i.e., different distribution of gains among the dynodes and amounts of cesiation of each dynode, were possible.

There is a variation in responsivity over the active area of the photocathode. The variation may exceed a factor of 2 to 1. When the PMT is being used with an image-forming system, it is common practice to use a Fabry lens. This lens images the pupil of the image-forming element onto the photocathode, which has the effect of spreading the radiation uniformly and reducing variations in the output caused by vibrations and image motion.

C. Speed of Response

There is a delay from the time at which a photoelectron is generated at the cathode to the time at which the corresponding pulse containing many electrons is detected at the anode. This transit time is strongly dependent upon tube geometry and overall voltage. The intrinsic width of the pulse is also determined by tube geometry and overall voltage but is usually much less than the transit time itself. Manufacturers make special efforts to keep the pulse width as narrow as possible in order that individual pulses will not overlap at high counting rates.

Care must be taken not to degrade the intrinsic pulse width by the $C_S R_L$ time constants in the anode circuit, where C_S is the stray capacity between the anode and everything else and R_L is the anode load. R_L is often chosen to be as low as 50 Ω in order to maximize the count rate. An additional benefit of this particular value is the ease of matching the detector to the following electronics with standard coaxial line techniques.

With analog circuitry, a PMT resembles an almost perfect current generator with shunt capacitance C_S. As such it resembles a photodiode, and the discussion on frequency response given previously may be applied to the PMT with only notational changes being necessary.

D. Practical Considerations

For dc operation it is desirable to ground the anode end of the dynode chain. This keeps the output circuit at safe and convenient voltage levels.

The unwanted complication is that high electric fields are established in the glass envelope in the region of the photocathode and, because of this, there may be an increase in the dark level count rate and other deleterious effects. To prevent this it is normal to surround the cathode end of the PMT with an electrostatic screen connected to the cathode pin. To eliminate the electric shock hazard it is wise to shield the electrostatic screen mechanically so that it cannot be touched. The screen should not extend too far toward the tube base, otherwise the electric field in the intermediate space may cause luminescence within the envelope and corona discharge with a corresponding increase in dark current.

In pulse-counting operations, it is normal to ground the cathode side of the power supply. This obviates the need for the additional screen with its attendant problems. The output pulses are coupled to the following circuit by means of a dc blocking capacitor.

PMTs are sensitive to magnetic fields. A magnetic shield should be employed to prevent noise from being introduced from nearby power transformers and the like.

Surface leakage currents may also be a problem. They can be reduced to an insignificant level by paying proper attention to cleanliness (including fingerprints) and by controlling the humidity with dessicants. Phenolic base materials are particularly trouble-prone in humid conditions.

The cathode responsivity and the dynode gain are often significantly dependent upon temperature. Coefficients of 1% per degree are not uncommon. The tube temperature should be stabilized at the operating temperature to prevent drift from these effects.

In order to minimize gain drift it is common practice to stabilize the photomultiplier tube by operating it in the dark for several hours (up to 24 h) before use.

As a general rule it is best to choose a photocathode with the highest possible quantum efficiency at the wavelength(s) of interest. At the same time, it is also beneficial to have the long-wavelength cutoff as short as possible. The former increases the signal electrons and decreases the relative signal noise, while the latter minimizes the dark noise. If red-sensitive photocathodes are necessary, they should be as small as possible. In addition, it is often necessary to cool the PMT in order to reach an acceptable noise level. There is a wide variety of equipment available for this purpose.

Finally, the collection efficiency at the first dynode and the gain of the first dynode should be maximized. The optimum voltage between the photocathode and the first dynode (or focusing electrodes, if appropriate) can be determined by experiment, or the manufacturer's recommended operating voltages should be adhered to closely. This is especially important

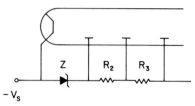

Fig. 20. The use of a zener diode (Z) to maintain a constant voltage between the photo-cathode and the first dynode.

for tubes with large-area photocathodes. The penalty for allowing the voltage to get too low is that the collection efficiency of the first dynode decreases. The result is a rapid, often dramatic, decrease in the signal-to-noise ratio that occurs as the overall supply voltage is lowered. When it is intended to vary the supply voltage over a wide range, a zener diode of appropriate rating can eliminate this problem. The diode replaces the resistor R_1 between the cathode and first dynode (see Fig. 20).

REFERENCES

The following articles, most of which have not been cited in the text, have been selected for further reading. They are representative of the large amount of literature in the field.

Baker, D. J., and Wyatt, C. L. (1964). Irradiance linearity corrections for multiplier photo-tubes, *Appl. Opt.* **3**(1), 89–91.

Budde, W. (1979). Multidecade linearity measurements on SI photodiodes, *Appl. Opt.* **18**(10), 1555–1558.

Eppeldauer, G. (1973). Some problems of photocurrent measurement of photovoltaic cells, *Appl. Opt.* **12**(2), 408–409.

Hamstra, R. H., Jr., and Wendland, P. (1972). Noise and frequency response of silicon photodiode operational amplifier combination, *Appl. Opt.* **11**(7), 1539–1547.

Havens, W. H. (1974). Measurement of low level photodiode noise currents, *Appl. Opt.* **13**(10), 2209–2211.

Keyes, R. J. (ed.), (1977). "Optical and infrared detectors" (Topics in Applied Optics, Vol. 19), Springer-Verlag, New York.

Kingston, R. H. (1978). "Detection of Optical and Infrared Radiation," (Optical Sciences Series, No. 10), Springer-Verlag, New York.

Lucovsky, G., and Emmons, R. B. (1965). High frequency photodiodes, *Appl. Opt.* **4**(6), 697–702.

Mohan, K., Schaefer, A. R., and Zalewski, E. F. (1973). Optical radiation measurements: Stability and temperature characteristics of some silicon and selenium photodetectors, NBS Technical Note 594-5, U.S. Department of Commerce, National Bureau of Standards, Washington, D.C.

Morton, G. A. (1968). Photon counting, *Appl. Opt.* **7**(1), 1–10.

Neiswander, R. S., and Plews, G. A. (1975). Low-noise extended-frequency response with cooled silicon photodiodes, *Appl. Opt.* **14**(11), 2720–2726.

RCA Corporation (1970). "RCA Photomultiplier Manual," RCA Corporation, RCA Electronic Components, Harrison, New Jersey.

Robben, F. (1971). Noise in the measurement of light with photomultipliers, *Appl. Opt.* **10**(4), 776–796.

Rodman, J. P., and Smith, H. J. (1963). Tests of photomultipliers for astronomical pulse-counting applications, *Appl. Opt.* **2**(2), 181–186.

Ruffino, G. (1971). "Comparison of photomultiplier and Si photodiode as detectors in radiation pyrometry, *Appl. Opt.* **10**(6), 1241–1245.

Sauerbrey, G. (1972). Linearitätsabweichungen bei Strahlungsmessungen mit Photovervielfachern, *Appl. Opt.* **11**(11), 2576–2583.

Seyrafi, K. (1978). "Electro-Optical Systems Analysis," 2nd ed., Electro-Optical Research Company, Los Angeles, California.

Schaefer, A. R., and Geist, J. (1979). Spatial uniformity of quantum efficiency of a silicon photovoltaic detector, *Appl. Opt.* **18**(12), 1933–1936.

Yariv, A. (1976). "Introduction to Optical Electronics," 2nd ed., Holt, New York.

Young, A. T. (1969). Photometric error analysis. IX: Optimum use of photomultipliers, *Appl. Opt.* **8**(12), 2431–2447.

Young, A. T. (1974). Photomultipliers: their cause and cure, *In* "Astrophysics. Part A: Optical and Infrared" (N. Carleton, ed.), Chap. 1, Academic Press, New York.

Youngbluth, O., Jr., (1970). Fatigue effects on the area sensitivity, dynode gain, and anode output for several end-on photomultipliers, *Appl. Opt.* **9**(2), 321–328.

Zatzick, M. R. (1972). How to make every photon count, in "Electro-Optical System Design," Milton S. Kiver Publications, Inc., Los Angeles, California.

CHAPTER 8

Arrays and Charge-Coupled Devices

JAMES A. HALL

Advanced Technology Laboratories
Westinghouse Defense and Electronic Systems Center
Baltimore, Maryland

I. INTRODUCTION

A solid-state image sensor is a photosensitive device for converting an irradiance image into a corresponding video signal. An approximate analog of the image can be reconstructed from this signal. Three basic functions are included in this process. First, the photon flux at each image point is absorbed in a corresponding elemental area of the material of the sensor (usually a silicon chip), which then generates free electrons and holes. Second, either the electrons or the holes (usually only the minority carriers) are collected at localized sites in the sensor that correspond to a one- or two-dimensional lattice of discrete image points. Third, after the conversion and collection process has proceeded for a designated integration time, the lattice of collection sites is interrogated in a scanning pattern so that a signal corresponding to the charge at each collection site in turn is developed and delivered to the output terminal of the device. Usually the scanning process also erases the stored charge pattern so that the sensor is ready to sense a new radiation image.

Because the scanning or commutating function takes place within the silicon wafer, rather than through the use of a separate electron beam as in a television camera tube, these sensors are known as self-scanned arrays. Two examples of solid-state image sensors are the Fairchild CCD221, a 380 × 488 element, charge-coupled imaging device (see Fig. 1) and the Reticon RA 100×100 photodiode array (see Fig. 2).

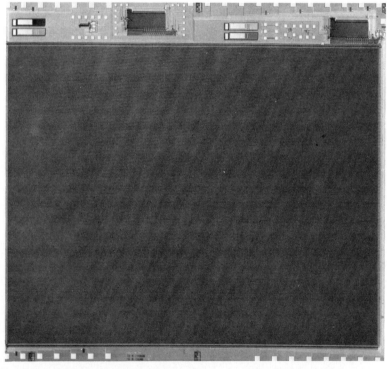

Fig. 1. The Fairchild CCD221 is a buried-channel charge-coupled area-imaging device providing an array of 380 × 488 elements for operation with normal U.S. broadcast television scanning standards. (Courtesy of Fairchild.)

FIG. 2. The Reticon RA 100×100 photodiode area array provides 10,000 light-sensitive elements on 60-μm centers, with MOSFET address switches controlled by an on-chip vertical shift register and two on-chip horizontal bucket brigade transport devices for signal output. (Courtesy of Reticon Corporation.)

A. SPECTRAL RESPONSE

In silicon solid-state image sensors, photons in the image are absorbed by exciting electrons in the silicon from the filled valence band across the gap into the conduction band. Thus, these devices respond only to light with a photon energy greater than that of the silicon band gap (about 1.1 eV) and a wavelength shorter than about 1.13 μm. Typical spectral quantum efficiency curves for silicon image-sensing devices are shown in Fig. 3. Curve A applies to a linear photodiode array with an antireflection coating, and curve B to a linear buried-channel charge-coupled device (CCD) time delay and integration (TDI) array with a carefully designed transparent tin oxide gate structure. Both arrays are sensitive over their entire exposed areas, and quantum efficiency values are referred to these areas.

In both cases, the quantum efficiency is high throughout the visible and very near-infrared wavelengths. The response falls at longer wavelengths because the silicon absorbs less strongly, the light is absorbed

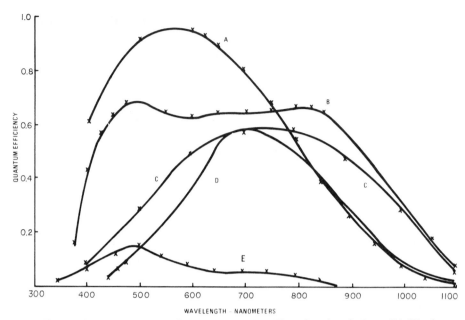

FIG. 3. Spectral quantum efficiency curves for silicon imaging devices. (A) Westing-house silicon diode array with antireflection coating peaked at 600 nm. (B) Westinghouse CCD TDI array with tin oxide transparent gate structure. (C) Typical TI thinned back-illuminated 100 × 160 CCD. (D) RCA SID52501 CCD silicon imaging device. (E) S25 trial-kali photocathode, 300 μA.

deeper in the structure, and more of the resulting charge carriers recombine at bulk centers before reaching the accumulation sites. The CCD response, curve B, is higher beyond about 800 nm, because the buried-channel collecting field (the depletion region) extends farther into the silicon and the minority carrier lifetime is longer in these devices because of processing differences. The response falls essentially to zero at 1130 nm where the silicon chip becomes transparent, and the response falls for wavelengths shorter than about 450 nm because the light is absorbed more strongly and many carriers generated near the first silicon surface are lost at surface recombination centers. A lower response below 450 nm is also due to a rising silicon index of refraction that increases reflection losses at shorter wavelengths.

Deviations from these essentially ideal curves are found in other typical devices for three reasons. First, in some charge-coupled imaging devices and two-dimensional diode arrays the devices and interconnecting structures required for commutation represent nonphotosensitive areas within the array. The net response over the image drops below that in Fig. 3 to the fraction of the area which is sensitive, typically 25–40% for interline transfer charge-coupled imaging devices. Curve D for the RCA SID52501 shows such a reduction, caused in part by operation in an anti-blooming mode, which reduces the effective sensing area, especially for shorter wavelengths. This device also uses polysilicon gate electrodes that absorb some of the incident light in the blue and violet wavelength region.

Second, in most CCDs and charge injection devices (CIDs), the gate structure is formed of interleaved thin films of conducting silicon or tin oxide and insulating silicon dioxide, perhaps with layers of insulating silicon nitride. For collimated light, the optical interference pattern resulting from multiple reflections at the layer interfaces can modulate the transmission into the silicon as a function of wavelength. Such modulation produces a fine structure in the spectral response curves shown in some data sheets, but barely seen in curve B in Fig. 3 for a Westinghouse tin oxide transparent gate CCD, perhaps because of averaging over the converging incident beam. If such optical interference effects exist, they can be eliminated by the use of a thinned chip in which the image illumination is incident on the back surface instead of passing through the gate structure. Some Texas Instruments (TI) CCDs use such rear-surface illumination. For best performance, this surface should be accumulated by adding doping after thinning so that photogenerated minority carriers are repelled from the surface away from the surface traps, enhancing the blue and near-ultraviolet response. When combined with antireflection coatings, this design can give a uniformly high, smooth spectral response throughout the visible spectrum. Curve C in Fig. 3 is described by TI as typical. As shown, the quantum efficiency falls below the antireflection-coated photodiode line array and the front-surface tin oxide gate buried-channel CCD array at short wavelengths, apparently because the rear-surface doping profile is not ideal and carriers are combining there. The response in the near infrared is slightly lower than that for the transparent gate CCD because more long-wavelength photons pass through the thinned wafer without being absorbed.

With these qualifications, relatively high quantum efficiency over the visible and near infrared is a desirable characteristic found generally in the class of silicon-based solid-state image sensors. For comparison, curve E shows that a good S25 photoemissive cathode used in some low-light-level image sensors averages only 9% quantum efficiency through the visible spectrum.

B. Charge Storage Operation

The photogenerated current reaching each collection site is very small, typically on the order of picoamperes. Thus an attempt to develop a signal by measuring the instantaneous current at each collection site in turn would fail because the noise of scanning and amplifying circuits is substantially higher. Instead, the sensor is designed to collect and store the charge carriers reaching each collection site during the entire integration time between interrogations, and the signal corresponds to this charge. This is known as charge storage operation and corresponds to full-frame integration in a television camera tube. Each charge collection site should collect only carriers generated nearby and should be capable of storing them for a period long compared to the integration time. Any thermally generated leakage current reaching each site should be small compared to the current of photon-generated carriers. Collection structures used are either a lattice of back-biased diodes at one surface of the silicon chip or a lattice of induced depletion regions under insulated gates in a metal insulator semiconductor (MIS) structure in a CCD or CID. In either case, the minority photocarriers are stored in potential minima. The charge storage medium is the depletion layer capacitance associated with each real or induced diode. The leakage charge is kept below the signal charge by careful processing and by operation of the device below room temperature if integration times are longer than a few milliseconds.

C. RESOLVING POWER

The resolving power for either diode arrays or CCDs is normally specified in terms of the modulation transfer function (MTF). Because the discrete array structure of the sensing elements makes the output signal a function of the test pattern image position as well as the image spatial frequency, a signal from a single sensing element is often measured as the image of a slit or an edge is moved past that element. The image sensor MTF is then computed as the Fourier transform of the line spread function, after mathematically removing the spread function of the optical system. Typical measured values are shown in Fig. 4a, which gives the MTF versus wavelength of illumination for a Fairchild CCD121H 1728-element linear CCD with $18 \times 13 \ \mu m^2$ elements on 13-μm centers. Because any array samples the image over a finite number of discrete points, the maximum image spatial frequency that can be unambiguously transformed is the Nyquist frequency, at which each sensing element corresponds to either a maximum or a minimum in the image. It is therefore customary to express the MTF as a function of spatial frequency normalized as a fraction of the Nyquist frequency for each array.

Some data sheets express the resolving power as the contrast transfer function (CTF) or the square wave amplitude response. This CTF is the relative signal amplitude versus the spatial frequency obtained when

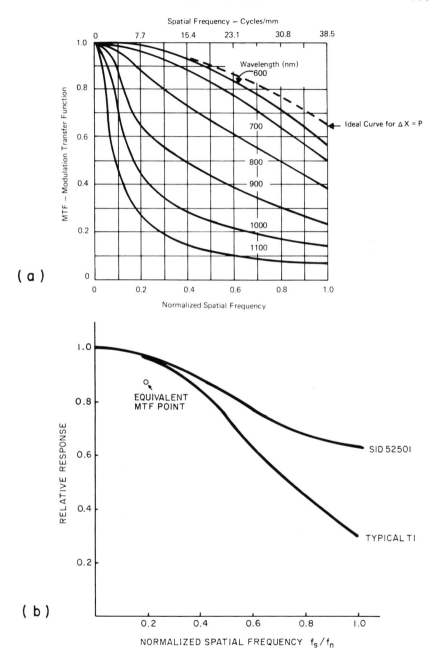

FIG. 4 (a) MTF data for the Fairchild CCD121H 1728-element linear array shows high MTF at shorter wavelengths below 800 nm. (Courtesy of Fairchild.) (b) CTF (square wave response) from published data for RCA SID52501 and TI back-side-illuminated array.

imaging a high-contrast bar chart test pattern like that normally used in evaluating television camera tubes. The bar pattern image is moved slowly past the array. The difference between the maximum and minimum signal values from a single detector element is related to the low-frequency amplitude and is recorded as a function of spatial frequency as in Fig. 4b. Unfortunately, the published bar chart data cannot readily be transformed to the MTF, because one needs correctly phased response data to several times the Nyquist frequency to perform the entire conversion. A single approximate value is shown at $f_s = 0.2 f_n$ to indicate that the MTF is significantly lower than the square wave response curve. The RCA and TI data are assumed to have been taken with illuminant A, a broad-band 2854-K tungsten source, but with an infrared attenuating filter for the RCA device. As shown, the relatively large elements of the SID52501 give a reasonably high square wave transfer function (CTF) at all frequencies up to the Nyquist frequency, but the MTF falls significantly below 100% at even 20% of the Nyquist frequency. The reason, suggested by the data on the Fairchild CCD121H, is primarily that the longer-wavelength photons from the source are absorbed deep in the silicon and the carriers can diffuse laterally before being confined in a CCD well. For all transparent gate arrays, normally fabricated on chips approximately 250 μm thick, a high MTF is found only for radiation wavelengths shorter than 600 or 700 nm, where most charge carriers are generated in or very near the potential wells in the depletion region. Even though the chip is thinned to about 10 μm, the CTF of the TI back-surface-illuminated array falls below the RCA results. This is probably because the TI array elements are smaller and because nearly all the carriers are generated in the bulk so that lateral diffusion is a factor at all wavelengths.

The problem of lateral charge diffusion of near-infrared-generated carriers is basic to the properties of the silicon material. The use of an infrared blocking filter to restrict sensor illumination to the visible or near visible improves the MTF performance significantly, without a great loss in response for many illuminants, since the sensor response falls rapidly beyond $\lambda = 900$ nm.

D. Speed of Response

Most solid-state image sensors are read and reset every integration time, so that each exposure cycle is essentially independent of the one that preceded it. Thus, there is no "lag" effect like that found in television camera tubes. Further, for a short-pulsed light source, the time required for charge carriers to reach a collection site is typically less than a few microseconds. A reciprocity law relates exposure irradiance and exposure time for light pulses shorter than the time between reading cycles. The speed of response to moving scenes is therefore limited simply by the read-out rate. Each sensor element generates a signal that represents the average irradiance on that element over the preceding exposure time. Changes in intensity or position of an image within an exposure time are averaged, just as they would be with a film camera. Thus, speed of response is a property of the system, not of the solid-state image sensor. To follow rapid motion, one must read the sensor more frequently and therefore provide a larger video channel bandwidth to handle the higher data rate. For astronomical photon-counting applications, 10^4-element arrays have been read as frequently as 100 or 200 times/s, and line arrays have been read 1000 or more times per second. (Still higher rates are possible but, since signal bandwidths must then be wider, higher circuit noise may limit the attainable signal-to-noise ratio. See Sections III,C and VII,A.)

II. PHOTODIODE ARRAYS

The earliest form of solid-state image sensor used a matrix of photo-diodes or phototransistors to sense the radiation image.

A. COMMUTATION

To interrogate each collection site of a photodiode array sensor, one must form a conducting path from each back-biased diode in turn to the device output terminals, while using as little sensing area for interconnecting devices as possible. Consider one photodiode in an array, as shown in Fig. 5. This element is connected to a signal output bus by a metal-oxide-semiconductor field-effect transistor (MOSFET) address switch, and the bus is connected through a load resistor to the reset voltage supply. The switch is closed to back-bias the diode to the supply voltage and then is opened to isolate the diode that collects photogenerated charge carriers during the following integration time. These reduce the voltage across the diode. When the address switch is again closed, current to recharge the diode flows through R_L, and the peak signal voltage approximately equals the collected charge divided by the output bus capacitance C_S. Figure 5 also shows a variant of this scheme in which the address switch is a second smaller diode (D) back to back, with the photo-diode. A negative pulse on the reset line forward-biases the isolation diode, recharging the photodiode essentially to the reset supply voltage.

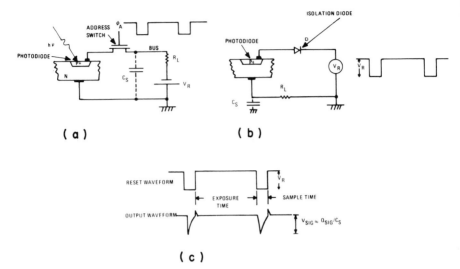

FIG. 5. Read–reset cycle for a photodiode operating in the charge storage mode. In (a) the address switch is a MOSFET and the peak signal voltage V is approximately Q_{SIG}/C_S, decaying as shown in (c). In (b) the sampling pulse provides the reset potential through the isolation diode, but the diode serves the same function as the MOSFET in circuit (a).

Removal of the pulse leaves both diodes back-biased essentially to the reset voltage, since the isolation diode capacitance is small compared to that of the photodiode, and photogenerated charge carriers are collected on the back-biased diode combination. The next address–reset pulse recharges the photodiode, and the reset current peak amplitude appearing as a voltage pulse across R_L is the signal from that element. In either case, the signal pulse is likely to be contaminated by switching transients capacitively coupled through the switch, since peak signal occurs essentially in synchronism with the reset pulse. However, the general scheme is widely used because of its simplicity and because for many applications available light is adequate to provide signal pulses 100 or more times larger than the switching transients.

Individual photodiode sensor elements are combined into line or area arrays as shown in Figs. 6 and 7. Figure 6 shows a Reticon RL256 linear diode array in which each diode is addressed by a series MOSFET, as indicated in the simplified schematic diagram. Each address switch is turned on and off in turn by a pulse from an on-chip shift register at a rate set by an external clock. Each scan of the array is initiated by a start pulse, which may be derived from the shift register output if continuous scanning is desired. This type of array is usually used to view moving

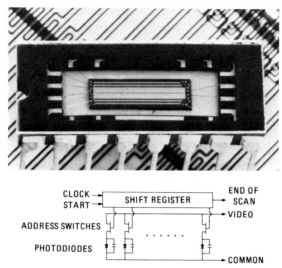

FIG. 6. RL-256 photodiode line array with 25 μm × 25 μm elements. (Courtesy of Reticon Corporation.)

images. For example, similar but longer arrays are used to televise pages of typed information in facsimile systems, where mechanical motion of the image across the array provides the second dimension for area scanning. As shown, the address switch MOSFET array is placed beside the photodiode array so that the sensing elements are contiguous and the entire sensitive area defined by the on-chip light shield is effective. For the array shown, there are actually two shift registers, one on each side of the photodiode array, with two sets of address switches and two output buses, one set addressing the even sensor elements and the other the odd elements. This and other similar schemes for subdividing the output data stream permit use of narrower bandwidth amplifiers for lower noise.

An area array is more complicated, since each address component must be placed near its sensing element, hence within the photosensitive area. Much ingenuity has been used in devising address structures for minimum complexity and minimum use of area on the chip. Unless a separate control lead is provided to each element, two address switches are required for each photodiode. With two switches per element, each row can be enabled as a unit and each column then addressed in turn. Only the single element where both row and column switches are closed is read into the output signal bus. All other elements have at least one switch open and are isolated and active in collecting photogenerated charge carriers. Figure 7 shows a photograph and a simplified circuit of a 50 × 50

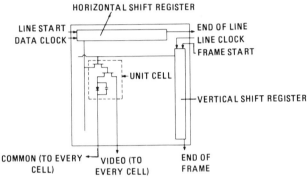

(b)

FIG. 7 (a) RA 50×50 matrix photodiode array. (b) Schematic diagram of photodiode matrix array. (Courtesy of Reticon Corporation.)

element photodiode array. The vertical shift register provides a pulse which turns on the MOSFET switch gates of an entire row of photodiodes at a time through a single control bus per row. The horizontal shift register, operated at higher speed but in synchronism with the first, biases a column of enable switch drains at a time, turning on the single address switch at the selected intersection. Thus, every element in an $m \times n$ element array can be addressed with $m + n$ control buses, rather than $m \times n$, at the cost of providing two control elements at each diode location. A similar design is used in all photodiode area array sensors.

Commercially available photodiode area arrays are relatively small, 50×50 or 100×100 elements, although a 400×500 element array was made experimentally several years ago. On the other hand, photodiode linear arrays are commercially available in sizes up to about 2000 elements per chip, and linear arrays can be made effectively as long as desired by placing several chips end to end if the sensitive area extends nearly to the chip edge. In addition, solid-state linear array sensors can provide the option of multiple-output data streams, each representing only a short segment of the array, so that low-level signal amplification can be accomplished through relatively narrow video bandwidths for an improved signal-to-noise ratio.

B. Output Circuits

Photodiode array sensors as described so far have reasonable image quality at comparatively high exposure levels. There are three limitations on image quality at lower exposures. First, transients produced by turning address switches on and off couple into the signal output through gate-to-source, gate-to-drain, or bus-to-bus stray capacitances. With the peak amplitude output circuit described above, video and address pulses occur simultaneously, and transients from both the turning on of the present element and the turning off of the previous element are mixed with the signal. (Averaging and compensating circuits are often used to minimize their amplitudes.) Second, the signal voltage is limited to the ratio of the signal charge to the output bus capacitance, including interconnection capacitance and the input capacitance of the first off-chip amplifier. Third, the amplifier bandwidth must be comparatively wide to maximize the peak signal at its output.

The signal circuits in Fig. 8a and b were devised to solve these problems and are now widely used. The first amplifier stage is a MOSFET on the sensor chip that measures a signal voltage from each element. One amplifier serves only one or a few elements, minimizing shunt capacitance and maximizing signal voltage. The gate node is reset before each new element is addressed. However, address, read, and reset operations are separate, and transients can be essentially eliminated from the signal if amplifier bandwidths are wide enough so that the output settles between the reset, address, and sample operations. Image quality is then limited primarily by random noise and by element-to-element nonuniformities.

Noise is discussed in Section II,C. Nonuniformities are of two classes. The first occurs in the dark and is due to differences in dark current from element to element, or to differences in the dark signal output resulting from differences in amplifier transconductance or in the effective reset voltage from element to element; both are usually related to MOSFET threshold differences. The second occurs when the array is illuminated and is usually signal level dependent. This indicates differences in response from element to element and may be due to dif-

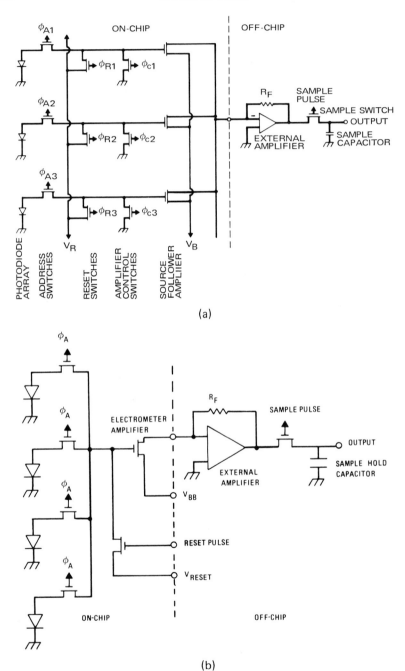

(a)

(b)

ferences in minority carrier lifetimes resulting from local impurity concentrations in the silicon, to differences in amplifier gains, to optical property variations which can produce element-to-element variations in spectral response, or to variations in element size due to photolithographic variables in manufacturing. These nonuniformities are usually repetitive in time and can in most cases be calibrated out in the signal-processing equipment if the sensor is used for a critical application where the cost of individual element calibration is warranted. Dark-level variations due to leakage currents can usually be reduced by operating the array at a lower temperature. Leakage currents in silicon devices are halved when the device is cooled 8°C for temperatures near 300 K. The remaining nonuniformities are minimized by the imaging array manufacturer to achieve better quality. Typical response uniformity values are given in Section V.

C. Random Noise

When the address switch in a diode array is conducting, the charge on the photodiode is transferred to the bus until the same voltage appears across each. The equivalent circuit is shown in Fig. 9, where the switch is represented by an equivalent resistance R_A in parallel with the series combination C_D and C_S. At equilibrium, the voltage across R_A is zero, except for the randomly varying Johnson–Nyquist noise that appears across the resistor. The shunting capacitor combination attenuates the higher-frequency noise components. It can be shown that the effective rms noise voltage appearing across R_A in this circuit is $V_n = \sqrt{kT/C}$ or the rms charge fluctuation on each capacitor is $Q_n = \sqrt{kTC}$, where C is the capacitance of the series combination of C_D and C_S. For $C_D \ll C_S$, the usual case,

$$Q_n = [kT\, C_D C_S/(C_D + C_S)]^{1/2} \approx \sqrt{kTC_D}$$

Since this expression does not involve R_A, use of a lower-resistance transistor switch does not reduce the effective noise charge. However, the expression does not include excess white or $1/f$ noise in the MOSFET, which can increase Q_n by up to $\sqrt{2}$. For a near-optimum 0.015-pF diode capacitance, the address switch noise at room temperature is 48–70 elec-

---

FIG. 8 (a) Each photodiode has its own amplifier for minimum shunt capacitance. Sensing signal voltage across each photodiode with an on-chip MOSFET amplifier can greatly increase the signal-to-interference ratio. (b) A short segment of the photodiode array is read into a single amplifier by address switches.

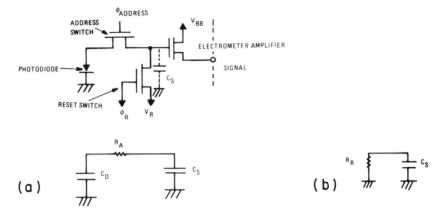

Fɪɢ. 9. Address and reset switches introduce Johnson–Nyquist noise which is sampled and added to the signal. (a) Address switch when closed is a resistor paralleling combination of series C_D and C_S introducing Johnson–Nyquist noise $Q_n \approx \sqrt{kTC_D}$ for $C_D \ll C_S$. (b) Reset switch introduces Johnson–Nyquist noise of $\sqrt{kTC_S}$.

trons per sample, and any switch-addressed array can be expected to have address noise of this order.

The reset switch also introduces Johnson–Nyquist noise as illustrated in Fig. 9b. For simplicity, assume that only the bus is reset. By an ingenious design and the use of a small transistor, serving one or a very few diodes, the bus shunt capacitance can be reduced to about 0.2 pF. Reset noise is then about $Q_n = \sqrt{kTC_s} = 177$ electrons per sample if no excess transistor noise is assumed. While nearly optimum for diode arrays, these noise levels are not insignificant. Yet there are many applications where this performance is perfectly acceptable and where diode arrays are preferred to CCDs, since a diode array can have important advantages in higher quantum efficiency and larger signal-handling capacity, hence a larger dynamic range. Further, use of correlated double sampling (CDS) described in Section III,C,1, can nearly eliminate reset noise, leaving address switch noise as the principal random noise source characteristic of a diode array. A more complete discussion of noise sources in solid-state imagers is given in Section III,C and D.

III. CHARGE-COUPLED DEVICES

A. Basic Principles

Charge-coupled image sensors not only are commutated differently but also use a different means for forming charge-integration sites. These

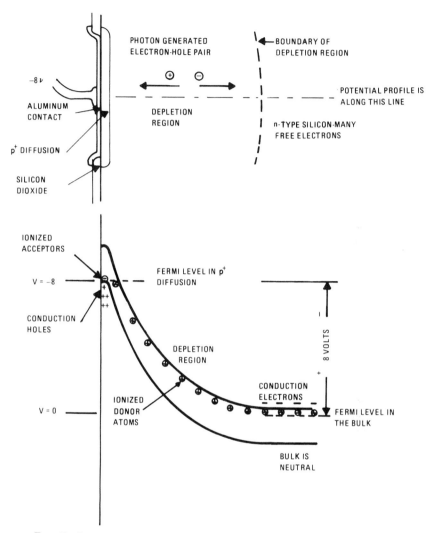

FIG. 10. Structure and potential distribution of back-biased p^+–n photodiode.

are described by contrasting them with the diode array structure. (Examples are given for devices formed on n-type silicon where holes are the minority photon-generated carriers. For p-type silicon devices the diagrams in Figs. 10, 11, and 13 would be inverted.) In a back-biased diode array, the structure of an integration site is as shown in Fig. 10. The p^+ diode diffusion is biased negatively to repel the free conduction electrons in the surrounding n-type silicon. The diffusion is surrounded by a depletion region with essentially no free electrons but with a space charge of positively ionized donor atoms so that the potential energy for holes varies parabolically from a minimum at the conducting p^+ diffusion up to the neutral silicon. Light absorbed in the depletion region or in the surrounding n-type silicon creates electron–hole carrier pairs. The electrons are urged outward and merge with the free electron population in the bulk silicon beyond the depletion region. Some of the free holes diffuse to the depletion region, and they and the holes generated in this region are swept toward the negative p^+ diffusion, making it slightly less negative. Once they reach the diffusion, the photon-generated holes are confined there, surrounded by an essentially insulating region with a strong field pre-

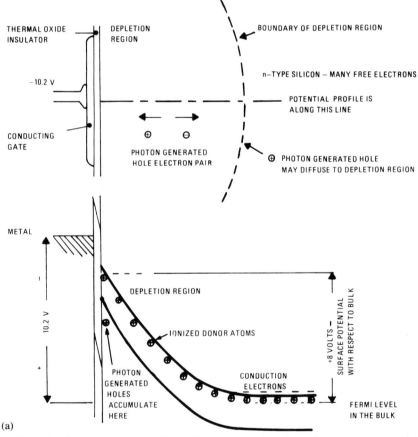

FIG. 11a. Structure and potential profile of a collection site under a gate in a surface-channel CCD. For the example shown, the silicon surface is 8 V more negative than the bulk and the photon-generated charge is small. A larger charge makes the surface less negative.

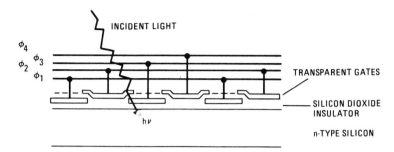

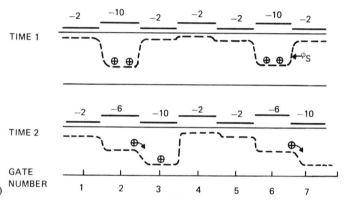

Fig. 11b. CCD shift register structure and surface potential distribution. This CCD electrode structure uses four transparent gates per cell. One or two phases are negative at any time, and their neighbors are less negative to localize the collected charge. Surface potential distribution is varied as shown to move collected charge packets along the array.

venting them from leaving. This is the structure and mechanism for each of the collection sites in every diode-sensing array. Addressing each site to measure the charge, as described above, is accomplished by closing an amplifier reset switch and then a switch which creates a conducting path between the diode diffusion and a signal bus and first amplifier, although Johnson noise is introduced by the switches.

A completely different method of forming and addressing collection sites in a solid-state image sensor was described by Boyle and Smith (1970).* The basic structure, shown in Fig. 11a, consists of a silicon chip with a thin film of insulating oxide on its surface. On the oxide are depos-

---

* An excellent basic treatment from the device viewpoint is Barbe (1975).

ited a series of conducting gate electrodes, each forming a small metal oxide semiconductor (MOS) capacitor.

Figure 11a shows the potential and energy band structure beneath one such gate electrode. As the gate is biased more negatively, the potential at the silicon surface beneath the oxide becomes more negative. For the assumed n-type silicon this repels the free electrons in the conduction band, creating a depletion region essentially empty of free electrons at and just below the surface. If light falls on the chip, passing through the transparent gate and insulator structure, electron–hole carrier pairs will be created in the depletion region and beyond. The electrons are urged away from the surface and merge with the cloud of conduction electrons away from the depletion region. The holes, which are the minority charge carriers in the n-type material, diffuse through the chip, with a lifetime of many microseconds. Those that diffuse into the depletion region and those generated originally in the depletion region are urged toward the more negative silicon surface and collect there, much as they did on the p^+ diffusion in the diode structure. Here there is no conducting diffusion—the holes are simply confined to the region under the gate by the field in the insulating depletion region. As they accumulate, the positive charge of the holes makes the silicon surface less negative. In a sense the charge is stored jointly on the depletion layer capacitance and on the MOS capacitance between the surface and the gate, with the latter usually dominant.

Now assume that the gate in Fig. 11a is gate 2 in a series of such gates, isolated from each other but placed very close together or overlapping, as shown in Fig. 11b. As long as gate 2 is more negative than its neighbors, holes will be isolated under it (unless the stored charge becomes large enough to alter the potential distribution). If gate 3 is now made more negative and gate 2 less so, the charge carriers will diffuse over to the potential minimum or "well" under gate 3. In the transfer they will be urged on by the lateral fringing field in the silicon from least negative gate 1 to more negative gate 2 to most negative gate 3. If the gate voltages are then varied cyclically in groups of three or four, the charge packet can be moved along the chip under the gate structure by keeping a potential hill behind it, a deeper well in front, and another hill beyond the deep well to confine the charge. Finally the charge packet can be moved to an output structure, either a more negative diffused charge-collecting diode or a charge-sensing gate, to be described later. If a diode is used, the same gated charge integrator output circuit can be used as was used for a diode array, except that the diode can be permanently connected to the gate of the first on-chip MOSFET amplifier, eliminating the address switch and its Johnson noise contribution. Further, the diode and transistor gate structure can be very small, hence low-capacity, and a small signal charge will produce a comparatively large voltage swing.

The concept of the CCD is significantly different. Charge is transferred from collection site to collection site not through a conducting switch and not by establishing a voltage equilibrium between two capacitors, but by changing potentials within what is essentially an insulator, the depletion region in a semiconductor. And since the charge is "untouched," because it never enters a conductor until it reaches the output-collecting diode, charge transfer is essentially complete. The essence of the process is that equilibrium is never established. As charge accumulates, from either photoexcitation or thermal excitation, it is transferred out to the collection diode. A good analogy is the process of lifting a bucket and pouring all the water from it into a following lower bucket. Except for a few drops clinging to the container walls, liquid transfer is complete. In a charge transfer device, except for a few carriers left in traps in the semiconductor, charge transfer is complete. Thus, to a first approximation there is no transfer noise. There is definitely no Johnson–Nyquist noise within the array. It is this feature that makes image-sensing CCDs attractive for possible use at low exposure levels.

B. TRANSFER EFFICIENCY

The use of charge coupling to read the charge signal from a large array at television rates could involve many transfers, effected at high speed, to provide output data rates of 5–10 megasamples per second. Under these conditions the number of charge carriers left behind in trapping states in the semiconductor at each transfer could be significant, because these trapped carriers would be reemitted into following charge packets. Thus if one were imaging an edge between a bright area and a dark area, charge left behind from the large, bright charge packets would appear in the following, nearly empty charge packets as a trailing edge, acting to reduce the effective MTF as shown in Fig. 12. Similarly, after a dark-to-light transition the first large packets would lose part of their signal to filling traps, reducing the sharpness of the transition.

The seriousness of this trapping-and-release mechanism depends on the concentration of trapping states in the silicon and on the time constant for reemission of trapped carriers. Unfortunately, localized states capable of trapping carriers are especially numerous at the silicon surface where the periodic crystalline silicon lattice stops and the mismatched silicon dioxide lattice begins. In the surface-channel CCD just described, the potential minimum is at the silicon surface, the signal charge is confined at or very close to the surface, and trapping is very likely. As the size of the charge packets varies, so does the effective surface potential, causing different local states to function as traps.

The first means of minimizing trapping and release of carriers in

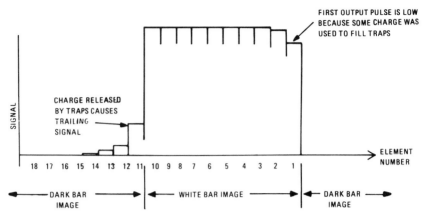

FIG. 12. Transfer inefficiency in a CCD can reduce the sharpness of transitions, hence reduce the MTF.

surface-channel devices is keeping the traps full. This can be approximated by never allowing the CCD wells to empty, which requires a continuous source of charge carriers whether the image is bright or dark. This bias charge, known colloquially as "fat zero" from digital terminology, has been commonly supplied by general low-level illumination of the sensor surface by bias lights placed within the lens housing (as is done for the RCA SID52501). Alternatively, charge can be injected into the imaging sensor through a series of diodes at the beginning of each row or column, but this approach requires great uniformity in the injector structures to avoid line-to-line background signal differences.

A second approach in avoiding the effect of surface traps is keeping the signal charge packets away from the surface. This requires a revision of the structure of the CCD so that the potential minimum occurs within the silicon with the silicon surface at a higher potential, the so-called buried-channel or bulk-channel design. The desired potential profile is shown in Fig. 13 and is accomplished by forming a surface layer of opposite doping type in the silicon before growing the insulating oxide. As shown, an originally n-type wafer is doped with a p-type acceptor impurity, usually ion-implanted boron, to a concentration 100 or more times higher than that of the underlying n-type wafer. A thin oxide insulating layer is grown on the surface, and transparent gate electrodes are deposited as in the surface-channel device. When the gates are grounded to the wafer, the band structure in the silicon is as shown in Fig. 13a. The p-type surface layer forms a large-area $p-n$ junction with a narrow depletion region separating it from the n-type wafer. Both the wafer and the surface layer are conducting. The p^+ charge collection diode is then biased negatively by 10 or 15 V. The conducting p-type layer becomes more negative

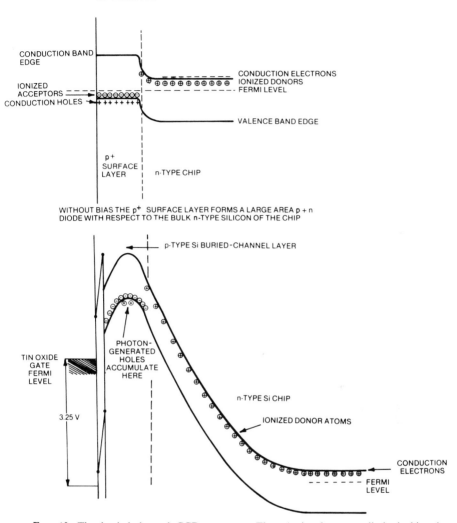

FIG. 13. The buried-channel CCD structure. The p^+ signal output diode is biased 10–15 V negatively, draining conduction holes until the surface layer is depleted. Band bending toward the less negative bulk silicon and gate electrode structure forms the buried channel away from the surface where photon-generated holes accumulate. Varying gate voltages keeps the charge confined in packets and moves them into the diagram toward the output diode.

as holes are conducted to the output collection diode, and the junction depletion region broadens into the substrate and to a lesser extent into the surface layer. Since the gates are near ground potential, the surface of the silicon is now more positive than the interior of the p-type layer, and a depletion layer extends from the surface further into the silicon as the output diode is made more negative. At a certain voltage the surface and junction depletion regions meet; the p-type surface layer is entirely depleted of free holes and ceases to be a conductor. The potential minimum, the channel, is now away from the surface. At this point making the diode more negative has no effect on the channel potential, which is determined by the gate and substrate potentials and by the amount of signal charge in the channel. The gates are now cycled between voltages such as 0 and −10 or +5 and −5, urging hole-type carriers toward the collection diode. Where the gates are more negative, the channel is deeper, and where more positive, the channel is shallower, providing a barrier to charge carrier flow along the channel. Thus the buried-channel charge-coupled image sensor functions like the surface-channel device, except that the potential minimum is provided by the output diode and the gates are more positive than the channel.

The buried-channel CCD structure is now used by virtually all device manufacturers. It can provide effective transfer efficiencies greater than 0.99999 with no bias charge, partly as a result of greater fringing field effects in the deeper channel. Hence it is suited for low-signal-level applications. Its principal disadvantages are somewhat lower charge-handling capacity and potentially higher leakage currents that might require lower operating temperatures.

C. Low-Noise Signal Output Circuits

A typical output circuit for a charge-coupled imaging device is the gated charge integrator in Fig. 14. The operating cycle is:

(a)   Close and then open the reset switch to reset the collection diode and MOSFET gate to the reset bias potential V_R.

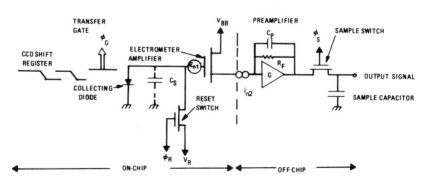

Fig. 14. For a simple gated charge integrator, the principal noise source is Johnson noise in the reset switch. (Courtesy of Westinghouse.)

(b) Cycle the CCD gate voltages to transfer a charge packet to the collection diode. The storage capacitance C_S is primarily that of the MOS gate. Hence the resulting voltage excursion is approximately linear with charge.

(c) Close and then open the sample switch to hold at the amplifier output a signal proportional to the charge. The succession of such signals is the video signal from the CCD register.

This circuit has no address switch, hence no address switch noise, but the Johnson–Nyquist noise of the reset switch is $Q_N = \sqrt{kTC_S}$. For $C_S = 0.15$ pF, $Q_N = 153$ electrons. When the reset switch opens, this noise charge is sampled and held on the gate node capacitance. The assembly of such samples will therefore form a video dark noise whose standard deviation is also 153 electrons.

1. Correlated Double Sampling

To eliminate reset noise, McCann (1973), and White *et al.* (1974) of Westinghouse devised CDS, shown in Fig. 15. They added a series capacitor in the video signal path and a shunting clamp switch. The operating cycle is:

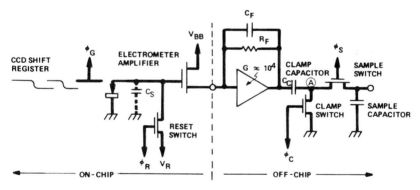

FIG. 15. CDS removes reset noise at the CCD output. A reset switch replaces the load resistor for CCD. The CDS shown removes most of reset switch Johnson noise, leaving CCD noise, electrometer amplifier noise, and off-chip amplifier noise. C_S is still a performance-determining factor. CDS operations:

1. Reset diode node to $V_R \pm \sqrt{kTC_S}$.
2. Close clamp switch so negative of reset noise sample appears across C_C. Open clamp switch.
3. Transfer next signal charge packet to gated collection diode.
4. Signal at A is $f[(q_s + q_n) - q_n] = f(q_s)$ alone. Operate sample switch to present corresponding signal output. (Courtesy of Westinghouse.)

(a)   Close and then open the reset switch. This samples and holds a noise charge $Q_N = \sqrt{kTC_S}$ on the gate node.

(b)   Close and then open the clamp switch. This stores on the clamp capacitor the negative of the amplified reset noise.

(c)   Cycle the CCD gate voltages to transfer a charge packet to the collection diode. The effective charge on the diode is now the sum of the signal packet and the sampled reset noise, and a corresponding voltage appears at the amplifier output. But the voltage at the output of the clamp capacitor is signal alone. The reset noise has been removed.

(d)   Close and then open the sample switch to hold the signal at the amplifier output.

CDS works because the resistance from the gate node to the chip substrate is high, so the storage time constant is long compared to the time T_0 between the opening of the clamp and sample switches, and also because sufficient gain is provided so that the noise of the clamp and sample switches may be neglected. The frequency response of the CDS circuit, shown in Fig. 16, is zero at dc and at frequencies of n/T_0. Thus, CDS cancels out reset switch noise, strongly attenuates $1/f$ noise in the on-chip amplifier, and makes the output signal independent of dark-level drifts in circuits ahead of the clamp switch. This reduction in noise has a price,

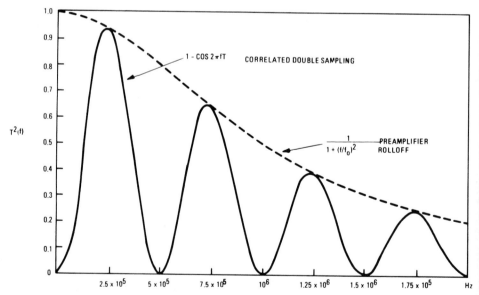

FIG. 16. Effective frequency response of a CDS system with 2 μs between opening of clamp and sample switches. (Courtesy of Westinghouse.)

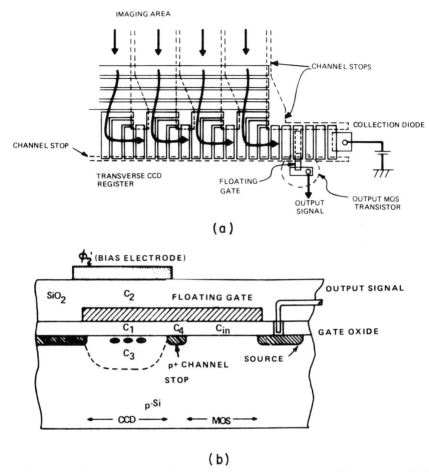

FIG. 17. (a) For floating gate output the gate extends under a gate of the extended CCD transverse output register. (b) Cross-sectional view of the floating gate structure. Amplifier drain diffusion is not shown. (Courtesy of Fairchild.)

however. Each signal sample includes two samples of the noise of the preamplifier, including the first on-chip stage, which doubles the amplifier noise power. Further, because transients must be allowed to settle before both clamp and sample operations, double sampling is difficult to mechanize at data rates above a few megahertz in a single channel. Nevertheless, this circuit is widely used and can provide noise equivalent signals as low as 25 electrons per sample for data rates of 1 or 2×10^5 samples/s.

2. *Floating Gate Amplifier Output*

A different approach to a low-noise signal output circuit is a floating gate amplifier (FGA), which reduces the noise bandwidth by removing the need for CDS. Since noise in the sensor output circuit is produced by address and reset switch action, and shunt capacitance is partly the junction capacitance of the output collecting diode and reset switch, workers at Fairchild (Wen, 1974; Wen and Salisbury, 1973) have dispensed with both switches and charge collection and sense the signal charge capacitively while it is being clocked along the CCD. Figure 17a shows how the output CCD register is extended beyond the sensing array, while Fig. 17b shows a section through the floating gate looking along the register. As can be understood from Fig. 11 or 13, signal charge in either a buried- or surface-channel CCD changes the potential at the semiconductor–insulator interface, hence changes the field in the insulator. If a well-insulated floating gate electrode were buried beneath a CCD gate in the oxide close to the silicon surface, its potential would change as a function of the amount of signal charge in the well. Figure 17a shows how the floating gate extends under the phase-2 CCD gate electrode, and is also the gate for a MOS transistor immediately adjacent. Signal electrons in the n-channel CCD make the floating gate more negative, decreasing the current in the n-channel MOSFET. A smaller charge in a following packet would increase the MOSFET current. Thus the output signal (the current in the MOSFET) simply varies with the signal charge without a need for collecting or resetting. The principal noise source is shot noise in the MOSFET channel current. To minimize the effect of noise in both the MOSFET and the following stages, the shunt capacitance from the floating gate should be as small as possible, whereas the MOSFET transconductance should be high but with a low channel current. Fairchild workers report designs in which C_S is only 0.05 pF and have demonstrated noise equivalent signals of 30–35 electrons per Nyquist sample in television applications at rates to 7×10^6 samples/s when the array is cooled to minimize leakage shot noise. Because there is only one sample per information packet rather than a difference of two samples, there is less need to wait for complete settling and the analog signal bandwidth in hertz can equal the data rate. This reduces the rms noise to 40 or 50% compared to that in a CDS output circuit and makes the FGA circuit better suited for higher data rates. Sometimes the "floating gate" is not allowed to float indefinitely but is reset once each line or frame time to keep the MOSFET amplifier at its optimum operating point. After being sensed with the floating gate, the signal charge in the CCD is clocked through more CCD stages to a following back-biased diode and is collected.

Astronomers Marcus, Nelson, and Lynds of Kitt Peak National Observatory report (Marcus *et al.*, 1979) random noise of only 17 electrons per sample on a Fairchild CCD211 with FGA output in a slow-scan camera. Operated at only 10,000 samples/s, 5–6 s per frame, this chip was cooled to $-112°C$ to reduce dark current greatly. This is the lowest noise reported so far on this CCD structure, although this relatively new circuitry may still not be optimized. A CDS integrator is used to limit both high-frequency and $1/f$ noise. (The CCD211 is described in Section V,D.)

Fairchild workers have attempted further performance gains by sensing the same signal charge packet consecutively with several FGAs distributed along a longer CCD register and combining their outputs in another CCD. This so-called distributed floating gate amplifier (DFGA) was built experimentally. However, variations in threshold voltage from MOSFET to MOSFET prevented optimum conditions from being achieved on all amplifiers in the group, and results were only marginally improved. Hence no sensing arrays are available commercially with DFGAs, but Fairchild offers many of their arrays with single-stage FGA outputs.

D. CCD NOISE MODEL

CCD imaging arrays are used for scientific applications with widely differing operating conditions. A noise model is given below to show how noise performance is affected. Noise is expressed in electrons per sample at the CCD output. Using an output circuit with either CDS or a FGA:

$$Q_N = \left[\left(\begin{array}{c} \text{signal} \\ \text{shot noise} \end{array} \right)^2 + \left(\begin{array}{c} \text{leakage} \\ \text{shot noise} \end{array} \right)^2 + \left(\begin{array}{c} \text{CCD transfer} \\ \text{noise} \end{array} \right)^2 \right.$$
$$\left. + \left(\begin{array}{c} \text{on-chip} \\ \text{MOSFET noise} \end{array} \right)^2 + \left(\begin{array}{c} \text{off-chip} \\ \text{amplifier noise} \end{array} \right)^2 \right]^{1/2}$$

For buried-channel CCDs, transfer noise is negligible compared to the other terms, and

$$Q_N = \left(\frac{HRt}{q} + \frac{I_l t}{q} + \frac{E_{n1}^2 \, \Delta f_n C_S^2}{q^2} + \frac{I_{n2}^2 \, \Delta f_n C_S^2}{g_{lm}^2 q^2} \right)^{1/2}$$

For a linear array in the dark where the irradiance H is zero, at a modest 1.5×10^5 samples/s data rate,

$$Q_{ND} = [0 + (13.7e)^2 + (16e)^2 + (13.4e)^2]^{1/2} = 25 \quad \text{electrons}$$

In the equation H is the sensor irradiance (W m^{-2}), R the sensor elemental response (approximately 2×10^{-10} Am^2/W for 25×25 μm^2 elements with sunlight illumination), t the exposure time (1×10^{-3} s), q the electronic charge (1.6×10^{-19} C), I_l the element leakage current (3×10^{-14} A), E_{n1} the on-chip MOSFET noise spectral density (15×10^{-9} V Hz$^{-1/2}$), Δf_n the effective noise bandwidth (for CDS Δf_n is four to six times the sample rate, 1.3×10^6 Hz), C_S the diode and amplifier gate capacitance (0.15×10^{-12} F), I_{n2} the equivalent off-chip operational amplifier input noise spectral density (2.5×10^{-12} A Hz$^{-1/2}$), and g_m the transconductance of the on-chip MOSFET amplifier (2×10^{-4} mho). The numerical values are those used in the sample calculation, which agrees with the measured rms noise for such an array when operated at room temperature. The parameters are appropriate for a short linear array scanner or for a longer array where separate output circuits are provided for each subsegment of the array to limit the data rate required of each output circuit.

Application of the same analysis to an area array suitable for television with U.S. broadcast scanning standards yields dark noise per sample:

$$Q_{ND} = [0 + (56e)^2 + (59.5e)^2 + (49.9e)^2]^{1/2} = 95.7 \quad \text{electrons}$$

This result is based on reading a 320×240 element array 60 times/s in an interlaced format (a data rate of 4.7×10^6 samples/s) with CDS and a noise bandwidth of at least 18×10^6 Hz.

While a noise of 96 electrons per Nyquist sample is only about 10% that for a good vidicon or plumbicon camera, much effort has been devoted to lowering it. To lower the leakage current noise, the silicon sensor can be cooled with thermoelectric coolers. Cooling 16°C halves the rms leakage shot noise from 56 to 28 electrons per sample. Cooling the sensor to 0°C is feasible in many scientific applications, although this requires either a hermetically sealed enclosure around the cooled chip or continuous flushing with a dry atmosphere, typically nitrogen, to avoid moisture condensation on the sensor chip. To lower the third and fourth terms, the sensor designer strives to minimize C_S and cooperates with the system designer in reducing Δf_n, the noise bandwidth, for example, by forming two or more output data streams through separate amplifiers operating at reduced data rates, or by use of the FGA described in Section III,C,2.

The noise model shows how leakage shot noise increases at longer exposure times unless the sensor chip is cooled to reduce leakage current, and also how amplifier noise increases at higher data rates, which are equivalent to shorter exposure and readout times. These effects impact astronomical and other applications, limiting time exposures and the size of arrays that can be read at high speed in photon-counting applications. For further discussion of these applications, see Section VII.

More significantly, the model shows how the signal-to-rms-noise ratio varies with exposure. In general,

$$Q_N = [(HRt/q) + Q_{ND}^2]^{1/2}$$

where Q_{ND} is the dark noise calculated earlier in this section. Since the signal is $Q_{SIG} = HRt/q$ electrons,

$$SNR = \frac{HRt/q}{[(HRt/q) + Q_{ND}^2]^{1/2}}$$

For a buried-channel CCD, with 25×25 μm^2 elements, the well capacity limits Q_{SIG} to $2500a = 2500 \times 625 = 1.5 \times 10^6$ electrons, where a is the element area in square micrometers. Thus for either assumed CCD, $SNR_{max} = 1.5 \times 10^6/(1.5 \times 10^6 + 625)^{1/2} = 1250$ at the saturation exposure, where the dark noise is essentially negligible. The difference between the arrays is primarily important in making radiometric measurements at low exposures where the dark noise is significant. The dynamic range above dark noise is about four times larger for the conditions postulated for the linear array.

IV. CHARGE INJECTION DEVICES

Charge injection devices (CIDs), developed by workers at the General Electric Company, are more closely related to diode array image sensors than to CCDs but offer some advantages over both. Because the charge signal at each sensing site is transferred only between potential wells under adjacent gate electrodes, operation is not compromised by moderate transfer inefficiency. CIDs can be built effectively on materials such as germanium, indium antimonide, and mercury cadmium telluride whose surface properties are not yet well controlled. Because the signal can be stored, read, and erased at each site by properly pulsing one row and one column gate bus, a full rectangular sensing array can be constructed simply with no extra address switch components within the light-sensitive area. On the other hand, since the signal is sensed by current flow in a row or column bus or in the substrate connection, the signal circuit shunt capacitance cannot be as low as for a CCD, and the theoretical signal-to-noise ratio is somewhat higher at low exposures.

A simple 16-cell charge injection area image sensor (Michon *et al.*, 1978) is shown schematically in Fig. 18. Each sensing site is covered by two gate electrodes side by side. The right-hand gate in each pair is connected to the row bus at that location, and the left-hand gate to the column bus. Both gates and buses can be made of electrically conducting transparent tin oxide or polysilicon, or the buses can be of metal and placed between the sensitive areas. As long as either or both gates at a site are negative, photon-generated holes are collected in the resulting potential

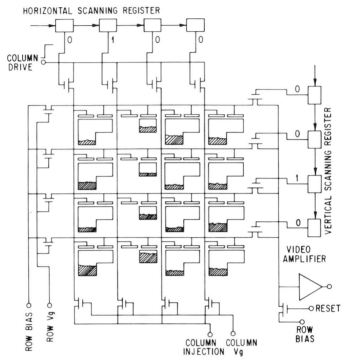

FIG. 18. Row readout array diagram illustrating sensing of the third row, second column. (Courtesy of General Electric.)

minimum, which is like that shown in Fig. 11a. When both gates are pulsed toward zero, the silicon surface becomes positive, the charge carriers are injected into the silicon, and a net current flows in a load resistor connected in series with the substrate connection or in the gate leads.

Reading the signal in the substrate lead was the first operating mode proposed for CID imagers, as shown in Fig. 19. Actually, there are four separate currents in the reading cycle, which must be separated. First, when a gate is pulsed more positively to read a charge packet or to transfer it to the other gate of the pair, a displacement current flows in the load resistor because of the gate-to-silicon capacitance, even if no signal charge is present. If a charge packet is present, the silicon surface is initially more positive, the packet is injected into the substrate, and the current pulse is somewhat larger. When the gate is returned to its normal negative potential, there is a negative-going displacement current pulse. The desired signal current can be separated from the displacement current transients by summing the positive- and negative-going transients in an averaging circuit, as shown. The fourth current, that of the collected car-

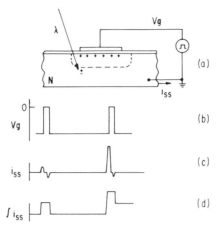

FIG. 19. MOS capacitor operated in the charge injection mode: (a) device cross section, (b) drive voltage, (c) substrate current, without and with injected charge, (d) substrate current with averaging. (Courtesy of General Electric.)

riers, changes the silicon surface potential during the exposure time and produces a displacement current that appears in the load circuit. However, this current is too small to be measured.

While reading the signal in the substrate lead is still done, a lower shunt capacitance and larger signal voltage can be realized by measuring the current flow in a row or column gate lead when the charge is injected into the substrate, especially if only one gate lead at a time is connected to the signal circuit. Further, the charge need not be injected into the substrate to develop an output signal. Just as charge in a CCD can be sensed nondestructively with a floating gate electrode, charge at each location in a CID can be sensed by monitoring the gate charging currents as the charge is moved from one gate to the other at each storage site. As suggested in Fig. 11a, the silicon surface potential under a gate is less negative when signal charge is present. Thus if the signal charge is initially under the left or column gate, and the right or row gate is then made more negative, the charging current will be larger as the signal charge is larger, because a larger potential difference must be produced across the capacitance represented by the insulating gate oxide.

In the row readout array, shown in Fig. 18, the row gates are biased to an intermediate negative voltage, but the signal charge is held under the more negative column gate. To read the signal from the elements in a row (or scanning line in television terms), the address switch transistor for the row is activated to connect the signal output bus to the selected row bus. The two buses are then reset to the row bias voltage by momentarily turning on the reset switch transistor. This leaves a sample of the reset switch noise on the buses. CDS is used to remove this noise, which is significant because the bus capacitance is several picofarads. (See Section III,C,1 for a description of CDS.) A clamp (or restore) switch is closed briefly in the following amplifier to store the negative of this sampled noise on a coupling capacitor. The column gates, which include the selected element, are then pulsed toward ground to place the signal charge under the right-hand row gates. The row bus potential change at the selected element is a measure of the signal charge, and the signal minus noise is read by sampling the output of the signal amplifier. When all elements in a row have been read, the row may be erased by pulsing both gates toward ground.

CIDs are available in sizes as large as 244×188, read in a television-compatible format. Dark nonuniformity is typically less than 3% full scale, and response nonuniformities are less than 0.5%. For row readout devices, random noise at television rate bandwidths of 7 MHz is approximately 2000 electrons per sample (compared to a maximum signal of 1.5×10^6) when CDS is used to remove reset noise. This is almost entirely Johnson–Nyquist noise in the address switch and buses and amplifier noise, all of which vary as the square root of the signal bandwidth. Thus in slow-scan operation noise can be significantly reduced to the order of a hundred electrons. This corresponds to a dynamic range above the random noise of about 15,000:1.

For applications requiring signal processing, the CID may be operated with nondestructive readout. By appropriately programming the column bus pulses and the address switches, one could, for example, read the signal from a block of elements around the selected element for subtraction of local background or for sensing the presence of edges. This use of the sensing array as memory, of course, is possible only if the input signal does not change the information content during the signal-processing time.

In summary, the fabrication simplicity of the CID, its usefulness in materials less well controlled than silicon, and its adaptability for random-access reading and signal-processing functions mean that it can satisfy applications that the diode array and CCD do not, despite its somewhat larger dark noise.

V. ARRAYS AND CCDs: SPECIFIC EMBODIMENTS

A. LINEAR PHOTODIODE ARRAY: RETICON RL1024S

The RL1024S is specially designed and optimized for applications in spectroscopy, with a linear array of 1024 elements on 25-μm centers, each element being 25 μm wide along the array and 2.5 mm high perpendicular to the array. The 100:1 aspect ratio elements are intended to be aligned with the slit image in a spectrum, where the 2.5-mm height can intercept much of the light for higher effective response and the large elemental area can store a charge as high as 1.4×10^{-11} C (nearly 10^8 electrons) per element in a single exposure with essentially linear response. The array can be read in times as short as 0.5 ms, which sets a lower limit on the exposure time for time-varying spectra unless a separate shutter is employed. Exposure times of 0.3 s at room temperature or of minutes to hours at a reduced array temperature can be used for sensing and measuring faint astronomical spectra. The typical relative spectral response shown in Fig. 20 (obtained with a model containing a quartz window) demonstrates useful sensitivity from 200 to 1100 nm.

The general appearance of the RL1024S is similar to that of the RL256 shown in Fig. 6. The RL1024S also has two sets of address switches and two output buses, one on each side of the sensing array, so that a signal from even elements is read from one bus and odd elements from the other. With appropriate circuitry the RL1024S has dark noise as low as 1000 electrons per sample, for a dynamic range above noise of about $10^5 : 1$. Element response of 2.2 C/(J/cm^2) is held to ±10% along the array, and a dark current of 5 pA per diode at 25°C (3×10^7 electrons/s) falls to 50 electrons/s at −100°C.

Other Reticon linear arrays are available from 64 elements, used for optical character recognition, to 1728 elements, used for page reading. The Reticon arrays described here are commercially available.

B. Photodiode Area Array: Reticon RA 100×100

This is the largest commercially available photodiode area array (Fig. 2); it has 10,000 light-sensitive elements located on 60-μm centers in both

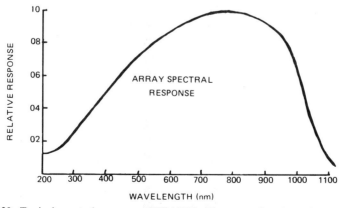

FIG. 20. Typical spectral response of RL1024S. (Courtesy of Reticon Corporation.)

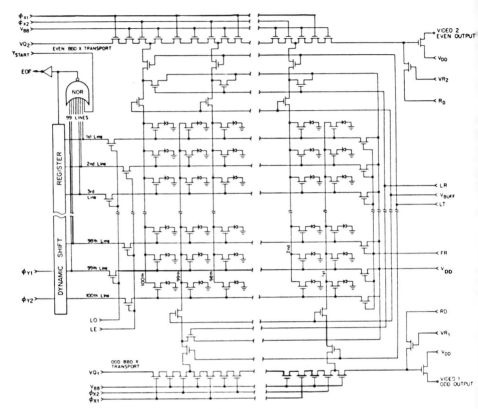

FIG. 21. Schematic diagram, RA 100×100. (Courtesy of Reticon Corporation.)

directions and covering a 6-mm square active area. As shown in Fig. 21, each sensing element has an associated MOS transistor address switch. The switch gates for each row of elements are connected in parallel, so that the charges from one row of diodes at a time can be connected to a set of column buses. These in turn transfer charge through switches to the elements of two horizontal bucket brigade analog shift registers, one serving the odd-numbered columns and the other the even-numbered columns. The bucket brigade is a relative of the CCD in which diffused diodes are connected by MOS transistor switches to effect charge transfer. The potentials of the diodes are controlled by the reset voltage applied to the charge collection diode at the output and by the clock voltages applied to the transistor gates, each of which overlaps the diode that becomes the drain when a switch is gated on. The bucket brigade device (BBD) is well adapted to an application where the sensing elements are diodes remote from the transfer register and charge must be injected through a diode at each stage by way of a metal or diffused interconnect. Transfer efficiency and noise performance tend to be inferior to those of a CCD, because transfers involve establishing equality between the potential under each gate and the potential in the preceding diode, as opposed to the completely nonequilibrium charge transfer assumed for CCDs.

The outputs of the RA 100 × 100 may be combined to provide a time sequential readout or may be clocked to provide interlacing by reading the odd field and then the even. Responsivity is a 2-V signal across a 2-kΩ load per μJ/cm^2 exposure. The saturation exposure is 0.2 μJ/cm^2, the dynamic range is 100:1, and the response uniformity is ±10%. Output data rates as high as 10^7 samples/s permit frame rates as high as 1000/s for following rapidly changing phenomena.

C. CCD LINEAR IMAGING DEVICES

The Fairchild CCD121H is a typical example of a linear charge-coupled imaging device intended principally for page-scanning applications. With 1728 elements, it can provide a 200 samples/in. resolving power across an 8½-in. page, or 16 samples in the width of a letterspace of elite type, more than adequate for good reproduction. For telefacsimile or similar applications the letter page is moved through the field of view perpendicular to the electrical scan of the CCD to scan the entire page area. At the maximum 10^7/s output bit rate, the line array is scanned 6000 times/s, and an 8½ × 11 in. page can be covered in 0.36 s.

For the application outlined, the CCD must be more complex than the conceptual designs suggested above so that exposure and scanning do not interact. If the photosensitive array were exposed and scanned continuously, the image would be smeared while the charge pattern was being moved out of the array, much as motion picture film would be smeared with no shutter and no fast pulldown mechanism. To avoid exposure mixing one could include a shutter, exposing only briefly while interrupting the electrical scan and then clocking the charge pattern out while the shutter was closed. To eliminate a mechanical shutter one could expose the CCD with the electrical scan interrupted and then very quickly clock the charge out of the array so that exposure mixing would be negligibly small. To keep the output data rate within feasible limits, this scheme is usually implemented by making the linear CCD twice as long as required and shielding the second half from light. When the exposure of a line of the image is complete, the charge pattern on the first half is transferred rapidly along the array until it is on the second half under the light shield.

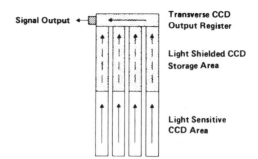

FIG. 22. Schematic diagram of a "frame transfer" charged-coupled area imaging device. (From Séquin and Tompsett, 1975.)

Electrical scan of the first light-sensitive half is then interrupted, and the next image line is exposed while the charge pattern from the original line is clocked out of the second half into the output amplifier. (This intermittent scan scheme is used principally for "frame transfer" area array sensors, such as the RCA SID52501 described in Section V,D,1, and is shown schematically in Fig. 22.)

The Fairchild CCD121H image sensor uses a different and more efficient organization in which sensing and electrical scanning functions are separated. As shown schematically in Fig. 23, the optical line image is defined by a slit in an opaque aluminum light shield. The light passes through a single transparent "photogate" electrode, producing charge carrier electrons in the p-type silicon. The electrons are confined in resolution element-sized bins by potential minima formed under the positive photogate, by "channel stops" (more heavily doped stripes in the silicon where the surface potential is little affected by the positive photogate), and by a transfer gate electrode at the silicon chip potential one each side of the photogate. At the end of the exposure time, the transfer gates are pulsed positive, transferring the charge from each sensor well laterally out under the light shield and into the light-shielded CCD shift registers. These perform the electrical scanning function by clocking the charge packets to the output circuit, a collection diode, and an on-chip MOSFET source follower amplifier and reset switch transistor. Thus the image charge packets are transferred in parallel into the serial output registers in a single transfer time that can be truly negligible compared to the exposure time. This scheme is used in most line array CCD sensors, since it has clear advantages. It is also used in interline transfer CCD area arrays like the Fairchild CCD 221 in Fig. 1, although the light-sensitive fraction of the image area must then be reduced to allow space for the transfer registers.

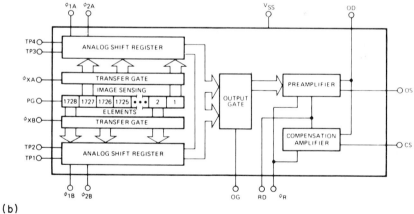

FIG. 23. (a) 1728-element linear image sensor (CDD121H). (b) Block diagram. (Courtesy of Fairchild.)

Characteristics of the CCD121H include a dynamic range of 2500 : 1 above rms noise, a response of 0.5 V across a 1000-Ω load resistor per $\mu J/cm^2$ of 2854-K tungsten light with element-to-element response uniformity within $\pm 25\%$, and a MTF at the Nyquist frequency of about 47% if a Corning 1-75 infrared cut filter is used with 2854-K tungsten light. Complete MTF data are shown in Fig. 4a. Rms noise is computed at 500 electrons without the use of CDS.

D. CCD AREA IMAGING DEVICES

1. *RCA* SID52501

This silicon imaging device is an excellent example of an area array sensor intended for compact television cameras. It provides 256 rows of sensor elements with 320 elements per row in a 7.31 mm high, 9.75 mm wide image area. Each element area is $30 \times 30 \ \mu m^2$.

Light from the image passes through transparent polycrystalline gate electrodes and excites charge carrier electrons in the p-type silicon chip. The electrons are collected in potential minima at each element formed by the horizontal gate electrode structure and by vertical channel stops.

The SID52501 is a frame transfer device, as shown schematically in Fig. 22. The CCD array is twice as high as the optical image, and the upper half is shielded from light. During the $\frac{1}{60}$-s exposure, scanning is interrupted in the light-sensitive lower half, with one gate in each elemental triad held positive to provide the potential minimum for the element. At the end of the exposure time, during the normal vertical retrace interval, the charge image is clocked rapidly up into the light-shielded upper half of the array. Scanning is again interrupted in the lower half, exposing the next field, but a different gate in each triad is held positive so that the second-field elements are interlaced with the first-field elements, as they should be for broadcast television.

During exposure of the second field, the charge pattern stored in the upper half array is shifted up by one line every 62.5 μs into the transverse horizontal output register and clocked out to a collection diode, MOSFET output amplifier, and reset switch located at the upper left corner of the array. Thus by moving the charge pattern the picture elements are read in the same order as they would be in a conventional television camera with electron beam scanning. With appropriate clocking signals, specified by the device manufacturer, the output data occur in groups corresponding to the active line times of the television scanning pattern, with appropriate gaps for insertion of the sync, blanking, and black-level reference signals.

The complete signal, including the interlacing, contains data from 320 elements along each of 512 scanning lines. This slightly exceeds the normal 488 active lines vertically, allowing for variations in the vertical retrace time in typical television receivers or display monitors. The 320 horizontal samples correspond only to 240 television lines, measured in equivalent half-cycles per picture height, but the CTF or bar chart response at this sample rate is 63%, far higher than that for a vidicon, when exposure is to 2854-K tungsten light through an infrared blocking filter. Thus picture quality is comparable to that of a vidicon camera with a $\frac{2}{3}$-in. vidicon.

The SID52501 uses a surface-channel CCD structure in the sensing and frame storage parts of the array and a buried-channel structure in the higher-speed output register. To achieve high transfer efficiency, a bias charge is provided by flooding the sensitive area with light from four light-emitting diodes placed within the camera behind the lens, so that the CCD wells are never empty. The resulting bias current is set to 60 nA, about 20% of the maximum signal. Becuase of the shot noise of this bias signal, dark noise is about 800 electrons per sample. The maximum signal-to-rms-noise ratio is only 316:1 for this device. However, this value is competitive with that of the best vidicon cameras and better than most. Response to 2854-K tungsten radiation is 65 mA/W incident on the 0.7-cm^2 area, or about 5×10^{-7} A m^2 W^{-1}. Spectral response data are shown in Fig. 3 curve D and the CTF in Fig. 4b.

2. *Fairchild* CCD211

This buried-channel interline transfer array has a sensing area 4.4 mm high and 5.7 mm wide and is intended for television camera applications. The array has 244 rows of sensing elements with 190 elements per row having sensing areas 14 μm wide and 18 μm high. The vertical element pitch is 18 μm (that is, the elements are contiguous), but the horizontal pitch is 30 μm, allowing a 16-μm space between columns for the interline CCD transfer registers that transport the charge packets to the top of the image and into a transverse output register, as illustrated schematically in Fig. 24. The array is designed for interlaced scanning. When the even-numbered rows have been exposed for a frame time, say $\frac{1}{30}$ s, the photogate is clocked low, while the vertical register phase 1 is high, thus transferring charge packets from the sensors in the even-numbered rows into the vertical registers. These packets are moved upward one row at a time during the horizontal blanking interval into the horizontal CCD shift register and transferred out an element at a time past the FGA, developing the video signal during the active horizontal line time. At the end of a field time, say $\frac{1}{60}$ s, the photogate is clocked low while the vertical register

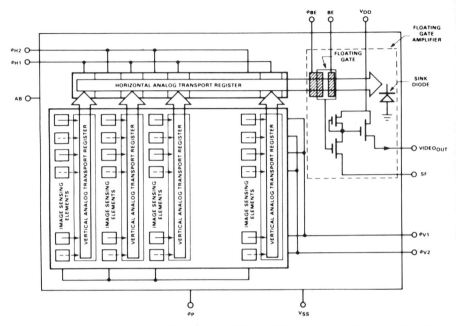

Fig. 24. Schematic diagram of Fairchild CCD211 or 221 charge-coupled area imaging device. (Courtesy of Fairchild.)

phase 2 is high, thus transferring charge packets from the sensors in the odd-numbered rows into the vertical registers. The transfer operation is repeated to develop the video signal for the odd field. Thus the sensing elements in each field are exposed for a full frame time, as contrasted to a field time in the RCA scheme, at the cost of having twice the number of (smaller) elements.

The CCD211 has a responsivity of 1 V/(μJ/cm^2) across a 500-Ω load versus a saturation output voltage of 200 mV, for a saturation exposure of 0.2 μJ/cm^2. The dynamic range above the dark signal is typically 300:1, with shading at half the saturation exposure less than $\pm$10% of signal. CTF (the response to a bar chart input) at the Nyquist frequency is 60% horizontal and 30% vertical. Noise performance data are not included in the bulletin for the CCD211 but, according to private communications, units have been measured with noise of 35 electrons when cooled to 0°C to remove dark current noise.

The CCD211 was the predecessor of the CCD221 shown in Fig. 1. The latter has 488 rows each containing 380 elements for broadcast television-compatible operation. Although the CCD221 is being manufactured regularly, data sheets were not yet available in late 1978. Performance of the CCD221 should be comparable to that of the CCD211, since the design was simply scaled up to form the larger chip, keeping the element size unchanged.

VI. PRESENT PERFORMANCE

The computed signal-to-noise capability of CCD imaging sensors promises a significant improvement in performance over that obtainable with television camera tubes. Noise of 20–25 electrons per sample and a maximum signal level without saturation of 1.25×10^6 electrons per sample for a buried-channel CCD with 25×25 μm^2 elements indicate a dynamic range of 50,000:1 above dark noise, compared to a range of only 250:1 for even a very good vidicon television camera. A 200-fold increase in dynamic range above noise would be a great step forward, because it would greatly increase the real information content of each picture (frame) and also would permit the reproduction of meaningful images of very low-contrast scenes. Most difficult imaging situations involve low contrast. For example, the visible range through the atmosphere is reached when light scattered to the viewer's eye by aerosol particles along the optical path reduces apparent contrast to the limit of visibility, about 2%. An imaging system capable of accepting information with lower contrast and reproducing clear imagery with higher contrast could greatly extend the range of human vision. Similar considerations apply in astronomy to detecting fainter stars against the unresolved sky background, in spectroscopy to detecting fainter spectral lines against a continuum, or in biology to the study of low-contrast tissue samples under the microscope, and so forth. But contrast can be stretched usefully only as long as the resulting image is not noisy; that is, as long as it does not appear "snowy" like a fringe area television picture. In every case system noise and maximum signal-handling capability limit the feasibility of electronic image enhancement.

Noise in an electronically reproduced image comes from two sources, from system noise (terms 2 through 4 in the noise model equation) and from the random arrival of photons in the image, or more accurately the random excitation of charge carriers in the image sensor (term 1 in the noise model equation). Because the signal-to-noise ratio against photon or charge carrier noise improves as the square root of the exposure per frame, and the signal-to-system-noise ratio improves directly as the exposure, image quality in low-contrast cases can be improved if the exposure is increased, as long as the sensor can handle the increased charge per frame without saturating. At maximum exposure a vidicon camera tube yielding highlight signals of 500 nA produces a camera signal-to-noise ratio of about 230:1, limited primarily by circuit system noise. This highlight current is an upper limit for most vidicon electron gun systems. A CCD operating just below saturation should have a signal-to-noise ratio of about 1100:1, limited by statistical fluctuations in the generation of carriers by photons. In infrared system terms, for moderately high exposures the CCD should have background-limited performance (BLIP), while the vidicon never approaches BLIP. Thus a CCD imager is predicted to give significantly improved imagery of low-contrast objects.

In fact, CCDs and other array sensors have not reached the imaging performance predicted from signal and noise data because nonuniformity in the reproduced image interferes with image visibility, as shown in Figs. 25 and 26. Nonuniformities are of two general types. The first, seen in Fig. 25, are variations in the signal level from element to element when the lens is capped or the system is viewing a uniformly dark area. These are caused primarily by dark or leakage current which varies from point to point in the silicon sensor chip because of microscopic variations in the density of impurities or the density of crystalline imperfections. Since leakage current can be reduced by cooling the sensor chip, (halving the current for each 8°C), the variations can be made negligible by operating a silicon sensor chip at 0°C or lower, although this may require a hermetically sealed enclosure to avoid moisture condensation on the sensor. The appearance of the reproduced imagery at low exposure levels can be significantly improved by cooling a silicon sensor to −10°C, as shown in Fig. 26.

The second type of uniformity is seen primarily at high exposures where object contrast is low and is due to differences in response from element to element in the sensor. This may also be due to microscopic impurity concentration variations that vary minority carrier lifetimes, hence the signal produced by a given exposure. More frequently it is caused by nonuniformity in the area of the sensor elements due to photolithography variations or to variations in surface state concentrations at the silicon–silicon dioxide interface that can change thresholds, hence affect response. While this type of nonuniformity is not apparent in the imagery in Fig. 26, manufacturer's specifications indicate response variations of ±10–25% can be expected across an array, and element-to-element response variations of 3–5% are normal.

This combination of a high signal-to-noise ratio but significant nonuniformity in dark level and response has caused CCD and diode array imagers to be candidates primarily for less critical applications. These include conventional television or telefacsimile line scanners and higher-cost applications such as National Aeronautics and Space Administration (NASA) imaging systems where the sensor can be calibrated element by

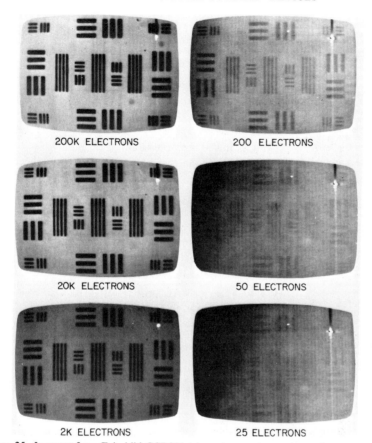

200K ELECTRONS · 200 ELECTRONS

20K ELECTRONS · 50 ELECTRONS

2K ELECTRONS · 25 ELECTRONS

FIG. 25. Imagery from Fairchild CCD221 taken at room temperature shows nonuniformities and blemishes at low exposures. (1975 photograph courtesy of Fairchild.)

element and the variations can be stored in a computer and removed from actual imagery, or in applications where conventional television with vidicon television cameras cannot do the job. An example of a high-technology NASA application is shown in Fig. 27. For LANDSAT applications, areas of the earth's surface are imaged from an orbiting satellite simultaneously in four "colors," including a band in the near infrared. Comparison of the signal in the four bands is made using a computer to assess the state of food crops, extent of drought or flooding, existence of mineral deposits, and so forth. The linear photodiode array shown is being considered for an advanced LANDSAT system in which each diode can be made sensitive to a different color by the use of filters. The large number of elements in the array will provide adequate system-resolving power even when four elements are used at each point for color sensing. The line

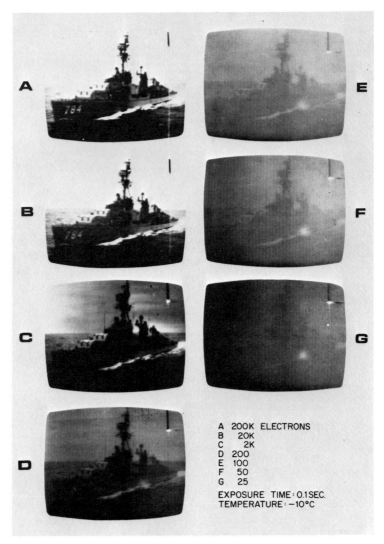

FIG. 26. Imagery at various exposure levels taken with a Fairchild CCD221 shows how array nonuniformities are reduced when the array is cooled to −10°C. (Courtesy of Fairchild.)

scan image shown was taken in the laboratory in black and white after calibration on a uniform field. It demonstrates that today's silicon diode arrays and silicon CCDs can produce excellent imagery as soon as the nonuniformities are removed.

Until recently, the removal of image nonuniformities with a computer

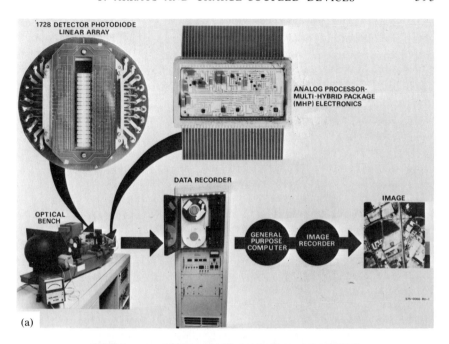

FIG. 27. (a) Linear photodiode array and (b) reproduced imagery. (Courtesy of West-inghouse.)

was thought to be limited to applications such as planetary exploration where the cost of computer correction was a minor part of the whole mission. However, recent accelerated development of large-scale memory chips and analog–digital and digital–analog converter chips now makes it probable that CCD and diode array imagers can be used in more modest applications in scientific, technical, or military applications where high-quality imagery can be obtained by the addition of an appropriate microprocessor.

VII. SOLID-STATE IMAGING APPLICATIONS

While RCA and Fairchild have produced charge-coupled area-imaging devices which operate at broadcast television rates in competition with vidicon cameras, the most significant applications for solid-state image sensors are those that are not well satisfied by cameras with television camera tubes. Some of these are briefly described because they may suggest other uses.

A. Photon Counting

For very low-contrast images with low photon fluxes, one must use long exposure times to improve the signal-to-statistical-noise ratio by accumulating as many photon-produced charge carriers as possible in each frame. (See the noise model in Section III,D.) But long exposures also integrate the nonuniform dark current in the sensor, forming a nonuniform pseudobackground, unless the sensor is cooled substantially. This background and its added shot noise make accurate radiometry difficult. To eliminate this problem and to remove the effects of system noise and the limitation on maximum exposure time set by the charge storage capacity of the device, one can use a photon counter. For photon counting, as suggested in Fig. 28, a charge-coupled area-imaging device is placed in a vacuum tube that includes a photocathode and a means for focusing and accelerating the photoelectrons through 10–20 keV to form an electron current image at the CCD. The electron energy is absorbed in the silicon, creating a hole–electron pair for every 3.5 eV, or about 3000 charge carriers per photoelectron with 10- or 11-kV acceleration. The CCD is then read out rapidly, usually at 100–1000 frames/s, so that on the average less than one photoelectron strikes each CCD element in a frame time. The video signal then consists of system noise, about 100 electrons, with a 3000-electron spike from each photoelectron event. The noise can be eliminated with a simple threshold circuit, set at the 400- or 500-electron

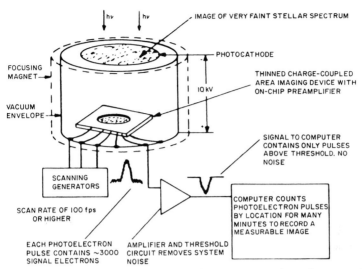

FIG. 28. Intensified CCD is near optimum for photon counting. (Courtesy of West-inghouse.)

level, and all events above threshold can be counted as logic events and stored by location in a computer memory. This makes a nearly ideal system. Dark current and system noise are eliminated by the combination of frequent scanning and the threshold circuit. The signal-to-noise ratio is determined only by signal statistics, and events can be counted indefi-nitely to improve the signal-to-noise ratio, which is limited only by the length of time available for exposure and the count-storing capacity of the computer.

This method is, however, restricted to relatively small arrays and very low photon fluxes. Scanning must be rapid compared to the rate of photo-electron arrival at each element to minimize the probability that two events will occur in the same frame time and be counted as one. A 100 × 100 element array can be read at 1000 frames/s with noise of 100 electrons or less, or a larger array can be read more slowly. The requirement is sim-ply that the noise be low enough that the threshold can be set at, say, 6σ so that noise spikes will exceed the threshold far less frequently than the rate for photon arrival, yet still have signal spikes well above threshold for easy counting. The low noise of the CCD imaging array makes it a useful candidate, while a normal intensified silicon diode array target camera tube requires an additional 15-kV intensifier stage so that signal spikes will sufficiently exceed the typical 1000- 2000-electron system noise.

Intensified CCDs are usually thinned arrays having the electron flux incident from the rear, since gate structures, while transparent to light, would absorb electron energy and reduce gain. The use of a photocathode for photon-to-electron conversion reduces quantum efficiency to a typical 10% or less (see Fig. 3, curve E), but the use of III–V crystalline negative electron affinity photocathodes, while more costly, raies quantum efficiency to 15–20%, at least over the visible spectrum.

B. Time Delay and Integration

In many applications, the image moves during exposure and either very short snapshot exposures or special image motion compensation schemes must be employed to yield blur-free reproductions. In early weather satellites, for example, the reproduced imagery was a set of individual photographic positives taken with a shuttered vidicon camera and correlated by hand after the fact. Current weather satellites, operating in polar orbits, use a "whisk broom" scanner in which the scene image is scanned across the vehicle track past a single silicon detector element by a rotating mirror, while vehicle movement provides along-track scan. This system provides a continuous 1600-mi-wide strip image of the earth's surface on a strip of film that covers three orbits, hence no matching is required. Further, the gain of the on-board signal amplifier is varied as a function of the instantaneous scan angle so that, where the image crosses the terminator, an exposure range of $10^5:1$ is compressed to provide a continuous-tone image within the contrast capabilities of the final photographic copy. In this case, the naturally moving image is made a virtue, since it allows for different treatment of different scene areas, and the use of a single detector element inherently eliminates any problem with nonuniformity of sensor response or of dark current.

The use of a single-element detector in a scanner is very wasteful of light. To improve system signal-to-noise ratio, one can use a linear array such as the Fairchild 121H (described in Section V,C) placed perpendicular to the direction of image motion in a "push broom" scanner. Charge is transferred into the light-shielded readout register every time the image moves across the width of the photosensing array, as is done in page-reading applications. For a given size detector element and rate of image motion, this increases element exposure time by the number of elements N in the line array and the signal-to-noise ratio for a given image irradiance by at least $\sqrt{N}$.

But for many applications more sensitivity is needed, and with CCDs one can take further advantage of the moving image. Consider a simple charge-coupled area-imaging device like the RCA 52501 described in Section V,D, but eliminate the light-shielded storage section, leaving the light shield over the transverse output register. If the direction of image motion is up the sensitive area, and if the vertical transfer gates are clocked continuously so that the charge image moves in synchronism with the optical image, the effective elemental exposure time can be as long as it takes a single image point to move from the bottom to the top of the sensing area. This type of moving image operation is known as TDI or "delay and add," because it was originally implemented with diode array sensors interconnected by an electrical delay line. The CCD implementation is much easier, since the rate of charge image motion can be varied by adjusting the frequency of the clock pulses.

The reader will recognize TDI operation of a CCD as the electrical analog of image motion compensation by film motion in a film camera, and of course it has similar limitations. The length of the sensitive array in the direction of motion is limited by the ability to match the optical and electrical image motion in both speed and direction, by effects of pin cushion or barrel distortion in the optical system, or by variations in keystoning where the optical axis is not perpendicular to the object surface or to the sensing array. But for a number of applications, TDI operation on a moving image makes new applications possible.

For example, NASA is considering the use of a CCD TDI array for a multispectral sensor for identifying crop areas and studying crop conditions for LANDSAT. The TDI array provides sensitivity for balancing the loss of light in the filter system and permits use of a smaller optical aperture. For an optical page reader application, TDI arrays with multiple outputs can permit much higher page speeds, making them candidates for the reading of addresses in a computerized high-speed mail sorter. Or the high sensitivity could be applied to computerized microscopic screening of biological samples on a continuous-flow basis. Or TDI array sensors could be used for panoramic imaging. Televising long, narrow images is poorly done with vidicon cameras, because the camera tube's sensitive area usually matches the round vacuum envelope, and resolving power is satisfactorily high in terms of picture information content only when most of this area is filled by the image. By using a scanning camera with a CCD TDI sensor positioned for horizontal image motion, one could provide high-quality, constant-scale imagery of the complete horizon or any part of the surrounding ocean from a ship or of the terrain around an observation post. The usefulness for automated or remotely monitored security surveillance or ocean search is obvious.

But TDI operation also has important advantages for the CCD imager. Because each image element is sensed consecutively by the several or many elements in a TDI column and their outputs are summed, variations in element response or in element dark current are averaged over these elements, significantly reducing their effective variability. Further, because the output signal now comes from n columns, rather than $m \times n$ elements, a comparatively small memory suffices for storing and canceling out the sensor variations and giving high-quality imagery.

In summary, while diode array and especially charge-coupled imaging devices can and will compete with television camera tubes in more compact cameras for conventional television applications, their real importance is in the new applications they make possible.

REFERENCES

Barbe, D. F. (1975). Imaging devices using the charge coupled device concept, *Proc. IEEE* **63**, 38–67.

Boyle, W. S., and Smith, G. E. (1970). Charge coupled semiconductor devices, *Bell Syst. Tech. J.* **49**, 587–593.

McCann, D. H. (1973). U.S. Patent 3,781,574.

Marcus, S., Nelson, R., and Lynds, R. (1979). "Preliminary evaluation of a Fairchild CCD-211 and a new camera system," Panoramic Detector Program, Kitt Peak National Observatory, Tucson, Arizona.

Michon, G. J., Burke, H. K., Vogelsong, T. L., and Wang, K. "Study of noise in CID array systems," Final Report on Naval Research Laboratory Contract N0013-77-C-0054.

Séquin, C. H., and Tompsett, M. F. (1975). "Charge Transfer Devices," Academic Press, New York.

Wen, D. D. (1974). Design and operation of a floating gate amplifier, *IEEE J. Solid-State Circuits,* **SC-9**, 410–414.

Wen, D. D., and Salisbury, P. J. (1973). Analysis and design of a single stage floating Gate Amplifier, *ISSCC Dig. Tech. Pap.,* 154–155.

White, M. H., Lampe, D. R., Blaha, F. C., and Mack, I. A. (1974). Characterization of surface channel CCD image arrays at low light levels, *IEEE J. Solid-State Circuits,* **SC-9**, 1–13; See also U.S. Patent 3,781,574.

Index

401